PROGRESS IN BRAIN RESEARCH

VOLUME 110

TOWARDS THE NEUROBIOLOGY OF CHRONIC PAIN

EDITED BY

G. CARLI

Institute of Human Physiology, University of Siena, Siena, Italy

M. ZIMMERMANN

II Institute of Physiology, University of Heidelberg, Heidelberg, Germany

ELSEVIER
AMSTERDAM – LAUSANNE – NEW YORK – OXFORD – SHANNON – TOKYO
1996

© 1996 Elsevier Science B.V. All rights reserved.

No part of this publication may be reproduced, stored in a retrieval system or transmitted in any form or by any means, electronic, mechanical, photocopying, recording or otherwise without the prior written permission of the publisher, Elsevier Science BV, Copyright and Permissions Department, P.O. Box 521, 1000 AM Amsterdam, The Netherlands.

No responsibility is assumed by the publisher for any injury and/or damage to persons or property as a matter of products liability, negligence or otherwise, or from any use or operation of any methods, products, instructions or ideas contained in the material herein. Because of the rapid advances in the medical sciences, the publisher recommends that independent verification of diagnoses and drug dosages should be made.

Special regulations for readers in the U.S.A.: This publication has been registered with the Copyright Clearance Center Inc. (CCC), 222 Rosewood Drive, Danvers, MA 01923. Information can be obtained from the CCC about conditions under which the photocopying of parts of this publication may be made in the U.S.A. All other copyright questions, including photocopying outside of the U.S.A., should be referred to the copyright owner, Elsevier Science BV, unless otherwise stated.

ISBN 0-444-82149-x (volume)
ISBN 0-444-80104-9 (series)

Published by:
Elsevier Science B.V.
P.O. Box 211
1000 AE Amsterdam
The Netherlands

Printed in The Netherlands on acid-free paper

List of Contributors

A.M. Aloisi, Istituto di Fisiologia Umana, Università degli Studi di Siena, Via Aldo Moro, I-53100 Siena, Italy

A. Berthele, Max-Planck-Institute of Psychiatry, Clinical Institute, Clinical Neuropharmacology, Kraepelinstrasse 2, 80804 Munich, Germany

G. Carli, Istituto di Fisiologia Umana, Universita' degli Studi di Siena, Via Aldo Moro, I-53100 Siena, Italy

E. Carstens, Section of Neurobiology, Physiology and Behavior, University of California, Davis, CA 95616 USA

M. Cavazzuti, Dipartimento di Patologia Neuropsicosensoriale, University of Modena, Via del Pozzo 71, I-41100 Modena, Italy

F. Cervero, Departamento de Fisiología y Farmacología, Universidad de Alcalá de Henares, Facultad de Medicina, Campus Universitario, Alcalá de Henares, 28871 Madrid, Spain

C.R. Chapman, Department of Anesthesiology, RN-10, University of Washington, Seattle, WA 98195, USA

A.H. Dickenson, Department of Pharmacology, University College, Gower Street, London, WC1E 6BT, UK

A. Dray, Sandoz Institute for Medical Research, 5 Gower Place, London WC1E 5BN, UK

A.W. Duggan, Department of Preclinical Veterinary Sciences, Royal (Dick) School of Veterinary Studies, University of Edinburgh, Summerhall, Edinburgh EH9 1QH, UK

C. Dykstra, Department of Pharmacology, University of Iowa, Iowa City, IA 52242, USA

G.F. Gebhart, Department of Pharmacology, University of Iowa, Iowa City, IA 52242, USA

T. Herdegen, II. Physiologisches Institut, Universtät Heidelberg, Im Neuenheimer Feld 326, D-69120 Heidelberg, Germany

A. Herz, Max-Planck-Institute for Psychiatry, Department of Neuropharmacology, Am Klopferspitz 18a, D-82151 Martinsried, Germany

U. Hoheisel, Institut für Anatomie und Zellbiologie, Im Neuenheimer Feld 307, D-69120 Heidelberg, Germany

J.M.A. Laird, Departamento de Fisiología y Farmacología, Universidad de Alcalá de Henares, Facultad de Medicina, Campus Universitario, Alcalá de Henares, 28871 Madrid, Spain

R.H. LaMotte, Department of Anesthesiology, Yale University School of Medicine, New Haven, CT 06510, USA

S.T. Meller, Department of Pharmacology, University of Iowa, Iowa City, IA 52242, USA

S. Mense, Institut für Anatomie und Zellbiologie, Im Neuenheimer Feld 307, D-69120 Heidelberg, Germany

M. Petersen, Department of Physiology, University of Wurzburg, Wurzburg, Germany

C.A. Porro, Dipartimento di Scienze Biomediche, University of Udine, Via Gervasutta 48, I-33100 Udine, Italy

H. Rees, 67 Nant Talwg Way, Barry, South Glamorgan, CF62 6LZ, UK

A. Reinert, Institut für Anatomie und Zellbiologie, Im Neuenheimer Feld 307, D-69120 Heidelberg, Germany

R.C. Riley, Department of Preclinical Veterinary Sciences, Royal (Dick) School of Veterinary Studies, University of Edinburgh, Summerhall, Edinburgh EH9 1QH, UK

J. Sandkühler, II. Physiologisches Institut, Ruprecht-Karls-Universität Heidelberg, Im Neuenheimer Feld 326, D-69120 Heidelberg, Germany

J. Schadrack, Max-Planck-Institute of Psychiatry, Clinical Institute, Clinical Neuropharmacology, Kraepelinstrasse 2, 80804 Munich, Germany

H.-G. Schaible, Physiologisches Institut, Universität Würzburg, Röntgenring 9, D-97070 Würzburg, Germany

R.F. Schmidt, Physiologisches Institut, Universität Würzburg, Röntgenring 9, D-97070 Würzburg, Germany

K.A. Sluka, 2600 Steindler Building, Iowa City, IA 52242, USA

T.R. Tölle, Max-Planck-Institute of Psychiatry, Clinical Institute, Clinical Neuropharmacology, Kraepelinstrasse 2, 80804 Munich, Germany

K.N. Westlund, Marine Biomedical Institute, 301 University Boulevard, Medical Research Building, Galveston, TX 77555-1069, USA

Z. Wiesenfeld-Hallin, Karolinska Institute, Department of Medical Laboratory Sciences and Technology, Section of Clinical Neurophysiology, Huddinge University Hospital, Sweden

W.D. Willis, Marine Biomedical Institute, 301 University Boulevard, Medical Research Building, Galveston, TX 77555-1069, USA

X.-J. Xu, Karolinska Institute, Department of Medical Laboratory Sciences and Technology, Section of Clinical Neurophysiology, Huddinge University Hospital, Sweden

J.-m. Zhang, Department of Anesthesiology, Yale University School of Medicine, New Haven, CT 06510, USA

W. Zieglgänsberger, Max-Planck-Institute of Psychiatry, Clinical Institute, Clinical Neuropharmacology, Kraepelinstrasse 2, 80804 Munich, Germany

M. Zimmermann, II. Physiologisches Institut, Universtät Heidelberg, Im Neuenheimer Feld 326, D-69120 Heidelberg, Germany

Preface

Pain research in the past was mostly related to the sensory aspects of acute experimental noxious stimuli in animals and humans and the clinical reality was not sufficiently taken into account. One of the main goals of the foundation of the International Association for the Study of Pain in 1973 was to concentrate the attention of the scientific community on chronic conditions which elicit enduring suffering and pain in patients.

The last few years have witnessed a major change in paradigms related to pain. With the development of a variety of animal models, it is now possible to focus on slow changes in the nervous system, at systemic, cellular, biochemical, molecular and genetic levels that occur in response to noxious events such as trauma, inflammation or nervous system lesions. It has become clear that plasticity of neural functions is a common denominator of what is seen following a noxious condition, and the end point is the increased sensitivity of the nervous system towards pain. This plasticicty of function bears similarities to the neurobiological mechnaisms of learning and therefore we may see the ability to learn as a basic nervous system process involved in the chronicity of pain. Since in some patients the process of pain chronicity is going on with a progressive severity of the process and with a decreased efficacy of therapy, the prevention and the early treatment of chronic pain represent the new tasks to be set out in clinical medicine.

It was, therefore, with the aim of providing a well-balanced review of activity-dependent modifications of synaptic efficacy which follow noxious events that the symposium "Towards the Neurobiology of Chronic Pain" was organized in Siena, at the Certosa di Pontignano. The title of the book was selected hoping that it will contribute to the continuing interest in understanding pain reinforcement by the brain. This emerging new concept in turn can be expected to have a great impact on the development of clinical paradigms of pain prevention and rehabilitation.

G. Carli and M. Zimmermann
1996

Acknowledgements

The conference "Towards the Neurobiology of Chronic Pain", which was organized in Siena, Certosa di Pontignano, on March 4–5, 1994, was made possible by the financial support of Bayer AG, Leverkusen; Formenti, Milan; Goedecke AG, Freiburg; Gruenenthal GmbH, Stolberg; Mundipharma GmbH, Limburg; Deutsche Gesellschaft zum Studium des Schmerzes and Università degli Studi di Siena. We are greatly indebted to the Servizio VI Convegni and Congressi della Università di Siena, to the members of the Istituto de Fisiologia Umana della Università di Siena and particularly to Dott. Anna Maria Aloisi for taking care of the innumerable organizational details without which neither the conference nor the publication of its Proceeding would have been possible. The editorial expertise of Almuth Manisali in helping to edit this volume is gratefully acknowledged.

The Editors

Contents

List of contributors v

Preface vii

Acknowledgements ix

Section I.   Psychobiology of Acute and Chronic Pain

1. From acute to chronic pain: mechanisms and hypotheses
   F. Cervero and J.M.A. Laird (Madrid, Spain) 3

2. Quantitative experimental assessment of pain and hyperalgesia in animals and underlying neural mechanisms
   E. Carstens (Davis, CA, USA) 17

3. Nociceptive, environmental and neuroendocrine factors determining pain behaviour in animals
   A.M. Aloisi and G. Carli (Siena, Italy) 33

4. Functional imaging studies of the pain system in man and animals
   C.A. Porro and M. Cavazzuti (Udine and Modena, Italy) 47

5. Limbic processes and the affective dimension of pain
   C.R. Chapman (Seattle, WA, USA) 63

II. Neuropeptides, Inflammation and Neuropathic Injuries

6. Neurogenic mechanisms and neuropeptides in chronic pain
   A. Dray (London, UK) 85

7. Peripheral opioid analgesia - facts and mechanisms
   A. Herz (Martinsried, Germany) 95

8. Alterations in the functional properties of dorsal root ganglion cells with unmyelinated axons after a chronic nerve constriction in the rat
   R.H. LaMotte, Jun-ming Zhang and M. Petersen (New Haven, CT, USA and Wurzburg, Germany) 105

9. Plasticity of messenger function in primary afferents following nerve injury – implications for neuropathic pain
   Z. Wiesenfeld-Hallin and X.-J. Xu (Huddinge, Sweden) 113

10. The possible role of substance P in eliciting and modulating deep somatic pain
    S. Mense, U. Hoheisel and A. Reinert (Heidelberg, Germany) 125

11. Studies of the release of immunoreactive galanin and dynorphin $A_{(1-8)}$ in the spinal cord of the rat
    A.W. Duggan and R.C. Riley (Edinburgh) 137

III. Neurotransmission and Plasticity

12. Cooperative mechanisms of neurotransmitter action in central nervous sensitization
    W.D. Willis, K.A. Sluka, H. Rees and K.N. Westlund (Galveston, TX, USA) 151

13. Neurophysiology of chronic inflammatory pain: electrophysiological recordings from spinal cord neurons in rats with prolonged acute and chronic unilateral inflammation at the ankle
    Hans-Georg Schaible and Robert F. Schmidt (Würzburg, Germany) 167

14. Acute mechanical hyperalgesia in the rat can be produced by coactivation of spinal ionotropic AMPA and metabotropic glutamate receptors, activation of phospholipase A_2 and generation of cyclooxygenase products
    S.T. Meller, C. Dykstra and G.F. Gebhart (Iowa City, IA, USA) 177

15. Involvement of glutamatergic neurotransmission and protein kinase C in spinal plasticity and the development of chronic pain
    T.R. Tölle, A. Berthele, J. Schadrack and W. Zieglgänsberger (Munich, Germany) 193

16. Neurobiology of spinal nociception: new concepts
    J. Sandkühler (Heidelberg, Germany) 207

17. Balances between excitatory and inhibitory events in the spinal cord and chronic pain
    A.H. Dickenson (London, UK) 225

18. Plasticity of the nervous system at the systemic, cellular and molecular levels: a mechanism of chronic pain and hyperalgesia
    M. Zimmermann and T. Herdegen (Heidelberg, Germany) 233

Subject Index 261

Section I

Psychobiology of Acute and Chronic Pain

CHAPTER 1

From acute to chronic pain: mechanisms and hypotheses

Fernando Cervero and Jennifer M.A. Laird

Departamento de Fisiología y Farmacología, Universidad de Alcalá de Henares, Facultad de Medicina, Campus Universitario, Alcalá de Henares, 28871 Madrid, Spain

Introduction

Some years ago we proposed a conceptual framework and a series of models that addressed the mechanisms of both acute and chronic pain and the inter-relation between different pain states (Cervero and Laird, 1991a). We argued that the neurophysiological mechanisms responsible for all of the various pain states are different and that normal (nociceptive) and abnormal (neuropathic) pain represent the end-points of a sequence of possible changes that can occur in the nervous system. Normally, a steady state is maintained in which there is a close correlation between injury and pain. However, changes or oscillations induced by nociceptive input or by changes in the environment can result in variations in the quality and quantity of pain sensation produced by a particular noxious stimulus. Such changes would be temporary, unless there was further noxious input, as the system would always tend to restore the normal balance between injury and pain. However, long-lasting or very intense nociceptive input or the removal of a portion of the normal input would distort the nociceptive system to such an extent that the close correlation between injury and pain would be lost.

We considered three major stages or phases of pain and proposed that different neurophysiological mechanisms are involved depending on the nature and time course of the originating stimulus. These three phases are: (1) the processing of a brief noxious stimulus; (2) the consequences of prolonged noxious stimulation leading to tissue damage and peripheral inflammation; and (3) the consequences of neurological damage, including peripheral neuropathies and central pain states (Fig. 1). However, it is important to point out that these phases are not exclusive, and that at any given time several of the neurophysiological mechanisms that underlie these pain states may co-exist in the same individual.

We believe that the basic framework is still valid and useful, but we would like to consider in this chapter a number of points which were not covered sufficiently in our original article or that need to be modified due to new experimental evidence obtained in the last 5 years.

Phase 1 pain

The mechanisms subserving the processing of brief noxious stimuli (Phase 1 pain) can be viewed as a fairly simple and direct route of transmission centrally towards the thalamus and cortex and thus the conscious perception of pain, with possibilities for modulation occurring at synaptic relays along the way (see Phase 1 model in Fig. 1). The relative simplicity of this model reflects the experimental observation that in humans undergoing Phase 1 pain, there is a close correlation between the discharges in peripheral nociceptors and the subjective expression of pain. Phase 1 pain is the type which has been most studied experimentally, and on the basis of the large body of data from both humans and animals, it is reasonably easy to construct plausible and detailed neuronal circuits to explain the features of Phase 1 pain. In our original proposal we suggested that Phase 1 pain can best be explained by models based on the specificity interpretation of pain mechanisms, that is, the exis-

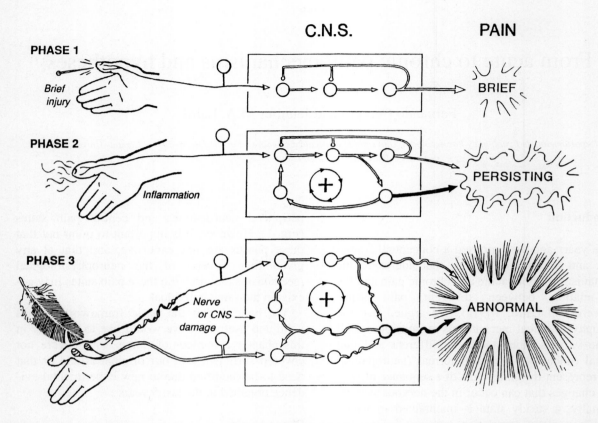

Fig. 1. Three models of pain processing for the three phases of pain discussed in this chapter. For a detailed explanation see text. From Cervero and Laird (1991a).

tence within the peripheral and central nervous systems of a series of neuronal elements concerned solely with the processing of these simple noxious events. This viewpoint remains unchanged.

Phase 2 pain

The situation changes to what we have called Phase 2 pain if a noxious stimulus is very intense, or prolonged leading to tissue damage and inflammation. The pain state under these conditions is different from that in Phase 1 pain, because the response properties of various components of the nociceptive system change. There is a greatly increased afferent inflow to the CNS from the injured area as a result of the increased activity and responsiveness of sensitised nociceptors. In addition, nociceptive neurones in the spinal cord modify their responsiveness in ways that are not merely an expression of the changes in their inputs from the periphery. These changes mean that the CNS has moved to a new, more excitable state as a result of the noxious input generated by tissue injury and inflammation. Phase 2 pain is characterised by its central drive, a drive that is triggered and maintained by peripheral inputs (see Phase 2 model in Fig. 1).

In Phase 2 pain, the subject experiences spontaneous pain, a change in the sensations evoked by stimulation of the injured area, and also of the undamaged areas surrounding the injury. This change in evoked sensation is known as hyperalgesia, defined as a leftward shift in the stimulus-response function describing the relationship between stimuli and pain sensation (Fig. 2, Treede et al., 1992). In this situation, normally innocuous stimuli like

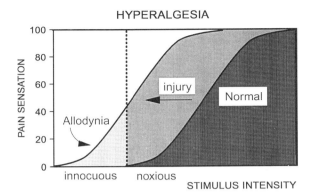

Fig. 2. Diagram illustrating the changes in pain sensation induced by injury. The normal relationship between stimulus intensity and the magnitude of pain sensation is represented by the curve at the right-hand side of the figure. Pain sensation is only evoked by stimulus intensities in the noxious range (the vertical dotted line indicates the pain threshold). Injury provokes hyperalgesia, defined as a leftward shift in the curve relating stimulus intensity to pain sensation. Under these conditions, innocuous stimuli evoke pain (allodynia), and stimulus intensities that normally evoke mild pain evoke more intense pain.

brushing and touch, are painful (allodynia), and normally mild painful stimuli like pinprick, are more painful (hyperalgesia; Lewis, 1936; LaMotte et al., 1991). Hyperalgesia in the area of injury is known as primary hyperalgesia, and in the areas of normal tissue surrounding the injury site, as secondary hyperalgesia.

Peripheral mechanisms contributing to Phase 2 pain

The sensitisation of peripheral nociceptors is characterised by two main changes in their response properties. Firstly, the appearance of spontaneous activity, which will provide a continuous nociceptive afferent barrage which probably contributes to spontaneous pain, and secondly a decrease in threshold, to an extent that normally non-noxious stimuli will activate the sensitised receptor. This drop in threshold is generally agreed to underlie the change in the sensation provoked by stimulation of an injured area known as primary hyperalgesia (Treede et al., 1992). A drop in threshold may also contribute to the development of continuous nociceptive input to the CNS, and thus spontaneous pain, depending on the site of injury. For example, the available evidence on the properties of visceral nociceptors indicates that a variety of noxious stimuli, evoke not only an acute excitation of the nociceptors, but also cause prolonged sensitisation of both the "normal" and the "silent" nociceptors (see Cervero, 1994 for review). In the sensitised state, visceral nociceptors respond to many of the innocuous stimuli that occur during the normal functioning of internal organs. As a consequence, after inflammation of an internal organ, the CNS will receive an increased barrage from peripheral nociceptors that is related initially to the extent of the acute injury, but for the duration of the inflammatory process initiated by the injury, will be dependent on the motor (smooth muscle) and secretory activity of that organ until the peripheral sensitisation has completely resolved.

Central mechanisms contributing to Phase 2 pain

The role of continuing afferent input

In normal human subjects, intense experimental pain produced by a burn injury or by chemical stimulation of nociceptors with capsaicin changes the responsiveness of an area of undamaged skin surrounding the injury site such that normally innocuous stimuli like brushing and touch, are painful (allodynia), and normally mild painful stimuli like pinprick, are more painful (hyperalgesia; Lewis, 1936; LaMotte et al., 1991). This phenomenon of "secondary hyperalgesia" is now known to be due to a central change, induced and maintained by input from nociceptors, such that activity in large myelinated afferents connected to low threshold mechanoreceptors provokes painful sensations as well as tactile sensations (Torebjörk et al., 1992), and input from nociceptors evokes enhanced pain sensations (Cervero et al., 1994).

It is becoming increasingly clear from both psychophysical and animal studies that the expression of both spontaneous pain and of hyperalgesia are very dependent on ongoing afferent activity from the injury site. The elegant studies of LaMotte and colleagues (LaMotte et al., 1991; Torebjörk et al.,

1992) have shown that after the appearance of an area of secondary hyperalgesia around an intradermal capsaicin injection in normal human subjects, the secondary hyperalgesia disappears within seconds if the injection site is cooled with a piece of ice (a procedure known to block the ongoing activity in the nociceptors), and re-appears if the ice is removed as the temperature returns to normal. Similarly, it has been shown that duration of secondary hyperalgesia after a burn injury (Moiniche et al., 1993) or the application of mustard oil (Koltzenburg et al., 1992) is dependent on the afferent input from the site of injury. The presence of allodynia and hyperalgesia in patients with peripheral neuropathy, a phenomenon very similar to and perhaps identical to secondary hyperalgesia can also be abolished by the injection of local anaesthetic into a presumed source of nociceptive ectopic activity (Gracely et al., 1992). In animal experiments, the magnitude of the changes seen in neurones at the first relay in the CNS also appears to be directly related to the amount of afferent input received. For example, in a recent study from this laboratory, the central changes in the excitability of trigeminal neurones in response to corneal injury were examined (Pozo and Cervero, 1993). In this study we were able to demonstrate a close correlation between the size of the nociceptive afferent barrage and the extent of the increase in excitability of the cells as measured by the increase in the size of their receptive fields (Fig. 3).

Central sensitisation

Experiments in laboratory animals have provided evidence for the concept that noxious input to the CNS sets off a central process of enhancement of responsiveness that continues independently of peripheral afferent drive. Short periods of electrical stimulation at noxious intensity produce increases in excitability with a time course similar to that of short term potentiation (from a few seconds to tens of minutes) in single dorsal horn neurones (Cervero et al., 1984; Cook et al., 1987), in C-fibre evoked potentials recorded in the superficial dorsal horn (Schouenborg, 1984) and in spinal reflexes (Wall and Woolf, 1984; Woolf and Thompson, 1991). Increases in excitability of dor-

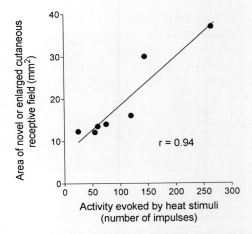

Fig. 3. Trigeminal neurones with input from the cornea, of which 6 responded only to stimulation of the cornea, and one to stimulation of the cornea and to noxious stimulation of periorbital skin were recorded in anaesthetised rats. After a series of noxious heat pulses were applied to the cornea, all six neurones that previously only had a corneal receptive field now had an additional periorbital receptive field that responded to noxious stimulation, and the neurone that had a cutaneous receptive field before heating showed an enlarged receptive field. The graph shows the relationship between the number of impulses evoked in the seven neurones by heat stimuli, and the size of the novel or enlarged cutaneous receptive field after the heat stimuli. A highly significant correlation between the two variables was found. Re-drawn from Pozo and Cervero (1993).

sal horn C-fibre evoked potentials with a time course similar to that of long term potentiation has also recently been described (Liu and Sandkühler, 1995). These increases in central excitability are also referred to as "central sensitisation" (Woolf, 1989; Woolf and Thompson, 1991).

The description of central sensitisation provided the rationale behind the procedure known as "pre-emptive analgesia" (Woolf, 1989; Woolf and Chong, 1993) whereby the administration of analgesics and/or neural blocks prior to a surgical procedure would reduce or eliminate postoperative pain. The idea was that blocking the initial injury-induced barrage from arriving at the CNS would prevent the development of central sensitisation, and therefore the resulting pain state. The results of some preliminary clinical trials were initially encouraging, but more recently many serious doubts have been raised about the clinical efficacy of "pre-emptive" analgesic measures (Dahl and Kehlet, 1993; Kehlet and Dahl, 1993).

The lack of success of this theoretical approach is probably due to the fact that outside the controlled conditions of the laboratory, noxious stimuli capable of evoking spontaneous pain and hyperalgesia produce not just central sensitisation but also sensitisation of nociceptors in the periphery. Whereas central sensitisation produces only an increase in the response to further noxious stimuli, sensitisation of nociceptors is characterised by the development of spontaneous activity, which will provide a continuous nociceptive afferent barrage, along with a drop in threshold, which, depending on the site of injury, may also contribute to the development of continuous nociceptive input to the CNS (see above). Since normal visceral nociceptors are not spontaneously active (Cervero, 1994), and the fact that "silent" nociceptors in viscera appear not to become activated until an inflammatory process is underway (Häbler et al., 1990), it is likely that the discharges from sensitised visceral nociceptors, both spontaneous and in response to innocuous stimuli during the inflammatory sequelae of an acute injury are greater in magnitude and duration that those produced during the injury itself. Thus in some circumstances, the central effects of post-injury discharges are likely to be greater that those of the acute insult.

States of "central sensitisation" maintained for a period of time independently of further afferent input can be demonstrated experimentally but it seems that the role of such a "self-sustaining" neurophysiological mechanism in the spontaneous pain and hyperalgesia produced by injury is relatively limited. On the one hand in the normal situation the nociceptive afferent barrage continues for as long as the injury takes to heal, and on the other hand, the available evidence from pyschophysical and clinical studies shows that the expression of hyperalgesia and spontaneous pain is highly dependent on this continuing afferent input. Any central mechanism that is a candidate for an important role in the generation and maintenance of spontaneous pain and hyperalgesia should logically therefore also be strongly dependent on afferent input.

Phase 3 pain

Phase 3 pains are abnormal pain states and are generally the consequence of damage to peripheral nerves or to the CNS itself. These are neuropathic pain states characterised by a lack of correlation between injury and pain. In clinical terms, Phase 1 and 2 pains are symptoms of peripheral injury, whereas Phase 3 pain is a symptom of neurological diseases that include lesions of peripheral nerves or damage to any portion of the somatosensory system within the CNS. Phase 3 pains are spontaneous, triggered by innocuous stimuli or exaggerated responses to minor noxious stimuli. These sensations are expressions of substantial alterations in the normal nociceptive system induced by peripheral or central damage (see Phase 3 model in Fig. 1). The particular combination of mechanisms responsible for each one of the various Phase 3 pain states is probably unique to the individual disease, or to particular sub-groups of patients. Phase 3 type syndromes are seen only in a minority of people, even patients with seemingly identical damage to their nervous systems may or may not complain of pain. Thus the development of Phase 3 pain may involve genetic, cognitive or emotional factors that have yet to be identified.

The loss of sensory function as a result of damage to the nervous system is conceptually easy to understand as the interruption of connections either within the CNS or between the CNS and the peripheral sensory receptors. However, the abnormal sensory symptoms that accompany many cases of neuronal damage require more explanation. Two groups of mechanisms probably account for these sensory symptoms; (i) pathological changes in the damaged neurones, and (ii) reactive changes in response to nociceptive (damage-related) afferent input, and to the loss of portions of the normal afferent input. Whereas the pathological changes in damaged neurones are almost certainly unique to Phase 3 pain, some of the reactive changes to nociceptive input may well be the expression of normal mechanisms also seen in Phase 2 type pain. In Phase 3 pain, the activation of these mechanisms may be abnormally prolonged or intense due to the abnormal input from damaged neurones, or simply the fact that since the regenerative properties of neurones are very poor, healing never occurs. Furthermore, even if successful regeneration occurs, the response properties of the regenerated receptors and the afferents to which they are connected may not be completely normal. For example, C-fibre nociceptors examined after successful regeneration of a peripheral nerve in the cat showed significantly lower response thresholds than those recorded in normal, undamaged nerves (Dickhaus et al., 1976).

Does the pathophysiology of Phase 3 pain include the expression of normal pain mechanisms?

The development of areas of secondary hyperalgesia is a normal consequence of injury and inflammation and, as such, a feature of Phase 2 pain. As described above, secondary hyperalgesia is a phenomenon induced mainly by changes occurring in the CNS whereby activity in low threshold mechanoreceptors evokes pain sensations. This kind of mechanism also operates in neuropathic pain (Campbell et al., 1988; Gracely et al., 1992) and it is therefore possible that the mechanisms of the allodynia seen in neuropathic pain patients include those normally present during secondary hyperalgesia.

Secondary hyperalgesia can even be evoked in the absence of the perception of pain per se in normal human subjects, as we showed in a recent study (Cervero et al., 1993). In these experiments, a long duration non-painful heat stimulus was applied to an area of normal skin on the back of the hand. The skin temperature was raised to 39°C for 30 min, then raised to 40°C for another 30 min, and so on in 1°C steps until the subject reported the sensation of pain, and the stimulus was stopped. The presence of flare and the extent of any hyperalgesia to punctate or stroking stimuli applied to the area of skin around that being stimulated was recorded every 15 min throughout the application of the heat stimuli, and at the end of the stimulation whenever this occurred. All subjects tested developed flare (an indication of nociceptor activation) and an area of hyperalgesia during the application of the heat stimulus. Flare and hyperalgesia always appeared well before the first report of pain, and the difference in time of onset was statistically significant (Fig. 4). After the discontinuation of the heat stimulus, the hyperalgesia disappeared within 5–6 min. These results show that cutaneous secondary hyperalgesia can be evoked in normal human subjects by prolonged thermal stimulation at intensities that are not perceived as painful.

Pathological activity in nociceptors as a result of nerve injury (ectopic discharge or responses to stimulation of sensitive nerve sprouts), would be expected to produce a secondary hyperalgesia-like central change. This mechanism could explain the observed allodynia and hyperalgesia, in the absence of an apparent peripheral injury, in neuropathic pain patients. In patients with painful peripheral neuropathy, allodynia has been shown to be mediated by large myelinated afferents (Campbell et al., 1988), and local anaesthesia of a presumed ectopic focus has been shown to abolish allodynia and hyperalgesia (Gracely et al., 1992). Pathological on-going activity in large myelinated afferents is also seen after certain types of experimental nerve injury (Kajander and Bennett, 1992), and the combination of the two types of activity via this central mechanism would presumably produce on-going "allodynia-like" pain independent of any peripheral stimulation.

Role of supraspinal components in the expression of Phase 2 and Phase 3 pain

As discussed above, central changes are very important in the expression of Phase 2 and 3 pain states. In many experimental studies in which spinalised animals have been used, central changes induced by noxious peripheral input have been reported. Thus it is clear that some of these central mechanisms are contained entirely within the spinal cord. However, it is also known that descending control from supraspinal centres can alter both the ascending nociceptive message and the reflex responses of animals with intact spinal cords. Although much work has been devoted to the study of the various descending pathways, the role of

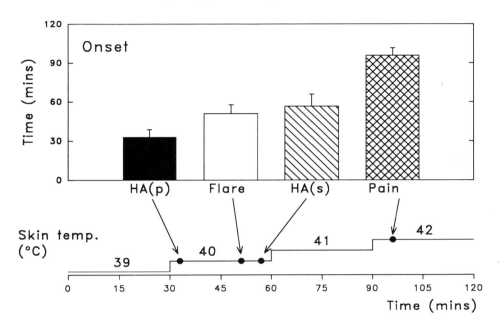

Fig. 4. The lower panel shows the time course and intensity of a heat stimulus delivered to normal skin of the back of the hand in 10 human volunteers. The upper panel shows the mean (± SEM) times after the start of the stimulation period at which flare was detected, or the presence of punctate hyperalgesia (HA(p)), stroking hyperalgesia (HA(s)) and pain reported for all of the subjects to whom the heat stimulus was applied. Taken from Cervero et al. (1993). Further experimental details are given in the text.

descending control in the changes in pain sensation that occur after either inflammatory pain or neuropathic pain states has received relatively little attention.

Work from this laboratory has shown that the magnitude of tonic descending inhibition is directly related to the afferent input to the spinal cord mediated by afferent C fibres. Some years ago we described that in animals treated at birth with capsaicin and that, as a consequence, had a reduced afferent C fibre input, the extent and magnitude of tonic descending inhibition was considerably reduced (Cervero and Plenderleith, 1985). More recently we have also described that in animals with an inflammation of the knee joint, tonic descending inhibition of spinal cord neurones increases progressively over time as the inflammation develops, which shows an increase in central inhibition that parallels the increase in afferent inflow due to the sensitisation of joint nociceptors (Cervero et al., 1991; Schaible et al., 1991).

It has also been demonstrated that dorsal horn neurones partially denervated by a dorsal rhizotomy are under considerably less tonic descending inhibition than similar neurones in normal animals (Brinkhus and Zimmermann, 1983), which further supports the idea that the degree of inhibition exerted by supraspinal centres over spinal nociceptive neurones depends on the amount of afferent input they receive. It is also known that DNIC, another form of descending control, is greater in arthritic as compared to normal animals (Calvino et al., 1987) which again indicates a direct relationship between noxious afferent inflow and the magnitude of central inhibition.

Descending inhibition thus depends on the in-

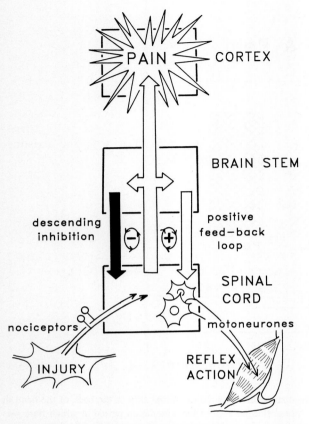

Fig. 5. Box diagram showing how the nociceptive message that results from a peripheral injury may be both enhanced by a positive feedback loop between the spinal cord and brain stem and suppressed by descending inhibition. Both processes are dependent on the incoming afferent barrage evoked in nociceptive afferents by the peripheral injury and also by the subsequent sensitisation of the nociceptive afferents. The nociceptive message may result in reflex action, the conscious sensation of pain, or both. Modified from Cervero and Laird (1991b).

coming afferent messages (see Fig. 5) and it is therefore very likely that supraspinal mechanisms would play a crucial role in the development and maintenance of persistent pain states, such as those that characterise Phases 2 and 3 pains. We suggest that the loss of a proportion of the normal input after peripheral nerve damage produces, in addition to purely spinal changes, a lifting of descending inhibition. Such dis-inhibition may well contribute to the exaggerated afterdischarges seen in dorsal horn neurones recorded in rats with a peripheral nerve injury (Laird and Bennett, 1993) and the high spontaneous discharge of dorsal horn neurones recorded in rats after dorsal rhizotomy (Brinkhus and Zimmermann, 1983). Equally, dis-inhibition could be responsible for the abnormally enhanced perception of painful stimuli in patients with neuropathy, in particular to the phenomenon known as "hyperpathia", whereby noxious stimuli evoked exaggerated pain responses in these patients.

Descending excitation of spinal neurones, both tonic excitation and excitation evoked by stimulation of brainstem sites has also been described (Cervero and Wolstencroft, 1984; Cervero and Plenderleith, 1985; Tattersall et al., 1986; Laird and Cervero, 1990), and is particularly prominent in visceral spinal pathways, where almost 50% of neurones with visceral input show evidence of excitatory descending influences (Tattersall et al., 1986). However, possible changes in descending excitation with changes in afferent input have received little study, since in spinal neurones with only somatic inputs they represent a small proportion (<10%) of the total population. The numbers and characteristics of these neurones did not appear to be changed when neurones from normal animals were compared with those recorded in rats treated with capsaicin at birth (Cervero and Plenderleith, 1985), or in cats with acute arthritis (Cervero et al., 1991; Schaible et al., 1991). However, since the total number of neurones showing descending excitation studied so far is small and since these neurones are present in much greater numbers in visceral pathways, in which the effects of phase 2 and 3 pain states on descending control have not yet been examined in detail, it is difficult to make any firm conclusions on the possible involvement of descending excitation in the expression of phase 2 and 3 pain.

Transmitter systems involved in the 3 phases of pain

The different pain states described by the three models in Fig. 1 are due to changes in the neurophysiological mechanisms involved in the nociceptive pathway, as described above. However, it is clear that the neurochemistry of the nociceptive system changes with noxious input, and there is increasing evidence that the pharmacology of pain changes as one moves from Phase 1 to Phases 2 and 3 as different transmitter systems are recruited as part of the different mechanisms that underlie the different pain states. These pharmacological differences are already reflected in clinical practice, and they have important implications not just for the treatment of the different pain states, but also for the development of novel analgesic agents. Until recently, both industrial drug discovery programs and academic pain pharmacologists routinely used very simple animal models of pain which mostly represent Phase 1 pain (e.g. tail-flick, hot-plate, etc.). However, in recent years there has been increased emphasis on the development of models of different pain states, and there is a growing body of literature showing that the pharmacology of several transmitter systems is different in the various different pain states. In the following paragraphs we have made a brief summary of some of these data.

Opiates

Opiates are extremely effective in animals in inhibiting pain in both Phases 1 and 2 type pain states. They are generally the first line treatment in the clinic for severe pain, and there is no theoretical reason why they should not be effective in treating Phase 3 or neuropathic pain (Hill, 1994). However, their usefulness in neuropathic pain patients is the subject of some debate (Arnér and Meyerson, 1988; Portenoy et al., 1990). It has been suggested that the dose-response curve for opiates in these patients may be shifted to the right, such that high doses are required to provide significant relief (Portenoy et al., 1990). This problem is likely to be compounded by the fact that neuropathic pain may be very severe. Thus the side-effects produced by an effective analgesic dose may well be intolerable, especially for long-term use. There are also varying reports of their effectiveness in animal models of neuropathic pain, (Attal et al., 1991; Yamamoto et al., 1994). The pre-clinical data indicate that the changes seen in the opiate system after nerve damage may vary depending on the type of input being studied, and the time elapsed after injury, which perhaps is a contributing factor in the clinical controversy.

The opioid receptor subtypes have recently been cloned, which will greatly assist the development of agents specific for each of the human opiate receptors. It is possible that agents selective for these receptor sub-types may have improved side-effect profiles compared with the existing compounds (Hill, 1994).

NSAIDs

The family of aspirin-like drugs, or non-steroidal anti-inflammatory drugs (NSAIDs) are non-opioid analgesic and anti-inflammatory agents which have actions at various sites in the nervous system, but are usually considered to exert their major analgesic actions in the periphery. Their mechanism of action was explained by the demonstration by Vane (1971) that all of the members of this large and chemically diverse family inhibit the cyclooxygenase (COX) enzyme, thus preventing the formation and release of prostaglandins. Prostaglandin concentrations increase at sites of injury and act on the nociceptive system both directly by sensitising nociceptors, and indirectly by stimulating the production of other substances which activate nociceptors. It has recently been discovered that the increased synthesis of prostaglandins at sites of inflammation is due to the de novo synthesis of a new COX enzyme (COX-2). The COX-1 enzyme is constitutively expressed in most tissues, and has a "house-keeping" function, producing prostaglandins as part of normal healthy tissue regulation. The COX-2 form is essentially absent from healthy tissues, but is rapidly induced under conditions of inflammation by tissue damage. In chronic inflammation, the levels of COX-2 parallel the increased levels of prostaglandins (see Vane and Botting, 1995 for review).

It is therefore to be expected that NSAIDs would be more effective in inflammatory conditions. The available experimental data shows that this is the case; although many NSAIDs can be shown to inhibit nociceptive processing in normal animals and humans, they are considerably more effective in experimental models that represent Phase 2 type pain. For example, acetylsalicylate (Pircio et al., 1975) and sodium diclofenac (Attal et al., 1988) are more effective in arthritic rats than in normal rats in inhibiting pain-related behaviour and the activity of thalamic neurones, and a recent study in human volunteers (Kilo et al., 1995) has shown that ibuprofen inhibits cutaneous mechanical hyperalgesia, but not mechanical pain from normal skin.

However, the analgesic potency of individual members of the NSAID class does not correlate with their efficacy as COX inhibitors in the periphery. It is possible that damage and inflammation induces COX activity in the CNS, or possibly even another isoform of the COX enzyme (COX-3 ?), or that some of these agents exert analgesic effects via other mechanisms which remain to be determined.

NMDA receptor antagonists

Glutamate is a major transmitter in the spinal cord, and the N-methyl-D-aspartate (NMDA) glutamate receptor subtype has been proposed as having a particular role in mediating persistent pain and hyperalgesia in the spinal cord (Dickenson, 1990). Consistent with this view, NMDA receptor antagonists are not very effective in Phase 1 type pain, but in behavioural tests of Phase 2 type pain they reverse the hyperalgesia evoked by local inflammation at doses which have no effect on the response to noxious stimulation of the contralateral, non-affected limb (e.g. Ren and Dubner, 1993). NMDA receptor antagonists have been shown to reduce some but not all types of abnormal pain behaviour in animal models of neuropathy (e.g. Tal and Bennett, 1993) and abnormal pain sensations in a small number of case reports in human pain patients (e.g. Kristensen et al., 1992). However, the psychotomimetic and motor side-effects of NMDA receptor blockade at present rule out widespread clinical use of these agents (see Kauppila et al., 1995).

The NMDA receptor antagonists currently available for clinical use act competitively at the glutamate recognition site, or by blocking the associated ion channel. However, the use of agents that act on the modulatory sites of the receptor may be more attractive clinically since such agents appear to have a much lower potential for adverse CNS side-effects (for review see Kemp and Leeson, 1993). Agents acting at the glycine modulatory site have been shown to have antinociceptive effects in animal behavioural experiments, for example in the formalin test (Millan and Seguin, 1993) and on hyperalgesia induced by inflammation (Laird et al., 1996) or by peripheral neuropathy (Mao et al., 1992). The development of such agents may provide an alternative therapy in the future for Phase 2 and Phase 3 pain.

Substance P receptor antagonists

Substance P has long been thought to be involved with nociceptive processing, since it is expressed in small diameter primary afferents, most of which are connected to peripheral nociceptors. However, the recent description of selective non-peptide receptor antagonists for the NK1 tachykinin receptor, at which substance P is the highest affinity endogenous ligand (see Maggi et al., 1993, for review), has allowed direct investigation of the role of the NK1 receptor in nociception. It seems that NK1 receptors are not involved in Phase 1 pain transmission, since NK1 receptor antagonists have no effect in simple behavioural analgesic tests in normal animals (tail-flick, paw pressure etc.; e.g. Rupniak et al., 1993). Similarly, NK1 receptor antagonists do not affect the responses of dorsal horn neurones or spinal reflexes to brief noxious stimuli (De Konick and Henry, 1991; Laird et al., 1993). However, NK1 receptors have a role in the processing of Phase 2 pain, since in electrophysiological experiments, NK1 receptor antagonists inhibit the responses to prolonged or intense stimulation (De Konick and Henry, 1991; Laird et al., 1993), and also inhibit enhanced responses provoked by inflammation in dorsal horn neurones (Neugebauer et al., 1994) and behavioural nociceptive responses (Rupniak et al., 1995). It is also known that peripheral inflammation increases the expression of both NK1 receptors and substance P in the spinal cord.

Little data have been published on the effects of NK1 receptor antagonists in Phase 3 pain. However, Yamamoto and Yaksh (1992) showed that intrathecal administration of NK1 receptor antagonists increased the latency of response to noxious heating of the paw in a rat model of painful peripheral neuropathy. This effect was seen both ipsilateral and contralateral to the nerve injury. These data indicate that NK1 receptor antagonists may also be effective in Phase 3 pain states.

Since this type of compound has been developed only recently, there is very little information (at least in the public domain) on their effect in experimental pain in humans, or in pain patients. Thus any judgement of their potential usefulness in the clinic must await the announcement of the results of the clinical trials of the NK1 receptor antagonists now in development.

References

Arnér, S. and Meyerson, B.A. (1988) Lack of analgesic effect of opioids on neuropathic and idiopathic forms of pain. *Pain*, 33: 11–23.

Attal, N., Kayser, V., Eschalier, A., Benoist, J.M. and Guilbaud, G. (1988) Behavioural and electrophysiological evidence for an analgesic effect of the non-steroidal anti-inflammatory agent, sodium diclofenac. *Pain*, 35: 341–348.

Brinkhus, H.B. and Zimmermann, M. (1983) Characteristics of dorsal horn neurones after partial chronic deafferentation by dorsal root transection. *Pain*, 15: 221–236.

Calvino, B., Villanueva, L. and LeBars, D. (1987) Dorsal horn (convergent) neurones in the intact anaesthetised arthritic rat. I. Segmental excitatory influences. *Pain*, 28: 81–98.

Campbell, J.N., Raja, S.N., Meyer, R.A. and Mackinnon, S.E. (1988) Myelinated afferents signal the hyperalgesia associated with nerve injury. *Pain*, 32: 89–94.

Cervero, F. (1994) Sensory innervation of the viscera: peripheral basis of visceral pain. *Physiol. Rev.*, 74: 95–138.

Cervero, F. and Laird, J.M.A. (1991a) One pain or many pains? a new look at pain mechanisms. *NIPS*, 6: 268–273.

Cervero, F. and Laird, J.M.A. (1991b) Neurophysiology of postoperative pain. In: J.H. McClure and J.A.W. Wildsmith (Eds.), *Conduction Blockade for Postoperative Analgesia*, E. Arnold, London, pp. 1–25.

Cervero, F. and Plenderleith, M.B. (1985) C-fibre excitation and tonic descending inhibition of dorsal horn neurones in adult rats treated at birth with capsaicin. *J. Physiol. (London)*, 365: 223–237.

Cervero, F. and Wolstencroft, J.H. (1984) A positive feedback loop between spinal cord nociceptive pathways and antinociceptive areas of the cat's brainstem. *Pain*, 20: 125–138.

Cervero, F., Schouenborg, J., Sjölund, B.H. and Waddell, P.J. (1984) Cutaneous inputs to dorsal horn neurones in adult rats treated at birth with capsaicin. *Brain Res.*, 301: 47–57.

Cervero, F., Schaible, H.-G. and Schmidt, R.F. (1991) Tonic descending inhibition of spinal cord neurones driven by joint afferents in normal cats and in cats with an inflamed knee joint. *Exp. Brain Res.*, 83: 675–678.

Cervero, F., Gilbert, R., Hammond, R.G.E. and Tanner, J. (1993) Development of secondary hyperalgesia following non-painful thermal stimulation of the skin: a psychophysical study in man. *Pain*, 54: 181–189.

Cervero, F., Meyer, R.A. and Campbell, J.N. (1994) A psychophysical study of secondary hyperalgesia: evidence for increased pain to input from nociceptors. *Pain*, 58: 21–28.

Cook, A.J., Woolf, C.J., Wall, P.D. and McMahon, S.B. (1987) Dynamic receptive field plasticity in rat spinal cord dorsal horn following C-primary afferent input. *Nature*, 325: 151–153.

Dahl, J.B. and Kehlet, H. (1993) The value of pre-emptive analgesia in the treatment of postoperative pain. *Br. J. Anaesth.*, 70: 434–439.

De Konick, Y. and Henry, J.L. (1991) Substance P-mediated slow excitatory postsynaptic potentials elicited in dorsal horn neurons in vivo by noxious stimulation. *Proc. Natl. Acad. Sci. USA*, 88: 11344–11348.

Dickenson, A.H. (1990) A cure for wind-up: NMDA receptor antagonists as potential analgesics. *Trends Pharmacol. Sci.*, 11, 307–309.

Dickhaus, H., Zimmermann, M. and Zotterman, Y. (1976) The development in regenerating cutaneous nerves of C-fibre nociceptors responding to noxious heating of the skin. In: Y. Zotterman (Ed.), Pergamon Press, Oxford, pp. 415–423.

Gracely, R.H., Lynch, S.A. and Bennett, G.J. (1992) Painful neuropathy: altered central processing maintained dynamically by peripheral input, *Pain*, 51: 175–194.

Häbler, H.-J., Jänig, W. and Koltzenburg, M. (1990) Activation of unmyelinated afferent fibres by mechanical stimuli and inflammation of the urinary bladder in the cat. *J. Physiol. (London)*, 425: 545–562.

Hill, R.G. (1994) Pharmacological considerations in the use of opioids in the management of pain associated with non-terminal disease states. *Pain Rev.*, 1: 47–64.

Kajander, K.C. and Bennett, G.J. (1992) Onset of a painful peripheral neuropathy in rat: a partial and differential deafferentation and spontaneous discharge in $A\beta$ and $A\delta$ primary afferent neurons. *J. Neurophysiol.*, 68: 734–744.

Kauppila, T., Grönroos, M. and Pertovaara, A. (1995) An attempt to attenuate experimental pain in humans by dextromethorphan, an NMDA receptor antagonist. *Pharmacol. Biochem. Behav.* 52: 641–644.

Kehlet, H. and Dahl, J.B. (1993) The value of "multimodal" or "balanced analgesia" in postoperative pain treatment. *Anesth. Analg.*, 77: 1048–1056.

Kemp, J.A. and Leeson, P.D. (1993) The glycine site of the NMDA receptor – five years on. *Trends Pharmacol. Sci.*, 14: 20–25.

Kilo, S., Forster, C., Geisslinger, G., Brune, K. and Handwerker, H.O. (1995) Inflammatory models of cutaneous hyperalgesia are sensitive to effects of ibuprofen in man. *Pain*, 62: 187–194.

Koltzenburg, M., Lundberg, L.E.R. and Torebjörk, H.E. (1992) Dynamic and static components of mechanical hyperalgesia in human hairy skin. *Pain*, 51: 207–219.

Kristensen, J.D., Svensson, B. and Gordh, T. (1992) The NMDA-receptor antagonist CPP abolishes neurogenic "wind-up pain" after intrathecal administration in humans. *Pain*, 51: 249–253.

Laird, J.M.A. and Bennett, G.J. (1993) An electrophysiological study of dorsal horn neurons in the spinal cord of rats with an experimental peripheral neuropathy. *J. Neurophysiol.*, 69: 2072–2085.

Laird, J.M.A. and Cervero, F. (1990) Tonic descending influences on the receptive field properties of nociceptive dorsal horn neurones. *J. Neurophysiol.*, 62: 854–863.

Laird, J.M.A., Hargreaves, R.J. and Hill, R.G., (1993) Effect of RP 67580, a non-peptide neurokinin$_1$ receptor antagonist, on facilitation of a nociceptive spinal flexion reflex in the rat. *Br. J. Pharmacol.*, 109: 713–718.

Laird, J.M.A., Mason, G.S., Webb, J., Hill, R.G. and Hargreaves, R.J. (1996) Effects of a partial agonist and a full antagonist acting at the glycine site of the NMDA receptor on inflammation-induced mechanical hyperalgesia in rats. *Br. J. Pharmacol.*, 117: 1487–1492.

LaMotte, R.H., Shain, C.N., Simone, D.A. and Tsai, E.-F.P., (1991) Neurogenic hyperalgesia: psychophysical studies of underlying mechanisms. *J. Neurophysiol.*, 66: 190–211.

Lewis, T. (1936) Experiments related to cutaneous hyperalgesia and its spread through somatic nerves. *Clin. Sci.*, 2: 373–423.

Liu, X.-G. and Sandkühler, J. (1995) Long-term potentiation of C-fiber-evoked potentials in the rat spinal dorsal horn is prevented by spinal *N*-methyl-D-aspartic acid receptor blockage. *Neurosci. Lett.*, 191: 43–46.

Maggi, C.A., Patacchini, R., Rovero, P. and Giachetti, A. (1993) Tachykinin receptors and tachykinin receptor antagonists. *Auton. Pharmacol.*, 13: 23–93.

Mao, J., Price, D.D., Hayes, R.L., Lu, J. and Mayer, D.J. (1992) Differential roles of NMDA and non-NMDA receptor activation in induction and maintenance of thermal hyperalgesia in rats with painful peripheral mononeuropathy. *Brain Res.*, 598: 271–278.

Millan, M.J. and Seguin, L. (1993) (+)-HA 966, a partial agonist at the glycine site coupled to NMDA receptors, blocks formalin-induced pain in mice. *Eur. J. Pharmacol.*, 238: 445–447.

Moiniche, S., Dahl, J.B. and Kehlet, H. (1993) Time course of primary and secondary hyperalgesia after heat injury to the skin. *Br. J. Anaesth.*, 71: 201–205.

Neugebauer, V., Schaible, H.-G., Weiretter, F. and Freudenberger, U. (1994) The involvement of substance P and neurokinin-1 receptors in the responses of rat dorsal horn neurons to noxious but not to innocuous mechanical stimuli applied to the knee joint. *Brain Res.*, 666: 207–215.

Pircio, A.W., Fedele, C.T. and Bierwagon, M.E. (1975) A new method for the evaluation of analgesic activity using adjuvant-induced arthritis in the rat. *Eur. J. Pharmacol.*, 31: 207–215.

Portenoy, R.K., Foley, K.M. and Inturrisi, C.E. (1990) The nature of opioid responsiveness and its implications for neuropathic pain: new hypotheses derived from studies of opioid infusions. *Pain*, 43: 273–286.

Pozo, M.A. and Cervero, F. (1993) Neurons in the rat spinal trigeminal complex driven by corneal nociceptors: Receptive-field properties and effects of noxious stimulation of the cornea. *J. Neurophysiol.*, 70: 2370–2378.

Ren, K. and Dubner, R. (1993) NMDA receptor antagonists attenuate mechanical hyperalgesia in rats with unilateral inflammation of the hind-paw. *Neurosci. Lett.*, 163: 19–21.

Rupniak, N.M.J., Boyce, S., Williams, A.R., Cook, G., Longmore, J., Seabrook, G.R., Caeser, M., Iversen, S.D. and Hill, R.G. (1993) Antinociceptive activity of NK_1 receptor antagonists: Non-specific effects of racemic RP67580. *Br. J. Pharmacol.*, 110: 1607–1613.

Rupniak, N.M.J., Webb, J.K., Williams, A.R., Carlson, E., Boyce, S. and Hill, R.G. (1995) Antinociceptive activity of the NK1 receptor antagonist, CP-99,994, in conscious gerbils. *Br. J. Pharmacol.*, 116: 1937–1943.

Schaible, H.-G., Neugebauer, V., Cervero, F. and Schmidt, R.F. (1991) Changes in tonic descending inhibition of spinal neurons with articular input during the development of acute arthritis in the cat. *J. Neurophysiol.*, 66: 1021–1031.

Schouenborg, J. (1984) Functional and topographical properties of field potentials evoked in rat dorsal horn by cutaneous C-fibre stimulation. *J. Physiol. (London)*, 356: 169–192.

Tal, M. and Bennett, G.J. (1993) Dextrorphan relieves neuropathic heat-evoked hyperalgesia in the rat. *Neurosci. Lett.*, 151: 107–110.

Tattersall, J.E.H., Cervero, F. and Lumb. B.M. (1986) Viscero-somatic neurones in the lower thoracic spinal cord of the cat: excitations and inhbitions evoked by splanchnic and somatic nerve volleys and by stimulation of brain stem nuclei. *J. Neurophysiol.* 56: 1411–1423.

Torebjörk, H.E., Lundberg, L.E.R. and LaMotte, R.H. (1992) Central changes in processing of mechanoreceptive input in capsaicin-induced secondary hyperalgesia in humans. *J. Physiol. (London)*, 448: 765–780.

Treede, R.-D., Meyer, R.A., Raja, S.N. and Campbell, J.N. (1992) Peripheral and central mechanisms of cutaneous hyperalgesia. *Prog. Neurobiol.*, 38: 397–421.

Vane, J.R. (1971) Inhibition of prostaglandin synthesis as a mechanism of action for aspirin-like drugs. *Nature*, 231: 232–235.

Vane, J.R. and Botting, R.M. (1995) New insights into the mode of action of anti-inflammatory drugs. *Inflamm. Res.*, 44: 1–10.

Wall, P.D. and Woolf, C.J. (1984) Muscle but not cutaneous

C-afferent input produces prolonged increases in the excitability of the flexion reflex in the rat. *J. Physiol. (London)*, 356: 443–458.

Wong, E.H.F. and Kemp, J.A. (1991) Sites for antagonism on the N-methyl-D-aspartate receptor channel complex. *Annu. Rev. Pharmacol. Toxicol.*, 31: 401–425.

Woolf, C.J. (1989), Recent advances in the pathophysiology of acute pain. *Br. J. Anaesth.*, 63: 139–146.

Woolf, C.J. and Thompson, S.W.N. (1991) The induction and maintenance of central sensitisation is dependent on N-methyl-D-aspartic acid receptor activation; Implications for the treatment of post-injury pain hypersensitivity states. *Pain*, 44: 293–299.

Woolf, C.J. and Chong, M.-S. (1993) Pre-emptive analgesia - treating postoperative pain by preventing the establishment of central sensitisation. *Anesth. Analg.*, 77: 362–379.

Yamamoto, T. and Yaksh, T.L. (1992) Effects of intrathecal capsaicin and an NK-1 antagonist, CP-96,345, on the thermal hyperalgesia observed following unilateral constriction of the sciatic nerve in the rat. *Pain*, 51: 329–334.

Yamamoto, T., Shimoyama, N., Asano, H. and Mizuguchi, T. (1994) Time-dependent effect of morphine and time-independent effect of MK-801, an NMDA antagonist, on the thermal hyperesthesia induced by unilateral constriction injury to the sciatic nerve in the rat. *Anesthesiology*, 80: 1311–1319.

CHAPTER 2

Quantitative experimental assessment of pain and hyperalgesia in animals and underlying neural mechanisms

E. Carstens

Section of Neurobiology, Physiology and Behavior, University of California, Davis, CA 95616 USA

Introduction

It is difficult to assess pain in animals in the absence of a verbal report, and a variety of methods have been used to make deductions about animal pain sensation based on nocifensive responses or spontaneous behavior. Many of these methods measure the latency of an animal's nocifensive behavioral response, such as limb withdrawal or vocalization, following delivery of an acute noxious stimulus. An example is the widely used rodent tail flick reflex. Such methods have been used to show that analgesics increase the nocifensive response latency, and that manipulations causing hyperalgesia shorten the latency.

In addition to animal models that assess acute pain-like responses, several new animal models of persistent neuropathic pain have been recently introduced. These involve partial ligation (Seltzer et al., 1990) or constriction of peripheral (Bennett and Xie, 1988) or spinal nerves (Kim and Chung, 1992). Each model produces its own pattern of altered behavior including abnormal posture, hyperalgesia and allodynia in the affected limb. Abnormalities in limb placement and guarding can be observed and rated. Hyperalgesia and allodynia are quantified in terms of a reduction in latency to withdraw the affected limb after application of a noxious or non-noxious stimulus. That is, methods of assessing acute pain are used to assess hyperalgesia in the chronic pain model.

The primary aim of this paper is to critically examine the use of acute noxious stimuli to assess pain in normal animals and in animal models of persistent pain. Using ideas and approaches stemming from my fruitful collaboration with Professor M. Zimmermann during the time that I was a post-doctoral fellow in his laboratory at the University of Heidelberg, new strategies are suggested to examine nocifensive behavior evoked by controlled stimuli over a wide range of stimulus intensities in assessing underlying neural mechanisms and treatments.

Quantitative animal models of acute pain

For animal models meant to assess acute pain, Wood (1984) has suggested the following criteria as desirable: simplicity, reliability, sensitivity, quantitation, and validity. These elements have been incorporated by Dubner (1985, 1989) into a more extensive list of criteria: (1) distinguish between innocuous versus noxious stimuli, (2) variation of response magnitude with stimulus intensity in the noxious range, (3) allow evaluation of responses over a range of stimulus intensities acceptable to the animal, (4) sensitivity to non-sensory factors such as attention, motivation and motor performance, (5) utilize multiple threshold and suprathreshold measures, and (6) sensitivity to behavioral and pharmacological manipulations that alter the sensory-discriminative capacities of the animal. The importance of the validity of the animal model was emphasized by Watkins (1989), who adds: (g) allow reliable comparison between data attained from humans and laboratory animals. Most of the commonly used tests, such as the tail flick reflex, assess the animal's nocifensive response threshold but do not allow assessment of responses to graded, suprathreshold stimuli across the noxious range. Thus, criteria (2), (3) and (5) of the preceding list are not fulfilled.

The relative absence of simple methods to assess nocifensive behavior over a wide stimulus range has prompted us to develop such methods. The models we have tested most rigorously to date are: (1) magnitude of hind limb withdrawal (assessed by flexor EMG recordings) evoked by graded noxious heating of the plantar surface of the paw in rats (Carstens and Ansley, 1993), (2) the magnitude (force vector) of tail flicks elicited by graded noxious thermal stimulation of the rat's tail (Carstens and Wilson, 1993), and (3) force and speed of an operantly conditioned response (bar press with the nose) to terminate graded noxious thermal stimulation of the rat's tail. Our rationale for choosing reflexes in the first two models is that the magnitude of flexion withdrawal evoked by acute graded noxious electrical or thermal stimuli in humans increases in parallel with magnitude ratings of the accompanying pain sensation (Willer, 1985; Carstens and Campbell, 1988, 1992; Campbell et al., 1991) to yield a linear or positively accelerating stimulus-response function. Noxious heat was chosen as a stimulus because it can be precisely controlled, and it selectively activates nociceptors but does not appreciably excite thermo- or mechanoreceptors as do electrical and mechanical stimuli.

Nocifensive responses assessed in each of the three models above can be characterized as follows:

(1) Response magnitude increases with stimulus intensity from a threshold of approximately 40–43°C. Fig. 1A shows an example of the approximately linear increase in limb withdrawal (flexor EMG) magnitude from 40–50°C, and Fig. 1B illustrates the mean population stimulus-response function. Limb withdrawal magnitude was better correlated with stimulus temperature than was withdrawal latency, supporting the utility of this magnitude measure in pain assessment (Carstens and Ansley, 1993).

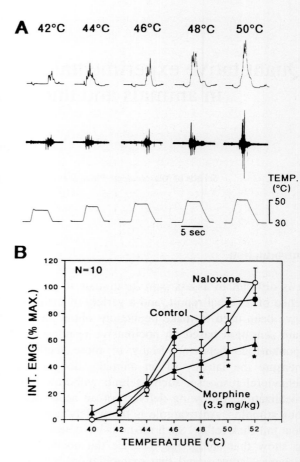

Fig. 1. Measurement of hind limb withdrawal magnitude in conscious rats. (A) EMGs recorded from biceps femoris in response to graded noxious heat stimuli. Columns show raw (middle trace) and integrated EMGs (upper trace) aligned vertically with the evoking heat stimulus (lowest trace; duration 5 s). Inter-stimulus adapting temperature was maintained at 35°C. (B) Stimulus-response function of limb withdrawal magnitude and suppression by morphine. Graph plots mean integrated EMGs versus temperature for 10 rats prior to (●, control) and following systemic administration of morphine (▲). Note the reduction in slope of the stimulus-response function following morphine, and reversal following naloxone (○). Error bars show S.E.M. in this and subsequent figures. From Carstens and Ansley (1993) with permission of the American Physiological Society.

To measure tail flick magnitude, forces of tail movements evoked by radiant tail heating were measured in three orthogonal planes and integrated to yield a vector (Fig. 2A). Fig.

2B shows that the mean tail flick force vector for a population of rats increased with stimulus temperature from 40–48 or 50°C and saturated at higher temperatures (Fig. 2B). However, tail flick force was no better correlated with stimulus temperature than was tail flick latency (Carstens and Wilson, 1993).

To obtain a quantitative measure of a more complex nocifensive behavior, we operantly conditioned rats to push a bar upward with the nose in order to terminate noxious stimuli delivered to the tail. We measured the speed and force of bar-presses evoked by graded noxious radiant tail heating. Fig. 3 illustrates that the mean integrated force of the operant response increased with stimulus temperature over a range that was higher compared to the tail flick reflex. The correlation between stimulus temperature and response force vs. speed was approximately equal (Carstens and Douglass, 1993; Douglass, 1993).

(2) Responses measured in each model were suppressed following systemic morphine in a naloxone-reversible manner (limb withdrawal: Fig. 1B; tail flick: Fig. 2B; operant response: Fig. 3). In each case there was an apparent reduction in the slope of the stimulus-response function. In our tail-flick model, however, a fraction of rats did not exhibit morphine suppression (upper-most curve in Fig. 2B), presumably reflecting the known quantal (all-or-none) nature of opioid suppression of the tail flick reflex (Levine et al., 1980; Yoburn et al., 1985; Carstens and Wilson, 1993). This is manifested as a complete suppression of the tail flick in a fraction of rats, with the remainder exhibiting normal tail flicks, following opiate administration. The incidence of analgesia (i.e., number of animals exhibiting total tail flick suppression) increases with dose of opiate.

(3) Responses in each model are sensitive to electrical stimulation through chronically implanted electrodes in midbrain areas associated with analgesia. Fig. 4A,B shows that stimulation in the medial midbrain periaqueductal

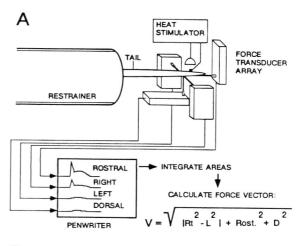

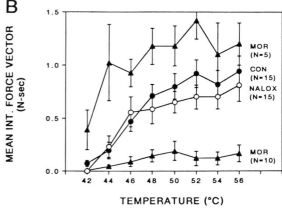

Fig. 2. Measurement of rat tail flick magnitude. (A) Method. The rat's tail is attached to 4 equidistant transducers to measure isometric forces in horizontal, vertical and longitudinal directions for tail movements elicited by noxious radiant heat pulses delivered to the tail. Adapting temperature, 30°C. Areas beneath force measurements were integrated to calculate the resultant vector. (B) Stimulus-response function of tail flick reflex magnitude, and suppression by morphine. Graph plots mean tail flick force vectors versus stimulus temperature prior to (Control, ●) and following systemic administration of 1 or 2 mg/kg morphine (▲) divided into groups showing no effect (upper-most curve) or suppression (lowest curve). Morphine effects were reversed by naloxone (○). From Carstens and Wilson (1993) with permission of the American Physiological Society.

gray (PAG) reduced the slope of the stimulus-response function for limb withdrawals and tail flicks, respectively, while stimulation at more lateral midbrain (LRF) sites shifts stimulus-response functions toward the right (Carstens and Douglass, 1995). These effects presumably reflect different inhibitory actions operating on interneurons in the spinal reflex circuits (Carstens et al., 1979, 1980), an observation that was originally made in collaboration with Prof. M. Zimmermann during my post-doctoral period in his laboratory. Operant responses evoked at all stimulus intensities were also markedly suppressed by midbrain stimulation (Douglass, 1993).

(4) Limb withdrawal and tail flick reflexes frequently decrement over repeated trials in a manner consistent with habituation. (Carstens and Ansley, 1993; Carstens and Wilson, 1993).

These quantitative behavioral models therefore fulfill many of the criteria noted above, particularly reliability of the response, use of threshold and suprathreshold measures over a range acceptable to the animal, and sensitivity to pharmacological modulation.

Neural mechanisms of acute pain and its modulation

An additional advantage of measuring the magnitude of a nocifensive behavioral response is that it can be more directly compared with the magnitude of responses, evoked by identical stimuli, of neurons in circuits that underlie the behavior. An understanding of the neurocircuitry mediating a nocifensive behavior, in turn, is important in developing pharmacological and other strategies to modify neural transmission and hence behavior.

The neural circuits mediating limb withdrawal and tail flick reflexes are located within the spinal cord, since these reflexes persist following cervical or mid-thoracic spinalization (Irwin et al., 1951). These are considered below. The circuitry underlying the operantly conditioned nose-push response is undoubtedly more complex, involving both spinal and supraspinal levels of organization associated with ascending transmission of the nocicep-

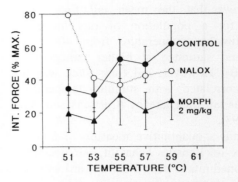

Fig. 3. Measurement of operant response magnitude and suppression by morphine. Rats were trained to push a bar upward with the nose to terminate noxious thermal stimuli delivered to the tail. The speed and force of the bar-press was recorded. Graph plots the mean area beneath force traces versus stimulus temperature prior to (Control, ●) and after systemic administration of morphine (▲). Note the reduction in slope by morphine and reversal following systemic naloxone (○, error bars omitted). From Carstens and Douglass (1993).

tive signal, and perceptual and decision-making processes to bar-press, respectively.

The afferent limb of the hind limb withdrawal reflex arc consists of thermal nociceptors with unmyelinated (C-) or thinly myelinated (A-delta) afferent fibers innervating skin on the plantar surface of the hind paw (Fleischer et al., 1983). These afferents travel in the tibial nerve to enter the spinal cord through dorsal roots L_3-L_6 (Swett and Woolf, 1985) and terminate in the medial superficial dorsal horn. The efferent limb of the reflex arc consists of motoneurons in segments L_3-L_6 innervating ipsilateral leg flexors including biceps femoris (Nicolopoulos-Stournaras and Illes, 1983; Woolf and Swett, 1984). Spinal interneuronal connections between nociceptor afferent terminals and motoneurons are presumed to be polysynaptic (see below).

The afferent limb of the tail flick reflex arc consists of thermal nociceptors innervating the tail (Necker and Hellon, 1978; Fleischer et al., 1983). Their unmyelinated afferents enter the spinal cord at segments S_4-Co_3 to terminate in the superficial dorsal horn (Grossman et al., 1982). The efferent limb consists of motoneurons in segments L_4-Co_3 that ipsilaterally innervate the three

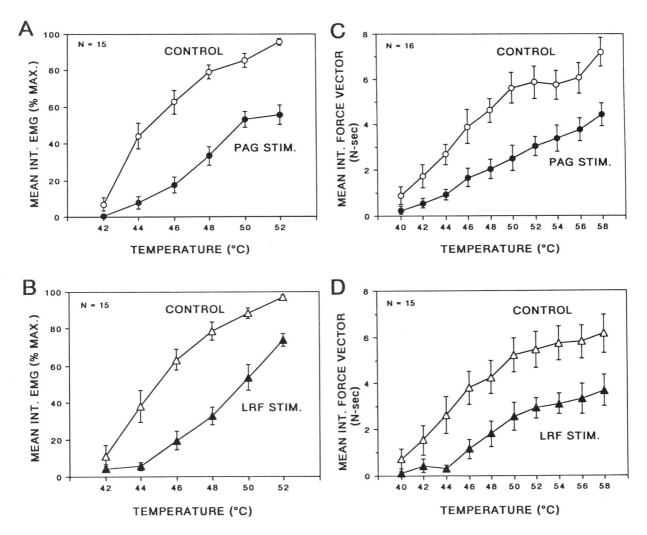

Fig. 4. Midbrain suppression of limb withdrawal and tail flick reflexes. (A) Graph plots mean limb withdrawal magnitude vs. stimulus temperature, without (○, control) and during stimulation in midbrain PAG (●). Note the apparent reduction in slope of the stimulus-response function with PAG stimulation. (B) Graph as in A, without (△) and during stimulation in the lateral midbrain reticular formation (LRF) (▲), which shifted the stimulus-response function toward the right with an increase in threshold. (C,D) Graphs as in (A,B) showing similar effects of PAG (C) and LRF (D) stimulation on mean tail flick force vectors. From Carstens and Douglass (1995) with permission of the American Physiological Society.

sets of back muscles that control tail movements (extensor caudae medialis and lateralis; abductor caudae dorsalis) (Brink and Pfaff, 1980; Grossman et al., 1982; Cargill, 1983). Interneuronal connections between nociceptor terminals in the dorsal horn and motoneurons in the ventral horn are thought to be polysynaptic, but have not been precisely delineated because of the inherent difficulty of tracing multisynaptic connections in the nervous system.

In an attempt to trace interneuronal connections of the tail flick reflex arc, we recently employed the method of retrograde transneuronal transport of pseudorabies virus (PRV) (Rotto-Percelay et al.,

1992). The rationale for this method is that PRV is picked up by synaptic terminals at its injection site and transported retrogradely to the cell body, where it replicates to increase the signal. PRV escaping the soma is picked up, in turn, by terminals synapsing on that cell, and retrogradely transported to the presynaptic neuron's cell body to replicate once more. In this manner, di- and possibly trisynaptic connections can be traced. Since it is an anatomical method, however, the functional nature of the synaptic connection cannot be determined. We found distinct changes in the pattern of PRV-labeled spinal neurons at increasing times after injection of PRV into the abductor caudae dorsalis muscle (Jasmin et al., 1994). After 1–2 days, cells were only in the ipsilateral motoneuronal pool at S_{2-4}, and the intermediolateral cell column of the thoraco-lumbar cord, with a few in the intermediate zone of the sacral cord. After 2–4 days, cells were in lamina I and in the intermediate zone bilaterally throughout the lumbosacral cord (Fig. 5A). After 4–5 days, numerous cells were distributed throughout the dorsal and ventral horn, including the substantia gelatinosa (Fig. 5B). These labeling patterns suggest several possible interneuronal connections that are schematically depicted in Fig. 5C,D. Similar labeling patterns (although with a more ipsilateral preponderance) were seen following injections of PRV into the biceps femoris, suggesting that the limb flexion reflex may have interneuronal connections similar to those shown in Fig. 5C,D.

Considerable evidence indicates that spinal dorsal horn neurons receiving nociceptor afferent input are an important nodal point in the modulation of pain and associated responses. For example, opiate suppression of withdrawal reflexes is thought to be mediated by inhibition of interneurons, rather than motoneurons or peripheral nociceptors. To establish the neural mechanisms underlying reflex modulation, much of our work has focused on candidate reflex interneurons in the dorsal horn. Such interneurons should be excited by the noxious stimulus in a manner preceding and overlapping the reflex. However, to actually demonstrate that a neuron polysynaptically excites a flexor motoneuron pool is difficult. We can only assume that some fraction of nociceptive dorsal horn neurons provide excitatory input to flexor motoneurons and therefore serve as interneurons in the reflex pathway.

Using this rationale, we have recorded in barbiturate-anesthetized rats from dorsal horn neurons in lumbar and sacral segments that respond to noxious heating of the hind paw and tail, respectively. A majority of lumbar dorsal horn neurons with cutaneous input from the hind paw generally give linearly increasing responses to graded noxious thermal stimuli from 40–52°C (Carstens and Watkins, 1986). In experiments simultaneously recording from a single dorsal horn unit, a single biceps femoris motor unit, and hind limb withdrawal force, dorsal horn unit responses to noxious heat preceded and overlapped those of the motor units and the reflex (Carstens and Campbell, 1992) consistent with an interneuronal role.

Similarly, a majority of sacral dorsal horn neurons with cutaneous tail input also responded to graded heating of tail skin in a manner that usually increased linearly from 40 up to 48–52°C and then plateaued at higher temperatures, much like tail flick force (Carstens and Douglass, 1995). An example is shown in Fig. 6.

Based on our behavioral results noted above, and on the descending modulatory effects operating on spinal dorsal horn neurons that were initially detailed in collaboration with Professor M. Zimmermann (Carstens et al., 1979, 1980), we were interested in determining if the sacral neurons were also under similar descending influences. We found, indeed, that stimulus-response functions of sacral dorsal horn neuronal responses to graded tail heating are similarly modulated, i.e., PAG stimulation reduced the slope, while LRF stimulation shifted the function in a parallel manner toward the right. An example of the effect of PAG stimulation is shown in Fig. 6. In a prior behavioral experiment using the same rat shown in Fig. 6, identical PAG stimulation also reduced the slope of the stimulus-response function for tail flick force. Furthermore, the slope of the population stimulus-response function of sacral neurons was reduced following systemic morphine (1 and 2 mg/kg)

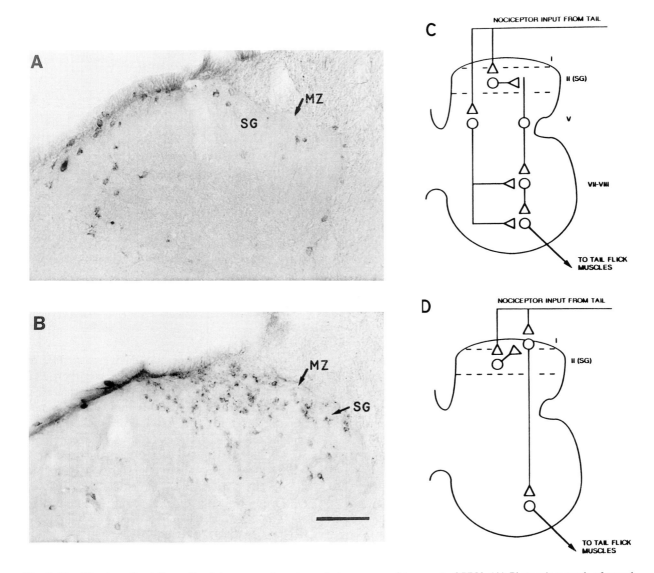

Fig. 5. Identification of putative reflex interneurons by retrograde transneuronal transport of PRV. (A) Photomicrograph of sacral dorsal horn 3 days following injection of PRV into the abductor caudae dorsalis (ACD) muscle. Note the presence of granular PRV-immunoreactive cells in the marginal zone (MZ; lamina I) but not in the substantia gelatinosa (SG; lamina II). (B) Photomicrograph of sacral dorsal horn 5 days after PRV injection into ACD. Note numerous PRV-immunoreactive cells in SG. Calibration bar: 100 μm. (C,D) Schematic drawings of spinal gray matter to show possible interneuronal pathways connecting tail nociceptor afferents to motoneurons supplying tail flick muscles.

(Douglass and Carstens, 1993) similar to our behavioral results (Fig. 2B). However, in a fraction of the sacral neurons noxious heat-evoked responses were not affected by morphine, consistent with the quantal effect of morphine on the tail-flick reflex mentioned earlier.

The main conclusion from these studies is that the responses of dorsal horn neurons, and their

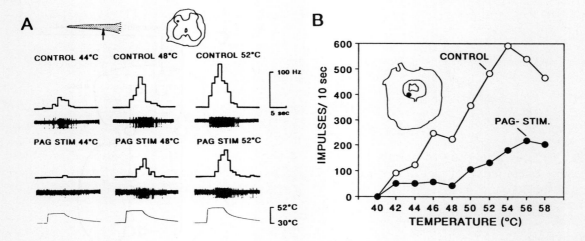

Fig. 6. Example of sacral dorsal horn neuronal responses to graded noxious heating of the rat's tail. (A) Columns show paired oscilloscope trace with peristimulus-time histogram above (bin width 1 s), of a dorsal horn neuron's response to a noxious radiant heat stimulus (lowest trace), without (control, upper pair) and during PAG stimulation (lower pair). Heat stimulus was delivered to the site indicated by the arrow in the upper left inset drawing of the tail. The recording site is indicated by the dot in the right-hand drawing of a section through the sacral spinal cord. (B) Graph plots responses of neuron from A versus stimulus temperature, without (○, control) and during PAG stimulation (●) at site indicated by dot on inset drawing of midbrain section. PAG stimulation reduced the slope of the linear part of the stimulus-response function.

modulation by analgesic brain stimulation or opiates, closely parallel the behavioral limb withdrawal and tail flick reflexes evoked by identical heat stimuli. Assuming that some of the dorsal horn neurons function as reflex interneurons, modulation of nocifensive reflexes can thus be accounted for largely by modulation of interneurons in the reflex circuit.

Animal models of persistent neuropathic pain

In this section, the argument will be presented that the quantitative models of acute pain described above may be useful in assessing hyperalgesia in animal models of persistent pain.

The animal models of persistent neuropathic pain mentioned at the outset have two features in common: hyperalgesia (increased pain evoked by a noxious stimulus) and allodynia (pain evoked by a normally innocuous stimulus). Hyperalgesia would be manifested as an alteration in the slope and/or threshold of an animal's stimulus-response function for nocifensive responses to graded noxious stimuli. The response function could be shifted in a parallel manner toward the left so that a given stimulus evokes a larger nocifensive response, or its slope could be reduced. In both cases there is a decrease in threshold stimulus intensity; that is, a previously ineffective stimulus now evokes a nocifensive response (allodynia). Allodynia and hyperalgesia might conceivably have pharmacologically distinct underlying neural mechanisms.

In most instances, hyperalgesia in neuropathic pain models is assessed by a decrease in latency of nocifensive responses such as limb withdrawal to heat. However, such latency measures would not reveal enhanced suprathreshold responses. It would therefore be desirable to use methods to assess nocifensive behavior over a range of stimulus intensities, as described above, to determine if suprathreshold as well as threshold responses are altered in animal models of neuropathic pain.

We have recently employed the Seltzer model (Seltzer et al., 1990), in which one-third to one-half of the rat's sciatic nerve is ligated on one side to create a neuropathy characterized by hind paw eversion, adduction of toes, decreased latency of limb withdrawal to heat, and reduced mechanical withdrawal threshold (Takaishi et al., 1996). Two to 4 weeks after nerve ligation, we tested neuropathic and sham-operated control rats in the limb withdrawal paradigm described earlier. The paradigm was modified so that the Peltier heating thermode, which contacted the plantar surface of the extended hind paw, was attached to a low-inertia lever system to directly measure the isometric force of attempted limb withdrawals. This method yields stimulus-response functions for paw withdrawal force as a function of stimulus temperature. There were no significant differences in slope or threshold (approximately 40°C) between ligated and unoperated sides of the neuropathic rats, or between neuropathic and sham-operated rats. This negative finding was surprising, since we observed a significant decrease in paw withdrawal latency to radiant heat on the ligated side using the method of Hargreaves et al. (1988). One possible explanation for this apparent discrepancy is that the partial sciatic nerve ligation may have damaged motor axons to flexor muscles, in addition to increasing the excitability of spinal nociceptive neurons. Thus, transmission over the spinal reflex pathway for limb withdrawals might be potentiated to result in a decreased withdrawal latency, but withdrawal force would be reduced because of paresis due to motor axon damage. Nonetheless, the reduced withdrawal latency should reflect a decrease in threshold (Ness and Gebhart, 1986) but this was not seen in our force measurement. A second possibility is that the affected hindpaw may have an elevated temperature which would also lower the withdrawal threshold to radiant heat (Hole and Tjolsen, 1993). In the withdrawal force paradigm, this problem was circumvented because we maintained a constant inter-stimulus adapting temperature of 35°C.

We also conducted electrophysiological experiments to determine if the responsiveness of lumbar dorsal horn neurons to mechanical and noxious thermal stimuli was altered in the neuropathic rats (Takaishi et al., 1996). The thermal responsiveness was unchanged, in agreement with a previous study of neurons recorded from rats receiving a loose ligation compression injury of the sciatic nerve (Laird and Bennett, 1993). We found cutaneous mechanosensitive receptive fields to be larger on the ligated compared to non-ligated side, also in agreement with a recent study of nociceptive dorsal horn neurons in rats with a partial sciatic nerve ligation (Behbehani and Dollberg-Stolik, 1994).

Our magnitude measure of limb withdrawal may therefore prove to be useful in assessing hyperalgesia in animal models of persistent pain. However, it is important to determine why this model did not reveal a hyperalgesia in neuropathic rats, while other measures of nocifensive response latency did.

Cellular mechanisms of hyperalgesia

Hyperalgesia that ensues following nerve damage, or in normal tissue surrounding an injury (secondary hyperalgesia), is thought to be mediated by an increase in the excitability of spinal nociceptive neurons (LaMotte et al., 1991; Simone et al., 1991; Dubner and Ruda, 1992; Willis, 1992; Coderre et al., 1993). Thus, after a skin injury nociceptors within the region of secondary hyper-

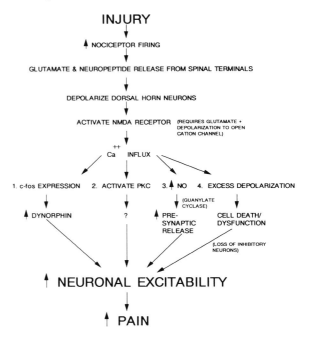

Fig. 7. Schematic diagram of the sequence of events that might lead to the development of hyperalgesia following tissue damage.

algesia do not become sensitized (Baumann et al., 1991; LaMotte et al., 1992). In addition, intraneural electrical stimulation of individual mechanoreceptor afferents, which normally evokes an innocuous tactile sensation referred to the cutaneous receptive field, will subsequently evoke pain if the receptive field lies within a region of secondary hyperalgesia that develops following intradermal injection of capsaicin (Torebjork et al., 1992).

Much recent evidence indicates that injury-induced sensitization of spinal nociceptive neurons is mediated by the sequence of events outlined in Fig. 7. A noxious stimulus excites nociceptors which release neuropeptides and excitatory amino acids (particularly glutamate) from their intraspinal terminals to excite dorsal horn neurons. The neuropeptides bind postsynaptic neurokinin receptor sub-types, and glutamate binds at least three postsynaptic receptor sub-types, including the N-methyl-D-aspartate (NMDA) receptor. This receptor appears to be the essential trigger for persistent increases in excitability of dorsal horn neurons. Thus, pretreatment with antagonists of the NMDA receptor or ion channel prevents: the increase in flexor reflex excitability following C-fiber stimulation (Woolf and Thompson, 1991); reduction in tail-flick reflex latency following intrathecal NMDA (Kolhekar et al., 1993); formalin-induced nocifensive responses (Coderre and Melzack, 1992; Vaccarino et al., 1993); development of neuropathic pain-like behavior following nerve constriction (Mao et al., 1993); and windup (Davies and Lodge, 1987; Dickenson and Sullivan, 1987), expansion of receptive fields (Ren et al., 1992) and development of hyperresponsiveness following intradermal capsaicin (Dougherty et al., 1992) or formalin (Haley et al., 1990) in spinal neurons.

The NMDA receptor requires both glutamate and sufficient postsynaptic depolarization (via neurokinin and non-NMDA receptors) to remove a Mg^{2+} block from its ion channel, thus allowing cation fluxes. Of particular importance is Ca^{2+} influx, which has several intracellular actions leading to a persistent increase in cell excitability including the following (Fig. 7). (1) Gene induction, leading to increased expression of neuromediators such as dynorphin (Noguchi et al., 1991; Dubner and Ruda, 1992) which may have an excitatory effect on dorsal horn neurons (Knox and Dickenson, 1987; Hylden et al., 1991) possibly via release from interneurons. (b) Activation of protein kinase C (PKC) to effect unspecified cellular changes associated with hyperexcitability. Thus, the increase in membrane-bound PKC levels, and development of neuropathic pain, following sciatic nerve constriction were prevented by GM1 gangliosides which interfere with PKC activation and translocation (Coderre, 1992; Mao et al., 1992, 1993; Price et al., 1994). (c) Activation of nitric oxide synthase (NOS), leading to increased production of nitric oxide (NO) (Meller and Gebhart, 1993) which can diffuse freely across cell membranes for short distances. NO is thought to bind to guanylate cyclase to increase intracellular cGMP. One possibility is that NO diffuses back into the presynaptic nociceptor terminal to enhance neurotransmitter release. Blocking NO production prevents NMDA-evoked hyperalgesia (Meller et al., 1992; Kolhekar et al., 1993; Malmberg and Yaksh, 1993) and prolonged formalin-evoked excitation of dorsal horn neurons (Haley et al., 1992). (d) Excitotoxicity from excessive glutamate release, causing dysfunction or death of dorsal horn neurons (Sugimoto et al., 1990; Dubner, 1991). If some of these normally function as inhibitory interneurons (Laird and Bennett, 1992), their loss could lead to a generalized increase in dorsal horn excitability via disinhibition.

Within this framework, we were interested to determine if our model of hind limb withdrawal magnitude is appropriate for studying mechanisms of hyperalgesia. We specifically tested the hypothesis that intrathecal NMDA enhances limb withdrawal magnitude, possibly by reducing the slope of the stimulus-response function, via the NMDA receptor and increased NO production. In one paradigm we recorded an initial control stimulus-response function of limb withdrawals evoked by a series of graded heat stimuli, followed 20 min later by a second heat series in which NMDA was administered intrathecally prior to each heat stimulus (Fig. 8A). Withdrawal magnitude was enhanced by intrathecal NMDA so as to reduce the slope and threshold of the population stimulus-response function (Fig. 8B). Surprisingly, control intrathecal injections of saline were found to have a similar albeit lesser effect. This suggests the possibility of a peripheral sensitization, triggered by the first heat series, that might confound any effect of NMDA in the second heat series. In a second paradigm employing only a single series of graded heat stimuli, NMDA lowered the threshold of the stimulus-response function (compared to a control group) in a manner that was prevented by intrathecal NMDA antagonists (APV, MK801) and the alternate substrate for NOS (L-NAME). These latter results are in agreement with previous studies cited above that relied on latency measures of nocifensive responses, and indicate that the quantitative animal models described here may prove useful in investigating pharmacological mechanisms of hyperalgesia.

Perspectives, interpretations and speculations

A major theme of this paper is that pain sensation in animals should ideally be inferred using quantitative measures of nocifensive behavior over a range of stimulus intensities from threshold to near-tolerance levels. To what extent can this be achieved using the approaches described here?

Two of our approaches utilized quantitative measures of withdrawal reflex magnitude. While nociceptive withdrawal reflexes commonly accompany pain sensation, they can occur in the absence of sensation (e.g., anesthesia, spinal transection) or can be absent in the presence of pain sensation (neuromuscular blockade). We found that midbrain stimulation suppressed limb withdrawals more effectively than simultaneously recorded dorsal horn neuronal responses to noxious heat (Carstens and Campbell, 1992). Because the reflex is a motor act, it is dangerous to infer that some manipulation suppressing the nociceptive reflex in a normal animal is "analgesic" without controlling for general motor inhibition. It is not inconceivable that reflex interneurons and ascending pain-transmitting (e.g., spinothalamic tract) neurons constitute separate populations (Jasmin et al., 1994) whose responses could be "uncoupled" under certain conditions. For example, it would be adaptive to suppress pain from injury yet maintain full motor function during escape behavior, whereas it would be more adaptive to suppress motor function (and also pain sensation?) in species that exhibit death feigning in the presence of a predator or during recuperation from injury. Normally, however, dorsal horn neurons and with-

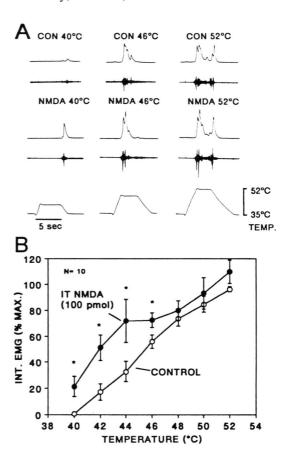

Fig. 8. Facilitation of hind limb withdrawal by intrathecal (IT) NMDA. (A) Individual example. Columns show pairs of raw and integrated EMGs aligned with the heat stimulus. Upper pair: control response; lower pair: enhanced response after IT NMDA. (B) Graph plotting mean integrated EMGs versus stimulus temperature, without (○, control) and following IT NMDA (●) Control responses to each temperature were recorded at 2 min intervals. The same stimulus series was then repeated, except that 4μ of IT NMDA was microinjected 15 s prior to each heat stimulus. Note facilitation by NMDA at lower temperatures.

drawal reflexes appear to be tightly coupled and subject to virtually identical modulatory influences. Thus, while caution should be exercised in extrapolating from reflexes to pain sensation, the accessibility of the spinal reflex circuitry nonetheless affords the distinct advantage of analyzing cellular and molecular mechanisms at the origin of the reflex behavior.

In animal models of chronic pain, hyperreflexia can be interpreted as increased pain sensation. Using our limb withdrawal model, however, withdrawal force was unaltered following partial sciatic nerve ligation even though paw withdrawal latency to radiant heat was reduced. As mentioned earlier, this may reflect conflicting consequences of the nerve injury: sensitization of spinal nociceptive processing, and paresis due to motor axon damage. It would be interesting to determine if similar results are obtained with other models of neuropathic pain (Bennett and Xie, 1988; Kim and Chung, 1992) which should also damage motor axons. It might also be worthwhile to develop a model that does not damage motor fibers, in an attempt to distinguish sensory from motor aspects of the neuropathy.

Since pain is normally a conscious perception, a more realistic animal model might incorporate a cognitive-evaluative component associated with nocifensive behavior. For this reason, we developed the operant response model that requires the rat to perceive and respond to a noxious stimulus with a bar-press. It was interesting that rats responded over a higher range of stimulus intensities in this paradigm compared to our measure of tail flick reflex force. One might speculate that the operant response is a measure of the motivation of the animal to work to avoid the noxious stimuli, or of the probability that the animal will or will not tolerate it. On the other hand, in human studies using identical noxious radiant heat stimuli, the threshold for "reflexive" limb withdrawal was higher compared to that for pain sensation (Campbell et al., 1991). However, the fact that subjects knew the stimuli would not cause tissue damage may have influenced their withdrawal threshold. These observations further underscore the need for caution in extrapolating from measures of nocifensive response to the actual pain sensation that an animal experiences.

Acknowledgments

I wish to express my gratitude to Professor Dr. M. Zimmermann for the friendship, interest, support and intellectual input that he has provided throughout my scientific career.

References

Baumann, T.K., Simone, D.A., Shain, C.N. and LaMotte, R.H. (1991) Neurogenic hyperalgesia: the search for primary cutaneous afferent fibers that contribute to capsaicin-induced pain and hyperalgesia. *J. Neurophysiol.*, 66: 212–227.

Behbehani, M.M. and Dollberg-Stolik, O. (1994) Partial sciatic nerve ligation results in an enlargement of the receptive field and enhancement of the response of dorsal horn neurons to noxious stimulation by an adenosine agonist. *Pain*, 58: 421–428.

Bennett, G.J. and Xie, Y.K. (1988) A peripheral mononeuropathy in rats that produces disorders of pain sensation like those seen in man. *Pain*, 33: 87–107.

Brink, E.E. and Pfaff, D.W. (1980) Vertebral muscles of the back and tail of the albino rat (*rattus norvegicus albinus*). *Brain Behav. Evol.*, 17: 1–47.

Campbell, I.G., Carstens, E. and Watkins, L.R. (1991) Comparison of human pain sensation and flexion withdrawal evoked by noxious radiant heat. *Pain*, 45: 259–268.

Cargill, C.L. (1983) *The Tail Flick Reflex in the Rat* (Master's thesis). University of Texas Medical Branch, Galveston, TX.

Carstens, E. and Watkins, L.R. (1986) Inhibition of the responses of neurons in the rat spinal cord to noxious skin heating by stimulation in midbrain periaqueductal gray or lateral reticular formation. *Brain Res.*, 382: 266–277.

Carstens, E. and Campbell, I.G. (1988) Parametric and pharmacological studies of midbrain suppression of the hind limb flexion withdrawal reflex in the rat. *Pain*, 33: 201–213.

Carstens, E. and Campbell, I.G. (1992) Responses of motor units during the hind limb flexion withdrawal reflex evoked by noxious skin heating: phasic and prolonged suppression by midbrain stimulation and comparison with simultaneously recorded dorsal horn units. *Pain*, 48: 215–226.

Carstens, E. and Ansley, D. (1993) Hindlimb flexion withdrawal evoked by noxious heat in conscious rats: magnitude measurement of stimulus-response function, suppression by morphine, and habituation. *J. Neurophysiol.*, 70: 621–629.

Carstens, E. and Douglass, D.K. (1993) Effect of morphine on the magnitude of an operant response to noxious thermal stimuli. *Abstr. 7th World Cong. Pain*, p. 204.

Carstens, E. and Wilson, C. (1993) Rat tail-flick reflex: magnitude measurement of stimulus-response function, suppression by morphine, and habituation. *J. Neurophysiol.*, 70: 630–639.

Carstens, E. and Douglass, D.K. (1995) Midbrain suppression of limb withdrawal and tail flick reflexes in the rat: correlates with descending inhibition of sacral spinal neurons. *J. Neurophysiol.*, 73: 2179–2194.

Carstens, E., Yokota, T. and Zimmermann, M. (1979) Inhibition of spinal neuronal responses to noxious skin heating by stimulation of the mesencephalic periaqueductal gray in the cat. *J. Neurophysiol.*, 42: 558–568.

Carstens, E., Klumpp, D. and Zimmermann, M. (1980) Differential inhibitory effects of medial and lateral midbrain stimulation on spinal neuronal discharges to noxious skin heating in the cat. *J. Neurophysiol.*, 43: 332–342.

Coderre, T.J. (1992) Contribution of protein kinase C to central sensitization and persistent pain following tissue injury. *Neurosci. Lett.*, 140: 181–184.

Coderre, T.J. and Melzack, R. (1992) The contribution of excitatory amino acids to central sensitization and persistent nociception after formalin-induced tissue injury. *J. Neurosci.*, 12: 3665–3670.

Coderre, T.J., Katz, J., Vaccarino, A.L. and Melzack, R. (1993) Contribution of central neuroplasticity to pathological pain: review of clinical and experimental evidence. *Pain*, 52: 259–285.

Davies, S.N. and Lodge, D. (1987) Evidence for involvement of N-methyl-D-aspartate receptors in "windup" of class 2 neurones in the dorsal horn of the rat. *Brain Res.*, 424: 402–406.

Dickenson, A.H. and Sullivan, A.F. (1987) Evidence for a role of the NMDA receptor in the frequency dependent potentiation of deep rat dorsal horn nociceptive neurones following C-fibre stimulation. *Neuropharmacology*, 26: 1235–1238.

Dougherty, P.M., Palecek, J., Paleckova, V., Sorkin, L.S. and Willis, W.D. (1992) The role of NMDA and non-NMDA excitatory amino acid receptors in the excitation of primate spinothalamic tract neurons by mechanical, chemical, thermal and electrical stimuli. *J. Neurosci.*, 12: 3025–3041.

Douglass, D.K. (1993) An operant response model for evaluating nocifensive behavior in rats: effects of morphine, diazepam, and midbrain stimulation and activity of the underlying dorsal horn neurons (Doctoral dissertation). University of California, Davis, CA.

Douglass, D.K. and Carstens, E. (1993) Effect of morphine on spinal dorsal horn neurons responding to noxious thermal stimulation of the rat's tail. *Abstr. 7th World Cong. Pain*, p. 204.

Dubner, R. (1985) Specialization in nociceptive pathways: sensory discrimination, sensory modulation and neural connectivity. *Adv. Pain Res. Ther.*, 9: 111–137.

Dubner, R. (1989) Methods of assessing pain in animals. In: P.D. Wall and R. Melzack (Eds.), *Textbook of Pain*, Churchill Livingstone, London, pp. 247–256.

Dubner, R. (1991) Neuronal plasticity and pain following peripheral tissue inflammation or nerve injury. In: M.R. Bond and C.J. Woolf (Eds.), *Proc. 5th World Cong. Pain*, pp. 263–276.

Dubner, R. and Ruda, M.A. (1992) Activity-dependent neuronal plasticity following tissue injury and inflammation. *Trends Neurosci.*, 15: 96–103.

Fleischer, E., Handwerker, H.O. and Joukhadar, S. (1983) Unmyelinated nociceptive units in two skin areas of the rat. *Brain Res.*, 267: 81–92.

Grossman, M.L., Basbaum, A.I. and Fields, H.L. (1982) Afferent and efferent connections in the rat tail flick reflex (a model used to analyze pain control mechanisms). *J. Comp. Neurol.*, 206: 9–16.

Haley, J.E., Sullivan, A.F. and Dickenson, A.H. (1990) Evidence for spinal N-methyl-D-aspartate receptor involvement in prolonged chemical nociception in the rat. *Brain Res.*, 518: 218–226.

Haley, J.E., Dickenson, A.H. and Schachter, M. (1992) Electrophysiological evidence for a role of nitric oxide in prolonged chemical nociception in the rat. *Neuropharmacology*, 31: 251–258.

Hargreaves, K., Dubner, R., Brown, F., Flores, C. and Joris, J.A. (1988) A new and sensitive method for measuring thermal nociception in cutaneous hyperalgesia. *Pain*, 32: 77–88.

Hole, K. and Tjolsen, A. (1993) The tail-flick and formalin tests in rodents: changes in skin temperature as a confounding factor. *Pain*, 53: 247–254.

Hylden, L.K., Nahin, R.L. Traub, R.J. and Dubner, R. (1991) Effects of spinal kappa-opioid receptor agonists on the responsiveness of nociceptive superficial dorsal horn neurons. *Pain*, 44: 187–193.

Irwin, S., Houde, R.W., Bennett, D.R., Hendershot, L.C. and Seevers, M.H. (1951) The effects of morphine, methadone and meperidine on some reflex responses of spinal animals to nociceptive stimulation. *J. Pharmacol. Exp. Ther.*, 101: 132–143.

Jasmin, L., Mendel, V.E., Carstens, E. and Basbaum, A.I. (1994) Interneurons presynaptic to rat tail-flick motoneurons as mapped by transneuronal transport of pseudorabies virus: few have long ascending collaterals. *Soc. Neurosci. Abstr.*, 20: 547.

Kim, S.H. and Chung, J.M. (1992) An experimental model for peripheral neuropathy produced by segmental spinal nerve ligation in the rat. *Pain*, 50: 355–363.

Knox, R.J. and Dickenson, A.H. (1987) Effects of selective and non-selective kappa-opioid receptor agonists on cutaneous C-fibre-evoked responses of rat dorsal horn neurones. *Brain Res.*, 415: 21–29.

Kolhekar, R., Meller, S.T. and Gebhart, G.F. (1993) Characterization of the role of spinal NMDA receptors in thermal nociception in the rat. *Neuroscience*, 57: 385–395.

Laird, J.M.A. and Bennett, G.J. (1992) Dorsal root potentials and afferent input to the spinal cord in rats with an experimental peripheral neuropathy. *Brain Res.*, 584: 181–190.

Laird, J.M.A. and Bennett, G.J. (1993) An electrophysiological study of dorsal horn neurons in the spinal cord of rats with an experimental peripheral neuropathy. *J. Neurophysiol.*, 69: 2072–2085.

LaMotte, R.H., Shain, C.N., Simone, D.A., and Tsai, E.-F.P. (1991) Neurogenic hyperalgesia: psychophysical studies of underlying mechanisms. *J. Neurophysiol.*, 66: 190–211.

LaMotte, R.H., Lundberg, L.E.R. and Torebjork, H.E. (1992) Pain, hyperalgesia and activity in nociceptive C units in humans after intradermal injection of capsaicin. *J. Physiol.*, 448: 749–764.

Levine, J.D., Murphy, D.T., Seidenwurm, D., Cortez, A., and Fields, H.L. (1980) A study of the quantal (all-or-none) change in reflex latency produced by opiate analgesics. *Brain Res.*, 201: 129–141.

Malmberg, A.B. and Yaksh, T.L. (1993) Spinal nitric oxide synthase inhibition blocks NMDA induced thermal hyperalgesia and produces antinociception in the formalin test in rats. *Pain*, 54: 291–300.

Mao, J., Price, D.D., Mayer, D.J. and Hayes, R.L. (1992) Pain-related increases in spinal cord membrane-bound protein kinase C following peripheral nerve injury. *Brain Res.*, 588: 144–149.

Mao, J., Mayer, D.J., Hayes, R.L. and Price, D.D. (1993) Spatial patterns of increased spinal cord membrane-bound protein kinase C and their relation to increases in 14C-2-deoxyglucose metabolic activity in rats with painful peripheral mononeuropathy *J. Neurophysiol.*, 70: 470–481.

Meller, S.T. and Gebhart, G.F. (1993) Nitric oxide (NO) and nociceptive processing in the spinal cord. *Pain*, 52: 127–136.

Meller, S.T., Dykstra, C. and Gebhart, G.F. (1992) Production of endogenous nitric oxide and activation of soluble guanylate cyclase are required for N-methyl-D-aspartate-produced facilitation of the nociceptive tail-flick reflex. *Eur. J. Pharmacol.*, 214: 93–96.

Neckar, R. and Hellon, R.F. (1978) Noxious thermal input from the rat tail: modulation by descending inhibitory influences. *Pain*, 4: 231–248.

Ness, T.J. and Gebhart, G.F. (1986) Centrifugal modulation of the rat tail flick reflex evoked by graded noxious heating of the tail. *Brain Res.*, 386: 41–52.

Nicolopoulos-Stournaras, S. and Illes, J.F. (1983) Motor neuron columns in the lumbar spinal cord of the rat. *J. Comp. Neurol.*, 217: 75–85.

Noguchi, K., Kowalski, K., Traub, R., Solodkin, A., Iadorola, M.J. and Ruda, M.A. (1991) Dynorphin expression and Fos-like immunoreactivity following inflammation induced hyperalgesia are colocalized in spinal cord neurons. *Mol. Brain Res.*, 10: 227–233.

Price, D.D., J. Mao, J. and Mayer, D.J. (1994) Central neural mechanisms of normal and abnormal pain states. In: H.L. Fields and J.C. Liebeskind (Eds.), *Progress in Pain Research and Management*, Vol. 1, IASP Press, Seattle, WA, pp. 61–84.

Ren, K., Hylden, J.L.K., Williams, G.M., Ruda, M.A. and Dubner, R. (1992) The effects of a non-competitive NMDA receptor antagonist, MK-801, on behavioral hyperalgesia and dorsal horn neuronal activity in rats with unilateral inflammation. *Pain*, 50: 331–344.

Rotto-Percelay, D.M., Wheeler, J.G., Osorio, F.A., Platt, K.B., and Loewy, A.D. (1992) Transneuronal labeling of spinal interneurons and sympathetic preganglionic neurons after pseudorabies virus injections in the rat medial gastrocnemius muscle. *Brain Res.*, 574: 291–306.

Seltzer, Z., Dubner, R. and Shir, Y. (1990) A novel behavioral model of neuropathic pain disorders produced in rats by partial sciatic nerve injury. *Pain*, 43: 205–218.

Simone, D.A., Sorkin, L.S., Oh, U., Chung, J.M., Owens, C., LaMotte, R.H. and Willis, W.D. (1991) Neurogenic hyperalgesia: central neural correlates in responses of spinothalamic tract neurons. *J. Neurophysiol.*, 66: 228–246.

Sugimoto, T., Bennett, G.J. and Kajander, K.C. (1992) Transsynaptic degeneration in the superficial dorsal horn after sciatic injury: effects of chronic constriction injury, transection and strychnine. *Pain*, 42: 205–213.

Swett, J.E. and Woolf, C.J. (1985) The somatotopic organization of primary afferent terminals in the superficial laminae of the dorsal horn of the rat spinal cord. *J. Comp. Neurol.*, 231: 66–77.

Takaishi, K., Eisele, Jr., H. and Carstens, E. (1996) Behavioral and electrophysiological assessment of hyperalgesia and changes in dorsal horn responses following partial sciatic nerve ligation in rats. *Pain*, in press.

Torebjörk, H.E., Lundberg, L.E.R. and LaMotte, R.H. (1992) Central changes in processing of mechanoreceptive input in capsaicin-induced secondary hyperalgesia in humans. *J. Physiol. (London)*, 448: 765–780.

Vaccarino, A.L., Marek, P., Kest, B., Weber, E., Keanam, J.F.W. and Liebeskind, J.C. (1993) NMDA receptor antagonists MK-801 and ACEA-1011, prevent the development of tonic pain following subcutaneous formalin. *Brain Res.*, 615: 331–334.

Watkins, L.R. (1989) Algesiometry in laboratory animals and man: current concepts and future directions. In: R.C. Chapman and J.D. Loeser (Eds.), *Adv. Pain Res. Ther.*, Vol. 12, *Issues in Pain Measurement*, Raven, New York, pp. 249–266.

Willer, J.C. (1985) Studies on pain: effects of morphine on a spinal nociceptive flexion reflex and related pain sensation in man. *Brain Res.*, 331: 105–114.

Willis, W.D. (1992) *Hyperalgesia and Allodynia*. Raven, New York, 400 pp.

Wood, P.L. (1984) Animal models in analgesic testing. In: M. Kuhar and G. Pasternak (Eds.), *Analgesics: Neurochemical, Behavioral and Clinical Perspectives*, Raven, New York, p. 175.

Woolf, C.J. and Swett, J.E. (1984) The cutaneous contribution to the hamstring flexor reflex in the rat: an electrophysiological and anatomical study. *Brain Res.*, 303: 299–312.

Woolf, D.J. and Thompson, S.W. (1991) The induction and maintenance of central sensitization is dependent on *N*-methyl-D-aspartic acid receptor activation: implications for the treatment of post-injury pain hypersensitivity states. *Pain*, 44: 293–299.

Yoburn, B.C., Cohen, A., Umans, J.G., Ling, G.S.F., and Inturrisi, C.E. (1985) The graded and quantal nature of opioid analgesia in the rat tailflick assay. *Brain Res.*, 331: 327–336.

CHAPTER 3

Nociceptive, environmental and neuroendocrine factors determining pain behaviour in animals

Anna Maria Aloisi and Giancarlo Carli

Istituto di Fisiologia Umana, Università degli Studi di Siena, Via Aldo Moro, I-53100 Siena, Italy

Introduction

Behaviour is not governed simply by reflex responses to immediate needs but depends upon complex factors, which include habits, incentives and rewards, learning and experience. Pain is a complex, multidimensional experience determined by the interactions among somatosensory, affective, motivational and social factors. In humans all these factors are integrated at the cognitive level and the experience of pain depends upon the manner in which the subject construes and interprets the events in the external and internal environments (Chapman and Turner, 1990). It is conceivable that a multitude of variables determines the behavioural expression of pain in animals, e.g. states of the brain, stress and situational factors of the experiments.

This article is an attempt to consider pain as an integrated rather than deterministic process. Woolf (1991) defines clinical (pathological) pain as being characterized by tissue damage, allodynia, hyperalgesia, persistent pain and referred pain. However these typical symptoms are not isolated but associated with a variety of modifications in other systems that mediate functions necessary for coordinated behaviour of the whole organism. We have adopted the formalin test in rats as a model of acute, persistent nociceptive stimulation to investigate three main topics: (a) the importance of the intensity of nociceptive stimuli and of the site of their application, not only on the typical responses to formalin (licking, flexing, jerking) but also on the standard measures of activity; (b) the interactions between some internal drives, such as hunger or the motivation to explore a new environment, and the need of the animal to react to the nociceptive stimulation and (c) the involvement of central state circuits and of the neuroendocrine system as indicated by the modifications of some physiological parameters, such as central and peripheral β-endorphin (β-EP), plasma ACTH and hippocampal choline acetyltransferase (ChAT) activity.

The description of our experiments will be preceded by a short summary of the main characteristics of formalin pain.

The formalin test

Pain induced by the formalin test is considered to be a reliable model for the study of acute, persistent pain conditions in animals and the physiological mechanisms underlying them (Tjolsen et al., 1992; Carli and Aloisi, 1993; Porro and Cavazzuti, 1993). The test is usually performed in freely-moving, unanaesthetized animals which are briefly restrained only at the moment of the injection. Different body regions have been selected as injection sites: face (Clavelou et al., 1989), shoulder (Takahashi et al., 1984) and paw (Dubuisson and Dennis, 1977). Each site displays typical pain-evoked behavioural responses that can be identified and categorized according to their temporal parameters (frequency and duration) and can be used to infer the intensity of pain experienced by the animal. The typical responses to formalin when injected into the paw are jerking, flexing and licking of the injected limb. Jerking consists of phasic, rapid flexion movements of the injected paw or even of the hind quarters. Usually it represents the first pain-evoked response. Jerking can be considered a spinal reflex since it has been observed also in spinalized animals (Coderre et al., 1994). Its frequency increases just before the occurrence of a licking episode (Aloisi, unpublished observations). Tonic flexion of the injected limb (flexing) represents the typical response to a sustained nociceptive input. Finally licking of the injection site consists of a series of coordinated, rhythmic movements involving the tongue, the head, the trunk and the injected limb. As summarized by Walters (1994), licking is a protective behaviour as it cleans the wound and the antimicrobial factors in the saliva accelerate wound healing. While flexing and jerking of the injected limb can be considered versions of the same spinal withdrawal reflex, licking requires the integration of spinal and supraspinal centres. In conclusion, in formalin-induced pain there is a simultaneous occurrence of signs of avoidance behaviour, such as flexing and jerking, and signs of recuperative behaviour, such as licking.

Different methods have been used to quantify the intensity of formalin-induced pain. Formerly, Dubuisson and Dennis (1977) introduced a rating scale based on weighted records of the durations of licking and flexing and this method has been validated recently by Coderre et al. (1993). Other authors have analyzed only licking duration or only jerking frequency (see Wheeler-Aceto and Cowan, 1991; Tjolsen et al., 1992). In recent years we have adopted the analysis of all pain-evoked responses (licking, flexing and jerking), but considered separately. This is based on the observation that the temporal aspects (latency, duration, time course) of each response may display different distributions (see later and Fig. 1).

Licking, flexing and jerking appear immediately after the formalin injection, reach a peak within 1–2 min, and then rapidly decrease (first phase). This phase is followed, after 10–15 min, by a resumption of all pain-evoked responses (second phase), which may reach even higher levels than in the first phase. During the second phase, there are also episodes of licking of the contralateral limb, although they are less frequent than licking of the ipsilateral, injected limb (Aloisi et al., 1993c). Finally, it has been demonstrated recently, by the tail flick test, that during the formalin test there is a prolonged hyperalgesia to heat stimuli. This extends beyond the body region in which the irritant has been injected and is supraspinally mediated (Wiertelak et al., 1994). All these characteristics of formalin-induced pain fulfill the criteria of pathological pain as defined by Woolf (1991). Electrophysiological recordings in unanaesthetized animals from afferent Aδ and C fibres have shown that, following formalin injection, nociceptors become sensitized and their pattern of activation follows a biphasic pattern which parallels the pain-evoked responses recorded in freely-moving animals (Heapy et al., 1987; Klemm et al., 1989). At the dorsal horn level, wide-dynamic-range neurons also display a biphasic pattern of increased discharges with a particular enhancement during the second phase (Dickenson and Sullivan, 1987).

The formalin test has been adopted to demonstrate in freely-moving animals the occurrence of plastic modifications in the central nervous system (Skilling et al., 1988; Coderre and Melzack, 1992a) previously described in anaesthetized preparations (Woolf and Thompson, 1991). There is evidence that prolonged electrical stimulation of Aδ and C fibres elicits the release of both substance P (Go and Yaksh, 1987) and glutamate (Skilling et al., 1988) and activates the NMDA receptors of spinal dorsal horn neurons (MacDermott et al., 1986). This results in a large influx of calcium ions, thereby activating both prostanoid and NO which diffuse out of the cell and may enhance the excitability of surrounding neurons through a variety of pre- and post-synaptic mechanisms (Wilcox, 1990; Malmberg and Yaksh, 1992; Meller and Gebhart, 1993). In the formalin test, it was found that there is a release of glutamate (Skilling et al., 1988) and of substance P (Coderre and Melzack, 1992a) from primary afferents. In addition, the application prior to formalin injection of antagonists of NMDA, AMPA, substance P, and prostaglandin E, as well as inhibitors of NO and voltage-sensitive calcium channels, does not affect the first phase but strongly reduces or abolishes the second phase of the pain-evoked reactions (Coderre and Melzack, 1992a,b; Yamamoto and Yaksh, 1992). This confirms the hypothesis that the release of substance P and of glutamate in the dorsal horn during the first phase activates the NMDA-mediated process of the second phase. All these studies reveal long-term modifications elicited by persistent nociceptive stimuli at the synapses and in the neural networks responsible for the codified transmission along sensory pathways. The persistent state of enhanced excitability in the dorsal horn seems to be crucial for the establishment of pathological pain and does not occur following acute, phasic nociceptive stimuli.

Effects of different formalin concentrations

In these experiments we studied the effects of three different formalin concentrations on the frequency and duration of pain-evoked responses and on standard measures of activity in rats. The separate analysis of licking, flexing and jerking reveals that the time courses of these behaviours are different for the same formalin concentration. For instance (Fig. 1, left), in animals treated with 50 μl of a 5 or 10% formalin concentration, licking duration gradually decreases after 40 min from the injection, while flexing and jerking remain elevated during the last part of the test. Cumulative analysis of the first (initial 10 min) and the second (the remaining 50 min) phases showed that (Fig. 1, right), in the animals treated with 5 and 10% formalin, there was no difference in either licking or flexing duration, which were both significantly ($P < 0.02$) higher than in the group treated with 0.1% formalin. As for jerking frequency (Fig. 1 right), during the first phase, only the injection of the highest concentration of formalin elicited a significant increase; during the second phase, the increase in formalin concentration elicited a corresponding significant increase in jerking frequency. Ceiling effects or even reductions in licking or rubbing duration have been observed by others (Rosland et al., 1990; Wheeler-Aceto and Cowan, 1991; Clavelou et al., 1995), thus confirming that a separate analysis of all pain-evoked responses is important.

Coderre and Melzack (1992a,b) noted that the hyperalgesic effects of intrathecal amino acid or calcium ionophore appeared when 1%, but not 5%, formalin was used. It would be interesting to know whether hyperalgesia (Rahman et al., 1994; Wiertelak et al., 1994) and all the related increases in excitability of dorsal horn neurons (Coderre and Melzack, 1992a,b; Yamamoto and Yaksh, 1992) occur also with the formalin concentrations used in our experiments.

An important observation of our experiments is that infliction of pain usually affects animal behaviour, and that the direction and magnitude of the modifications depend upon the formalin concentration. In one series of experiments (Aloisi et al., 1995a), we studied the effects induced by two different formalin concentrations (0.1%; 10%) on standard measures of activity by testing male rats in an open-field device (a transparent plexiglass box; 50 × 50 × 40 cm) for a 50-min period beginning 10 min after the injection (i.e. during the second phase of formalin pain). For the higher concentration (10%), rearing frequency (number of times the animal explored by lifting the fore paws) and exploratory duration (time spent exploring the environment) were reduced with respect to control animals, while locomotion (time spent walking around in the open-field) was inhibited to the point that it was virtually absent during the last 10–15 min of the test, when animals displayed sleeping-like posture. In contrast, the lower formalin concentration (0.1%) appeared to induce a general activation of behaviour, as suggested by the absence of sleeping-like episodes and a higher frequency of rearing (Aloisi et al., 1995a). The different arousal state of the two formalin-injected groups was confirmed also by the behavioural response to a new object (a small plastic cylinder) placed in the open-field at the end of the formalin test: only the animals receiving the lower formalin concentration displayed a high number of approaches (the animal moves and touches the object). These results are in agreement with a previ-

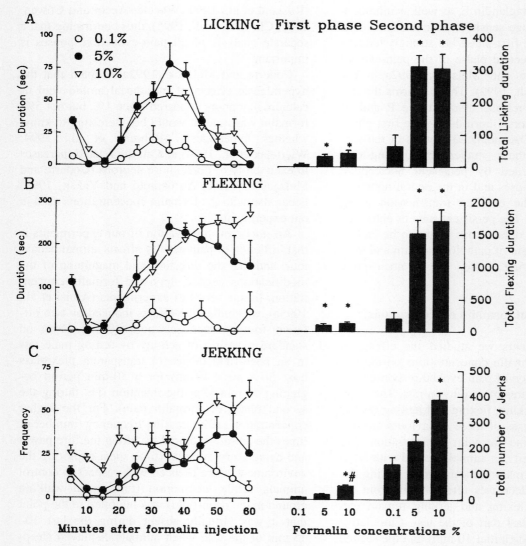

Fig. 1. Time course of nociceptive responses following s.c. injection of different concentrations of formalin (50 μl, 0.1, 5 and 10%) into the hind paw ($n = 8$ rats each group). Left: Formalin test: 60 min, values are the mean ± SEM of the behaviour during 5-min blocks. (A) Licking and (B) flexing duration; (C) frequency of jerking of the paw. Right: cumulative responses; (A,B) total time and (C) total number, of the pain-evoked responses recorded during the first (initial 10 min) and second (10–60 min) phases of the formalin test. *$P < 0.05$ versus 0.1% formalin group; #$P < 0.05$ versus 5% formalin group.

ous experiment in rabbits with a formalin concentration (100 μl, 5%) that induced both low-intensity short-lasting pain-evoked reactions and enhanced locomotion and exploratory durations in the open-field (Aloisi et al., 1993b). In conclusion, our results indicate that the two different concentrations of formalin elicit opposite effects on several behavioural parameters.

Role of the site of formalin injection

Most studies on formalin-induced pain (Tjolsen et al., 1992; Porro and Cavazzuti, 1993) do not address the question of injections in limb areas different from the paw. In one publication, nociceptive responses of rats to formalin injection in the subcutaneous tissue were shown to be maximal in the plantar side of the paw and to decrease when injections were made in the dorsum of the paw or in the wrist (Neoh, 1993).

It was the aim of our study to clarify further the issue of the site of the noxious stimulation. We used subcutaneous formalin injections (50 μl, 10%) into the dorsum of the hind paw and the lateral aspect of the thigh. Pain reactions were much more vigorous following injection into the paw (Fig. 2). As for the thigh, licking occurred sporadically at extremely low levels, while flexing was almost absent, except at low levels immediately after the injection. Jerking became substantial only 20 min after the injection. In contrast, the levels of locomotion and exploration were the same in both groups (Aloisi et al., 1995d).

Although these results suggest that the site of injection is critical for the expression of pain responses, it should not be inferred that the amount of pain perceived by the animal is different. In the case of formalin injection into the thigh, it is important to take into account that all the responses remain at extremely low levels throughout the test. This could be related to the difference in nociceptor density in the two sites. With regard to licking, the short duration could also be related to its reduced efficacy since in the thigh there is a thick layer of hairs which protects the surface of the skin.

Formalin test and food-hoarding behaviour

Animal behaviour depends upon many internal and external factors. When an animal is simultaneously exposed to different stimuli, the appearance of one behavioural "schema" may be associated with the inhibition of others.

Many animal species hoard food by carrying it either to their home-cage or to other protected places (Wolfe, 1939). In common laboratory rats, food-hoarding can be observed by providing a source of food at some distance, in an alley connected to the home-cage (Wishart et al., 1969; Whishow, 1990, 1993). In animals kept on a restrictive diet for some days, hoarding is associated with and/or followed by eating of the food pellets.

It was the aim of our experiments to study the

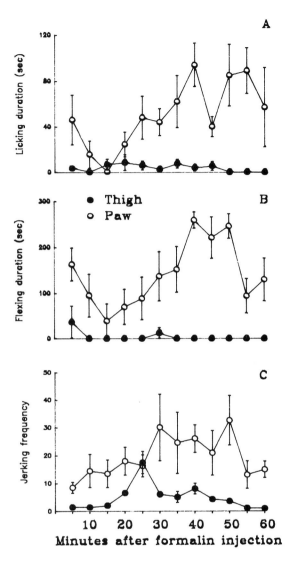

Fig. 2. Time course of nociceptive responses recorded in two groups of rats ($n = 8$ each) following formalin injection (50 μl, 10%) into subcutaneous tissue at two different sites. (A) Licking and (B) flexing durations; (C) jerking frequency. Formalin test (60 min): values are mean ± SEM recorded during 5-min blocks.

interactions among three different drives: exploratory activity, food-hoarding and pain avoidance (Aloisi and Carli, 1996).

The experimental apparatus consists of a home-cage (30 × 30 × 30 cm) connected by a guillotine door to an alley (50 cm), at the end of which ani-

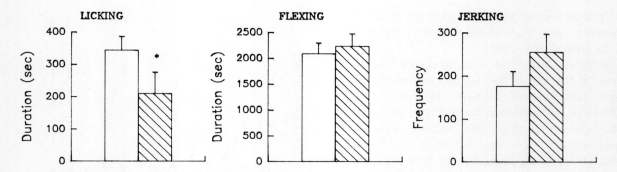

Fig. 3. Pain-evoked responses (licking and flexing duration, jerking frequency) recorded in food-deprived rats exposed to a food-hoarding apparatus for 60 min. All rats received a s.c. formalin injection (50 µl, 10%) immediately before the beginning of the test. One group ($n = 6$, hatched bars) was allowed to hoard food pellets, while the other group ($n = 6$, empty bars) did not receive food pellets. *$P < 0.05$, student's t-test.

mals allowed to hoard found a certain number of food pellets. Before the beginning of the experiments the animals were kept on a restricted diet (to 85% body weight) for one week. The time spent exploring the cage, eating the food pellets, licking and flexing the injected paw, and the frequency of jerking, were recorded during the test (60 min). The amount of pellets hoarded, i.e. transported from the alley to the home-cage, and/or eaten were measured at the end of the test. Three groups of animals were used, one of which was tested in the apparatus in the absence of food pellets. On the day of testing, the two groups allowed to perform food-hoarding were either formalin-injected (50 µl, 10%) or simply pricked with a syringe needle (sham-injected) in the dorsum of the hind paw; the third group, not allowed to hoard pellets, was also injected with formalin. In the animals treated with formalin, the availability of food reduced the licking duration but did not affect flexing and jerking (Fig. 3). In the same animals the amount of pellets eaten and the time spent eating were not affected by formalin-induced pain, while self-grooming and inactivity displayed shorter durations. Both the sham-injected and formalin-injected animals carried into the home-cage a number of pellets that exceeded the amount necessary for immediate consumption. Results indicate that, in animals in which the motivation to carry and eat food was increased by food deprivation, persistent nociceptive stimuli do not affect hoarding behaviour. The reduction in licking, but not in flexing and jerking, in animals allowed to perform food-hoarding can be explained by the simple hypothesis of response competition: the mouth has to be used either to lick the injected paw or to eat and to transport food pellets, while jerking and flexing can be performed at the same time as eating.

Formalin test and novelty

Exposing an animal to a new environment, such as a hole-board device (box with some holes in the floor), results in a series of exploratory behaviours (Terry, 1979). The time spent moving across the cage, the number of times the animal dips the head into the holes, and the frequency of rearing are well known indices of activity, which usually show higher scores during the initial minutes. The procedure of familiarization consists of keeping the animal in the hole-board for a certain period (10–20 min) for some days. During these exposures, the exploratory behaviours display a progressive decrease (Terry, 1979).

The formalin test was formerly used to investigate the effect of exposure to a novel environment on the behavioural response to persistent pain (Abbott et al., 1986). The results showed that the cumulative index of pain scores was reduced and that serotonin was involved in this decrease.

In recent experiments (Aloisi et al., 1994, 1995b), we used the formalin test to characterize further the interaction between novelty and pain. For this purpose, we analyzed both exploratory

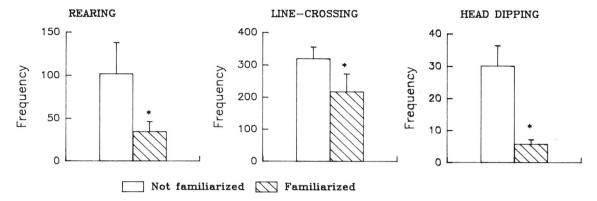

Fig. 4. Standard measures of activity (frequency of rearing, total number of line-crossings and of head dips) recorded during the formalin test (60 min) in rats familiarized and not familiarized with the open-field apparatus. Same animals as in Fig. 4. For all measures, the two groups differ significantly. *$P < 0.05$, student's t-test.

behaviours and all pain-evoked responses (licking, flexing and jerking) by providing separate records and not a cumulative index of pain. Rats both familiar and unfamiliar with the hole-board apparatus were used. The procedure of familiarization that we adopted consists of keeping the animal in the hole-board for 15 min on each of the three days prior to the day of testing.

Fig. 4 shows that, in formalin-injected animals, all the locomotor and exploratory parameters are lower in familiar animals than in the unfamiliar ones.

As regards the pain-evoked responses, familiarization elicited a significant increase, both in the first and second phases of formalin-induced behaviours, in flexing duration and jerking frequency, while licking duration was not affected (Fig. 5). In familiarized (but not in non-familiarized) animals, multiple regression analysis revealed negative correlations between jerking and rearing: $r = -0.81$, $P < 0.001$, and between jerking and line-crossing: $r = -0.77$, $P < 0.01$ (Aloisi et al., unpublished data). Besides analgesia, other factors related to novelty might have contributed to the decrease in jerking frequency and flexing duration: since exploratory activity is greater in animals unfamiliar with the test apparatus, a competition between the motivation to explore the unknown environment and the need to react to the painful stimulus could also be responsible for the selective effects on pain reactions.

Formalin test and central β-endorphin

Acute noxious conditions are known to induce activation of central opioid systems. Experiments concerning the effect of systemic or intracerebroventricular (i.c.v.) administration of an opioid receptor antagonist, naloxone, on responses evoked by formalin pain had led to conflicting results (North, 1978; Kocher, 1988). We have investigated the role played by β-EP during the formalin test. We first determined (Porro et al., 1991) the β-EP-like immunoreactivity levels (assayed by RIA) in discrete brain regions, both in controls and in rats injected with formalin. The results showed that there was an increase in this endogenous opioid in the ventro-medial hypothalamus, ventro-basal thalamus and periaqueductal gray matter 30, 60 and 120 min after formalin-injection. Our results have been confirmed by Facchinetti et al. (1992) by chromatography, which allowed the measurement of purified β-EP.

In another series of experiments, animals were i.c.v. injected with an anti-β-EP serum. Following such treatment, there was an increase in licking duration during the second phase of the formalin test (10–50 min) but no change during the first phase (Porro et al., 1991). This differential sensitivity to endogenous opiates in the first and the second phase is consistent with the finding that it takes time for β-EP to be released (Watkins and Mayer, 1982). Moreover, flexing duration was not modified by the anti-β-EP serum. These effects on flexing could be due to a ceiling effect already present in the control group. However, in contrast to licking, flexing has a predominant spinal mechanism. Hence the selective effect of β-EP on licking duration could be due to a major role of brainstem and limbic structures on opioid modulation of persistent pain. It would be interesting to know if β-EP effects are related to inhibitory mechanisms acting on the transmission of the nociceptive information to supraspinal structures and/or to selective inhibition of the motor output.

Formalin test and hormonal modifications

It is now well established that several hormones, in addition to producing their endocrine effects, participate centrally in the modulation of behaviour (Meyerson, 1979; Meyer and Meyer, 1993). There is evidence that ACTH and β-EP play important roles in the regulation of locomotor exploratory activities (Hoffman et al., 1991; Spanagel et al., 1991; Olson et al., 1993; Versteeg et al., 1993), whereas corticosterone, through its preferential binding in the hippocampal formation, plays an important role in the "behavioral inhibition system" (McEwen et al., 1986; Papolos et al., 1993). These hormones are released in response to aversive stimuli, including nociceptive ones (Rossier et al., 1977; De Souza and Van Loon, 1985; Harbuz and Lightman, 1992). Cytokines are released at the site of inflammation and exert their action both locally and centrally via the circulation. In particular, cytokines interact with the hypothalamic-pituitary-adrenal axis and with some structures of the central nervous system to participate in the modulation of animal behaviour (De Sarro et al., 1990; Plata-Salaman, 1991; Payne and Krueger, 1992; Gadient and Otten, 1993).

We have already shown that, in rats, two formalin concentrations (0.1% and 10%) produce opposite effects on standard measures of activity (Aloisi et al., 1995a). It was the aim of our research to evaluate the relationships between behaviour and the plasma levels of ACTH, β-EP, corticosterone and interleukin-6 (IL-6) in rats. The same two formalin concentrations were used

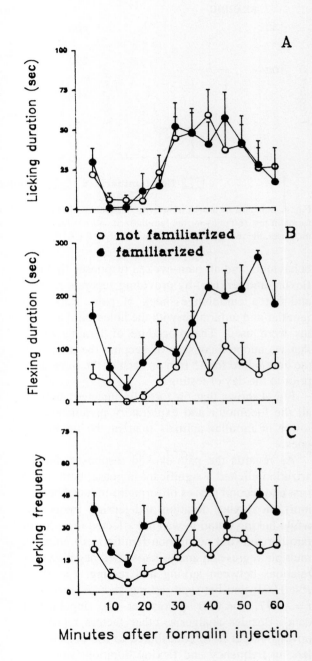

Fig. 5. Modifications of pain-evoked responses (A) licking and (B) flexing duration; (C) jerking frequency, recorded in rats familiarized ($n = 8$) and not familiarized ($n = 8$) with the open-field apparatus and treated with formalin ($50 \mu l$, 10%).

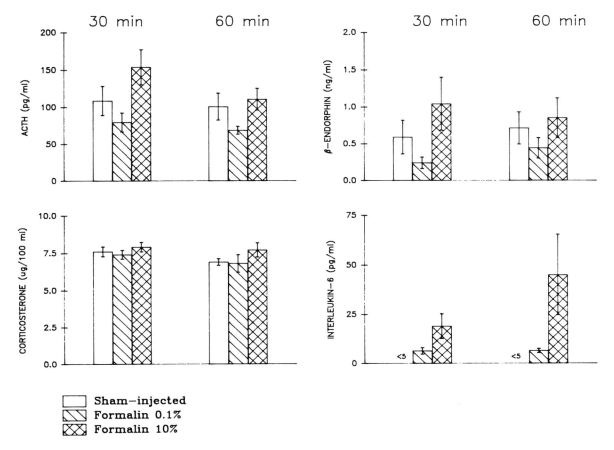

Fig. 6. Modifications of ACTH, β-EP, corticosterone and interleukin-6 plasma levels in sham- and formalin-injected (50 μl, 0.1% and 10%) animals. Rats (n = 8 each group) were sacrificed 30 and 60 min after treatment. Data are means ± SEM.

in animals previously familiarized with the environment and the experimenters (Aloisi et al., 1995c).

In confirmation of our previous experiments, (Aloisi et al., 1995a) all pain-evoked responses were greater in the group with the higher formalin concentration. Fig. 6 shows that, 30 min after formalin injection, ACTH plasma levels were increased only by the higher formalin concentration, while the lower concentration elicited a significant decrease in comparison with sham-injected animals. A similar trend was seen for β-EP (Fig. 6). IL-6, which could not be detected in controls, increased from min 30 to min 60 only in the animals receiving the higher formalin concentration. Interestingly, several negative correlations (Spearman-Rank test) were found in animals of the lower concentration group sacrificed 30 min after formalin injection: β-EP/rearing ($r = -0.80$, $P < 0.01$), β-EP/exploration ($r = -0.72$, $P < 0.004$); and in those of the higher concentration group sacrificed 60 min after the injection: IL-6/locomotion ($r = -0.75$, $P < 0.03$) and IL-6/licking ($r = -0.78$, $P < 0.02$).

There is evidence that administration of small doses of β-EP and of ACTH induces an increase in locomotor exploratory activities, while high doses induce opposite effects (Isaacson and Green, 1978; Fontani et al., 1987). In the present experiments (Aloisi et al., 1995c), the increase and the decrease found in the same exploratory parameters with the lower and the higher formalin concentrations, respectively, could be attributed to the different levels of β-EP and ACTH plasma levels.

Our observations suggest that rats are not stressed following injection of the lower formalin concentration. However, the ACTH levels suggest that even the higher concentration does not induce a prolonged stress. Indeed 60 min after formalin injection, although the pain reactions were still present, ACTH had already returned to basal levels (Fig. 6) and corticosterone was at the basal level both at 30 and 60 min after treatment. It seems that, in a situation in which rats are familiarized with the test environment, even the higher formalin concentration is not able to induce a typical stress response.

The significance of the reduction in the release of both ACTH and β-EP in response to mild nociceptive stimuli is not immediately clear since it has never been described in animal pain models. However, in studies of social interactions, Dijkstra et al. (1992) and Kaiser and van Holst (1993) found lower than control levels of ACTH and β-EP in dominant animals, and higher than control levels in subordinate animals. Since in animal groups the dominant animal controls social interactions much more extensively than subordinates, we suggest that a decrease, instead of an increase, of ACTH and β-EP release would occur in animals coping with mild stimuli that are perceived as controllable.

The increase in IL-6 plasma levels could be responsible for the modulation of the pain-evoked responses. In the present experiments (Aloisi et al., 1995c), as well as in previous ones (see Fig. 1), we have shown that, in the group treated with the higher formalin concentration, licking reached the highest levels between 30 and 40 min and then progressively decreased. Since IL-6 was negatively correlated with licking at min 60 in the same group of animals, the decrease in licking might be mediated by the progressive release into the circulation of IL-6 (Kishimoto et al., 1992), as suggested by the dose-dependent analgesia elicited by the application of IL-6 to inflamed tissue (Czlonkowsky et al., 1993). However, central nervous system effects cannot be excluded since these substances are known to facilitate the occurrence of light sleep and to delay REM sleep through their receptors in the CNS (Payne and Krueger, 1992). The high levels of circulating IL-6 during the second part of the test, following injection of the higher concentration of formalin, could have induced the sleep posture or sleep attempts that occurred during that period.

More in general, our results support previous observations concerning the relationship between inflammation, circulating cytokines, illness and pain (Watkins et al., 1994; Payne and Krueger, 1992).

Formalin test and hippocampal ChAT activity

The hippocampal formation, which plays a crucial role in learning and memory, is involved also in the regulation of several behavioural aspects concerning the adaptation to aversive situations and pain (Melzack and Casey, 1968). Electrophysiological studies have shown that septo-hippocampal neurons are driven either by mechanical or by thermal noxious stimulation (Dutar et al., 1985; Khanna and Sinclair, 1992). Moreover, injection of a local anaesthetic into the hippocampus results in a sustained analgesia in the second phase of the formalin test (McKenna and Melzack, 1992; Vaccarino and Melzack, 1992). The enzyme choline acetyltransferase (ChAT), which is the catalyst in the synthesis of acetylcholine in the cholinergic nerve terminals, is considered an indicator of presynaptic acetylcholine biosynthetic capacity (Fisher and Hanin, 1980). The presence of ChAT in the hippocampal formation is related to its localization in the cholinergic terminal fibres of the septo-hippocampal pathway (Kuhar et al., 1973). There is evidence (Fatranska et al., 1987) that ChAT activity may be increased by some environmental manipulations, such as 24 h of cold stress.

It was the aim of our experiments to detect modifications in hippocampal ChAT activity following injection of two different concentrations of formalin (50 μl, 0.1 and 10%). Fig. 7 shows that, 30 min after formalin injection, ChAT activity had decreased in the animals receiving both the low and high formalin concentrations (Aloisi et al.,

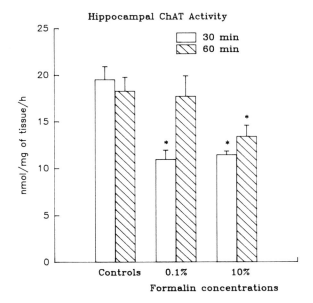

Fig. 7. Effect of two concentrations of formalin (50 μl, 0.1%, 10%) on the choline acetyltransferase (ChAT) activity in the hippocampus in rats sacrificed 30 or 60 min after treatment ($n = 8$ each group). *$P < 0.05$ versus controls, student's t-test.

1993a). After 1 h, however, while the animals receiving the lower concentration had recovered basal levels, the group with the higher concentration still displayed a lower ChAT activity than the controls (Fig. 7). In addition, in the 10% formalin group, ChAT was positively correlated with licking ($r = 0.68$, $P < 0.01$).

We have advanced the hypothesis that the reduction in ChAT activity is related to the activation of the opioid system elicited by formalin-induced excitation of nociceptors (Aloisi et al., 1993a). This hypothesis is supported by several findings: (a) there is an opioid pathway from the hypothalamus to the hippocampus (Beauvillan et al., 1983); (b) the β-EP system is activated by formalin-induced pain (Porro et al., 1991); (c) iontophoretic application of β-EP depresses the turnover of acetylcholine in cholinergic nerve terminals (Moroni et al., 1977); (d) there are large amounts of μ and κ receptors in the hippocampus (Monsour et al., 1988) and (e) naloxone, injected into the medial septal nucleus, enhances the release of acetylcholine in the hippocampus (Nishimura et al., 1992). In conclusion, our results suggest that hippocampal ChAT activity is reduced by persistent pain and that the duration of the effect is related to the intensity of peripheral nociceptive stimulation.

Concluding remarks

Our experiments provide some findings which may be critical for an understanding of the mechanisms of pathological pain. Nociceptive stimuli may have ambivalent effects on animal behaviour since at low levels they elicit activation and at high levels inhibition of behavioural activity. The first effect might be useful for detection of the source of the injury and the context in which the injury has occurred (Bolles and Fanselow, 1980), while inactivity could facilitate recuperative behaviour (Wall, 1979). These coping reactions are associated with hypothalamic-pituitary system responses (ACTH, β-EP) which have opposite directions according to the intensity of the noxious stimulus. It is commonly believed that nociceptive stimuli which produce tissue lesion are usually associated with inflammation, pain and stress. In the formalin test, various influences, such as familiarization with the experimenter and test apparatus and the freely-moving condition, might allow the animal to cope with the situation without the activation of 'typical' stress mechanisms. More in general, our results suggest that the central β-EP system, the septo-hippocampal ChAT activity and the hypothalamic-pituitary system are sequentially involved by persistent nociceptive stimuli.

The multiparametric recording of animal behaviour in all its aspects provides a complete picture of the animal's reactivity and the mechanisms involved. In particular, when experimental manipulations reduce the magnitude of some selected pain responses while leaving the others unmodified, influences other than analgesia or hypoalgesia should be taken into consideration.

Acknowledgements

The authors are grateful to Dr. Peter Christie for English revision. Experiments were supported by

40% and 60% grants from the Ministero Ricerca Scientifica e Tecnologica (MURST).

References

Abbott, F.V., Franklin, K.B.J. and Connell, B. (1986) The stress of a novel environment reduces formalin pain: possible role of serotonin. *Eur. J. Pharmacol.*, 126: 141–144.

Aloisi, A.M. and Carli, G. (1996) Formalin pain does not modify food-hoarding behaviour in male rats. *Behav. Proc.*, 36: 125–133.

Aloisi, A.M., Albonetti, M.E., Lodi, L., Lupo, C. and Carli G. (1993a) Decrease of hippocampal choline acetyltransferase activity induced by formalin pain. *Brain Res.*, 629: 167–170.

Aloisi, A.M., Lupo, C. and Carli, G. (1993b) Effects of formalin-induced pain on exploratory behaviour in rabbits. *NeuroReport*, 4: 739–742.

Aloisi, A.M., Porro, C.A., Cavazzuti, M., Baraldi, P. and Carli, G. (1993c) 'Mirror pain' in the formalin test: behavioral and 2-deoxyglucose studies. *Pain*, 55: 267–273.

Aloisi, A.M., Albonetti, M.E. and Carli, G. (1994) Sex differences in the behavioural response to persistent pain in rats. *Neurosci. Lett.*, 179: 79–82.

Aloisi, A.M., Albonetti, M.E. and Carli G. (1995a) Behavioural effects of different intensities of formalin pain in rats. *Physiol. Behav.*, 58: 603–610.

Aloisi, A.M., Albonetti, M.E. and Carli G. (1995b) Effects of formalin pain on standard hole-board parameters in male and female rats. *Med. Sci. Res.*, 23: 95–96.

Aloisi, A.M., Albonetti, M.E., Muscettola, M., Facchinetti, F., Tanganelli, C. and Carli, G. (1995c) Effects of formalin-induced pain on ACTH, beta-endorphin, corticosterone and interleukin-6 plasma levels in rats. *Neuroendocrinology*, 62: 13–18.

Aloisi, A.M., Decchi, B. and Carli, G. (1995d) Importance of the site of injection in the formalin test. *Med. Sci. Res.*, 23: 601–602.

Beauvillan, J.C., Poulain, P. and Tramu, G. (1983) Immunocytochemical localization of enkephalin in the lateral septum of the guinea pig brain. A light and electron-microscope study. *Cell Tissue Res.*, 228: 265–276.

Bolles, R.C. and Fanselow, M.S. (1980) A perceptual-defensive-recuperative model of fear and pain. *Behav. Brain Sci.*, 3: 291–323.

Carli, G. and Aloisi, A.M. (1993) Integrated complex responses following tonic pain. In: L. Vecchiet, D. Albe-Fessard, U. Lindblom and M.A. Giamberardino (Eds.), *New Trends in Referred Pain and Hyperalgesia*, Elsevier, Amsterdam, pp. 223–238.

Chapman, C.R. and Turner, J.A. (1990) Psychologic and psychosocial aspects of acute pain. In: J.J. Bonica (Ed.), *The Management of Pain*, 2nd edn., Vol. 1, Lea and Febiger, London, pp. 122–132.

Clavelou, P., Pajot, J., Dallel, R. and Raboisson, P. (1989) Application of the formalin test to the study of orofacial pain in the rat. *Neurosci. Lett.*, 103: 349–353.

Clavelou, P., Dallel, R., Orliaguet, T., Woda, A. and Raboisson, P. (1995) The orofacial formalin test in rats: effects of different formalin concentrations. *Pain*, 62: 259–274.

Coderre, T.J. and Melzack, R. (1992a) The contribution of excitatory amino acids to central sensitization and persistent nociception after formalin-induced tissue injury. *J. Neurosci.*, 12: 3665–3670.

Coderre, T.J. and Melzack, R. (1992b) The role of NMDA receptor-operated calcium channels in persistent nociception after formalin-induced tissue injury. *J. Neurosci.*, 12: 3671–3675.

Coderre, T.J., Fundytus, M.E., McKenna, J.E., Dalal, S. and Melzack, R. (1993) The formalin test: a validation of the weighted-scores method of behavioural pain rating. *Pain*, 54: 43–50.

Coderre, T.J., Yashpal, K. and Henry, J. (1994) Specific contribution of lumbar spinal mechanisms to persistent nociceptive responses in the formalin test. *NeuroReport*, 5: 1337–1340.

Czlonkowsky, A., Stein, C. and Herz, A. (1993) Peripheral mechanisms of opioid antinociception in inflammation: involvement of cytokines. *Eur. J. Pharmacol.*, 242: 229–235.

De Sarro, G.B., Masuda, Y., Ascioti, C., Audino, M.G. and Nistico, G. (1990) Behavioural and ECoG spectrum changes induced by intracerebral infusion of interferons and interleukin 2 in rats are antagonized by naloxone. *Neuropharmacology*, 29: 167–179.

De Souza, E.B. and Van Loon, G.R. (1985) Differential plasma β-endorphin, β-lipotropin, and adrenocorticotropin responses to stress in rats. *Endocrinology*, 116: 1577–1586.

Dickenson, A.H. and Sullivan, A.F. (1987) Subcutaneous formalin-induced activity of dorsal horn neurones in the rat: differential response to an intrathecal opiate administration pre- or post-formalin. *Pain*, 30: 349–360.

Dijkstra, H., Tilders, F.J.H., Hiehle, M.A. and Smelik, P.G. (1992) Hormonal reactions to fighting in rat colonies: prolactin rises during defence, not during offence. *Physiol. Behav.*, 51: 961–968.

Dubuisson, D. and Dennis, S.G. (1977) The formalin test: a quantitative study of the analgesic effects of morphine, meperidine, and brain stimulation in rats and cats. *Pain*, 4: 161–174.

Dutar, P., Lamour, Y. and Jobert, A. (1985) Activation of identified septohippocampal neurones by noxious peripheral stimulation. *Brain Res.*, 328: 15–21.

Facchinetti, F., Tassinari, G., Porro, C.A., Galetti, A. and Genazzani, A.R. (1992) Central changes of β-endorphin-like immunoreactivity during rat tonic pain differ from those of purified β-endorphin. *Pain*, 49: 113–116.

Fatranska, M., Budai, D., Oprsalova, Z. and Kvetnansky, R. (1987) Acetylcholine and its enzymes in some brain areas of the rat under stress. *Brain Res.*, 424: 109–114.

Fisher, A. and Hanin, I. (1980) Choline analogs as potential tools in developing selective animal models of central cholinergic hypofunction. *Life Sci.*, 27: 1615–1634.

Fontani, G., Grazzi, F. and Aloisi, A.M. (1987) Different effects of ACTH fragments on hippocampal EEG and behaviour. *Neuropsychobiology*, 17: 169–172.

Gadient, R.A. and Otten, U. (1993) Differential expression of interleukin-6 (IL-6) and interleukin-6 receptor (IL-6R) mRNAs in rats hypothalamus. *Neurosci. Lett.*, 153: 13–16.

Go, V.L.W. and Yaksh, T.L. (1987) Release of substance P from the cat spinal cord. *J. Physiol.*, 391: 141–167.

Harbuz, M.S. and Lightman, S.L. (1992) Stress and the hypothalamo-pituitary-adrenal axis: acute, chronic and immunological activation. *J. Endocrinol.*, 134: 327–339.

Heapy, C.J., Jamieson, A. and Russell, N.J.W. (1987) Afferent C-fibre and A-delta activity in models of inflammation. *Br. J. Pharmacol.*, 90: 164.

Hoffman, D.C., West, T.E.G. and Wise, R.A. (1991) Ventral pallidal microinjections of receptor-selective opioid agonists produce differential effects on circling and locomotor activity in rats. *Brain Res.*, 550: 205–212.

Isaacson, R.L. and Green, E.J. (1978) The effect of ACTH1–24 on locomotion, exploration, rearing and grooming. *Behav. Biol.*, 24: 118–122.

Kaiser, C. and van Holst, D. (1993) Physical and psychological stress and its effects on behaviour as well as endocrinological and immunological parameters in male *Tupaia belangeri. Int. Conf. on Hormone, Brain and Behaviour*, Tours, p. 109.

Khanna, S. and Sinclair, J.G. (1992) Responses in the CA1 region of the rat hippocampus to a noxious stimulus. *Exp. Neurol.*, 117: 28–35.

Kishimoto, T., Akira, S. and Taga, T. (1992) Interleukin-6 and its receptors: a paradigm for cytokines. *Science*, 258: 593–597.

Klemm, F., Carli, G. and Reeh, P.W. (1989) Peripheral neural correlates of the formalin test in the rat. *Eur. J. Physiol.*, 414: S42.

Kocher, L. (1988) Systemic naloxone does not affect pain-related behaviour in the formalin test in rats. *Physiol. Behav.*, 43: 265–268.

Kuhar, M.J., Pert, C.B. and Snyder, S.H. (1973) Regional distribution of opiate receptor binding in monkey and human brain. *Nature*, 245: 447–450.

MacDermott, A.B., Mayer, M.L., Westbrook, G.L., Smith, S.J. and Barker, J.L. (1986) NMDA-receptor activation increases cytoplasmic calcium concentration in cultured spinal cord neurons. *Nature*, 321: 519–522.

Malmberg, A.B. and Yaksh, T.L. (1992) Hyperalgesia mediated by spinal glutamate or substance P receptor blocked by spinal cyclooxygenase inhibition. *Science*, 257: 1276–1279.

McEwen, B.S., De Kloet, E.R. and Rostene, W. (1986) Adrenal steroid receptors and actions in the nervous system. *Physiol. Rev.*, 66: 1121–1188.

McKenna, J.E. and Melzack, R. (1992) Analgesia produced by lidocaine microinjection into the dentate gyrus. *Pain*, 49: 105–112.

Meller, S.T., and Gebhart, G.F. (1993) Nitric oxide and nociception processing in the spinal cord. *Pain*, 52: 163–168.

Melzack, R. and Casey, K.L. (1968) Sensory, motivational and central control determinants of pain: a new conceptual model. In: D. Kensalo (Ed.), *The Skin Senses*, C.C. Thomas, Springfield, IL, pp. 423–439.

Meyer, M.E. and Meyer, M.E. (1993) Behavioural effects of opioid peptide agonists DAMGO, DPDPE, and DAKLI on locomotor activities. *Pharmacol. Biochem. Behav.*, 45: 315–320.

Meyerson, B.J. (1979) Hypothalamic hormones and behaviour. *Med. Biol.*, 57: 69–83.

Monsour, A., Khachaturian, H., Lewis, M.E., Akil, H. and Watson, S.J. (1988) Anatomy of CNS opioid receptors. *Trends Neurosci.*, 11: 308–314.

Moroni, F., Cheney, D.L. and Costa, E. (1977) B-endorphin inhibits Ach turnover in nuclei of rat brain. *Nature*, 267: 267–268.

Neoh, C.-A. (1993) More about the formalin test. *Pain*, 53: 237–239.

Nishimura, J.I., Endo, Y. and Kimura, F. (1992) Increases in cerebral blood flow in rat hippocampus after medial septal injection of naloxone. *Stroke*, 23: 1325–1330.

North, M.A. (1978) Naloxone reversal of morphine analgesia but failure to alter reactivity to pain in the formalin test. *Life Sci.*, 22: 295–302.

Olson, G.A., Olson, R.D. and Kastin, A.J. (1993) Endogenous opiates: 1992. *Peptides*, 14: 1339–1378.

Papolos, D.F., Edwards, E., Marmur, R., Lachman, H.M. and Henn, F.A. (1993) Effects of the antiglucocorticoid RU 38486 on the induction of learned helpless behavior in Sprague-Dawley rats. *Brain Res.*, 615: 304–309.

Payne, L.C. and Krueger, J.M. (1992) Interactions of cytokines with the hypothalamus-pituitary axis. *J. Immunother.*, 12: 171–173.

Plata-Salaman, C.R. (1991) Immunoregulators in the nervous system. *Neurosci. Biobehav. Rev.*, 15: 185–215.

Porro, C.A. and Cavazzuti, M. (1993) Spatial and temporal aspects of spinal cord and brain stem activation in the formalin pain model. *Prog. Neurobiol.*, 41: 565–607.

Porro, C.A., Tassinari, G., Facchinetti, F., Panerai, A.E. and Carli, G. (1991) Central beta-endorphin system involvement in the reaction to acute tonic pain. *Exp. Brain Res.*, 83: 549–554.

Rahman, A.F.M.M., Takahashi, M. and Kaneto, H. (1994) Involvement of pain associated anxiety in the development of morphine tolerance in formalin treated mice. *Jpn. J. Pharmacol.*, 65: 313–317.

Rosland, J.H., Tjolsen, A., Maehle, B. and Hole, K. (1990) The formalin test in mice - effect of formalin concentration. *Pain*, 42: 235–242.

Rossier, J., French, E.D., Rivier, C., Ling, N., Guillemin, R. and Bloom, F. (1977) Foot-shock induced stress increases β-endorphin levels in blood but not brain. *Nature*, 270: 618–620.

Skilling, S.R., Smulling, D.H., Beitz, A. and Larsson, A.A.

(1988) Extracellular amino acid concentrations in the dorsal spinal cord of freely moving rats following veratridine and nociceptive stimulation. *J. Neurochem.*, 51: 127–132.

Spanagel, R., Herz, A., Bals-Kubik, R. and Shippenberg, T.S. (1991) β-endorphin-induced locomotor stimulation and reinforcement are associated with an increase in dopamine release in the nucleus accumbens. *Psychopharmacology (Berlin)*, 104: 51–56.

Takahashi, H., Ohkubo, T., Shibata, M. and Naruse, S. (1984) A modified formalin test for measuring analgesia in mice. *Jpn. J. Oral Biol.*, 26: 543–548.

Terry, W.S. (1979) Habituation and dishabituation of rats' exploration of a novel environment. *Animal Learn. Behav.*, 7: 525–536.

Tjolsen, A., Berge, O.-G., Hunskaar, S., Rosland, J.H. and Hole, K. (1992) The formalin test: an evaluation of the method. *Pain*, 51: 5–17.

Vaccarino, A.L. and Melzack, R. (1992) Temporal processes of formalin pain: differential role of the cingulum bundle, fornix pathway and medial bulboreticular formation. *Pain*, 49: 257–271.

Versteeg, D.H.G., Florijn, W.J., Holtmaat, A.J.G.D., Gispen, W.H. and de Wildt, D.J. (1993) Synchronism of pressor response and grooming behavior in freely moving, conscious rats following intracerebroventricular administration of ACTH/MSH-like peptides. *Brain Res.*, 631: 265–269.

Wall, P.D. (1979) On the relation of injury to pain. *Pain*, 6: 253–264.

Walters, E.T. (1994) Injury-related behavior and neural plasticity: an evolutionary perspective on sensitization, hyperalgesia, and analgesia. *Int. Rev. Neurobiol.*, 36: 325–427.

Watkins, L.R. and Mayer, D.J. (1982) Involvement of spinal opioid systems in footshock induce analgesia: antagonism by naloxone is possible only before induction of analgesia. *Brain Res.*, 242: 309–316.

Watkins, L.R., Wiertelak, E.P., Goehler, L.E., Smith, K.P., Martin, D. and Maier, S.F. (1994) Characterization of cytokine-induced hyperalgesia. *Brain Res.*, 654: 15–26.

Wheeler-Aceto, H. and Cowan, A. (1991) Standardization of the rat paw formalin test for the evaluation of analgesics. *Psychopharmacology*, 104: 35–44.

Whishow, I.Q. (1990) Time estimates contribute to food handling decisions by rats: implications for neural control of hoarding. *Psychobiology*, 18: 460–466.

Whishow, I.Q. (1993) Activation, travel distance, and environmental change influence food carrying in rats with hippocampal, medial thalamic and septal lesions: implications for studies on hoarding and theories of hippocampal function. *Hippocampus*, 3: 373–385.

Wiertelak, E.P., Furness, L.E., Horan, R., Martinez, J., Maier, S.F. and Watkins, L.R. (1994) Subcutaneous formalin produces centrifugal hyperalgesia at a non-injected site via the NMDA-nitric oxide cascade. *Brain Res.*, 649: 19–26.

Wilcox, G.L. (1990) Transmission and modulation of pain at the spinal cord level. *Pain*, Suppl. 5: S247.

Wishart, T., Brohman, L. and Mogenson, G. (1969) Effects of lesions of the hippocampus and septum on hoarding behavior. *Animal Behav.*, 7: 781–784.

Wolfe, J.B. (1939) An exploratory study of food-storing in rats. *J. Comp. Psychol.*, 28: 97–108.

Woolf, C.J. (1991) Central mechanisms of acute pain. In: M.R. Bond, J.E. Charlton and C.J. Woolf (Eds.), *Proc. VIth World Cong. Pain.* Elsevier, Amsterdam, pp. 25–34.

Woolf, C.J. and Thompson, S.W.N. (1991) The induction and maintenance of central sensitization is dependent on *N*-methyl-o-aspartic acid receptor activation: implications for the treatment of post-injury pain hypersensitivity states. *Pain*, 44: 293–299.

Yamamoto, T. and Yaksh, T.L. (1992) Comparison of antinociceptive effects of pre- and posttreatment with intrathecal morphine and MK801, an NMDA antagonist, on the formalin test in the rat. *Anesthesiology*, 77: 757–763.

CHAPTER 4

Functional imaging studies of the pain system in man and animals

C.A. Porro[1] and M. Cavazzuti[2]

[1]*Dipartimento di Scienze Biomediche, University of Udine, Via Gervasutta 48, I-33100 Udine, Italy* and
[2]*Dipartimento di Patologia Neuropsicosensoriale, University of Modena, Via del Pozzo 71, I-41100 Modena, Italy*

Introduction

Pain is a complex experience, which is likely to involve changes in activity of large populations of neurons in the spinal cord and brain (Willis, 1985; Price, 1988). A thorough understanding of the functional neural correlates of acute and chronic pain states in humans would probably require mapping electrical and metabolic events occurring in tens, or hundred of thousands of individual nerve cells, including their spatial and temporal correlations. Clearly, this is far beyond the possibilities of any single available technique for the study of nerve function. Nevertheless, methodological advances over the last two decades have greatly expanded our chances of investigating metabolic events in large arrays of central nervous system (CNS) cells. In this chapter, we will deal with functional imaging mapping studies based on tracer methods for the study of local glucose metabolism or blood flow rates, which are closely related to the electrical activity of the neuronal population of a given CNS area under physiological circumstances (Sokoloff, 1983; Raichle, 1987; Roland, 1993). Indeed, nerve cells rely almost exclusively upon glucose for their energetic requirements; glucose and oxygen are supplied by blood flow, which is finely and continuously adjusted in the CNS to meet regional metabolic demands (Raichle, 1987; Roland, 1993). The 2-deoxyglucose (2-DG) autoradiographic method (Sokoloff et al., 1977) yields images depicting the degree of local glucose uptake in virtually the entire CNS (Fig. 1), with a spatial resolution of 100–200 μm in experimental animals if the standard technique is employed; single-cell resolution may be achieved by more complex procedures (see Duncan, 1992). Local glucose utilization rates obtained by the standard 2-DG technique represent the integrated metabolic activity over a period of 30–45 min (Sokoloff, 1983); both basal values and stimulation-induced increases are related predominantly to synaptic function (see References in Roland, 1993). Autoradiographic methods for mapping regional blood flow rates are also available, which measure the local tissue concentrations of chemically inert, diffusible, radioactive tracers (Raichle, 1987). With the aid of computerized image processing systems, a quantitative estimate of both the spatial distribution and the intensity of labelling in individual spinal cord or brain areas may be achieved.

Regional cerebral metabolism or blood flow may be studied in humans by imaging techniques such as positron emission tomography (PET) and single photon emission computerized tomography (SPECT) (Phelps et al., 1979; Raichle, 1987; Roland, 1993). These techniques are based on computer-assisted identification of the sources of radioactivity produced by positron-emitting (^{11}C, ^{13}N, ^{15}O, ^{18}F) or gamma-ray-emitting (e.g., ^{131}I, ^{99m}Tc, ^{133}Xe) isotopes, respectively. Isotopes are linked to tracer molecules which are inhaled or injected into the blood stream, and thence enter the

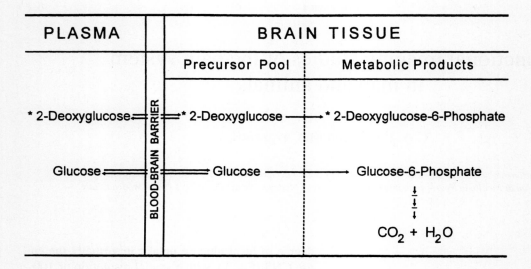

Fig. 1. Diagrammatic representation of the metabolic fate of 2-deoxyglucose (2-DG) and glucose in the brain. Glucose (G), an essential substrate for the production of high-energy phosphates, is continuously supplied to nerve tissues by cerebral circulation. It crosses the blood-brain-barrier by a carrier-mediated transport. After phosphorylation by hexokinase, it is further metabolized in nerve cells, mostly down the glycolytic pathway. 2-DG is a glucose analogue which competes with glucose for both the carrier-mediated transport across the blood-brain-barrier, and hexokinase-catalyzed phosphorylation. The rate constants of these reactions may be experimentally determined. The metabolic product of the latter step, 2-deoxyglucose-6-phosphate (2-DG-6P), unlike glucose-6-phosphate, can not proceed down the metabolic pathway, and it is therefore trapped inside the cells. In order to estimate local cerebral rates of glucose uptake, 2-DG is injected in tracer amounts into the peripheral circulation. The amount of 2-DG-6P accumulated in any cerebral tissue at any time after the injection is equal to the integral of the rate of 2-DG phosphorylation in that tissue, which is in turn related to the amount of glucose that has been phosphorylated over the same time interval. If 2-DG is labelled (*) with a radioactive isotope (^{14}C, ^{3}H, or ^{18}F), the total tissue concentration at the end of the experiment (almost exclusively *2-DG-6P, beginning from 25–30 min after the injection) may be quantified by autoradiographic techniques in experimental animals, and by positron emission tomography in humans. The absolute values of regional glucose consumption (expressed in μmoles/100 g/min) may then be calculated, provided that the time courses of the G and 2-DG plasma concentrations and specific rate constants are known (see the theoretical model and methodological details in Sokoloff et al., 1977; Phelps et al., 1979)

brain. PET and SPECT allow the simultaneous investigation of functional activity levels of several areas of the human brain in different pathophysiological conditions, such as sensory stimulation, motor activity, or cognitive tasks; furthermore, they are suitable for receptor mapping studies (Roland, 1993; Jones and Derbyshire, 1994). The recent development of noninvasive magnetic resonance imaging (MRI) techniques, sensitive to the local changes of blood flow, blood volume, and deoxyhemoglobin concentration which accompany neuronal activation, has provided a new powerful tool for the study of brain function (Kwong et al., 1992; Cohen and Bookheimer, 1994). Indeed, functional MRI (fMRI) does not require radioactive tracers, and it is characterized by a better spatial and temporal resolution than PET (Fig. 2).

During the last few years, functional imaging techniques for the study of regional metabolic or blood flow rates have been extensively applied to explore the functional organization of the pain system, both in rat models of acute prolonged or chronic pain, and during experimental or clinical pain in humans. Several issues related to pain mechanisms have been investigated by these approaches (Table 1). Available evidence suggests both similarities and differences between spatial distribution and intensity of CNS activity during acute or chronic pain states.

TABLE 1

Functional imaging studies of the pain system: current issues

- Stimulus-response relationships
- Somatotopy
- Time course
- Acute versus chronic pain states
- Effects of analgesic and anesthetic drugs
- Modulation by endogenous antinociceptive systems

Studies of regional glucose metabolism in rat models of prolonged and chronic pain

In our laboratory, a series of 2-DG experiments has been performed in the rat formalin model of acute prolonged (tonic) pain (Dubuisson and Dennis, 1977). This is characterized by prolonged activation of C-fibers and spinal neurons, and it may be considered a valid experimental model of injury-induced pain in humans. After a subcutaneous formalin injection, two distinct phases of pain-related behavioral changes (e.g., licking or flexion of the injected paw; see Chapter 3 by Aloisi and Carli) can be identified in rats. Animal behavior during the second and more prolonged phase (typically, 20–50 min after s.c. formalin in rats) depends at least in part on events occurring during the first phase (0–10 min), and it is related to plastic changes of the excitability of CNS units (see References in Coderre et al., 1993; Porro and Cavazzuti, 1993).

The main goals of this group of experiments were: (a) to provide a spatial map of CNS metabolic changes in the spinal cord and brain during acute prolonged pain and (b) to investigate the time profile of pain-related activation of different spinal cord and brain structures, in parallel with behavioral changes. To this end, a modification of the quantitative technique, or a semi-quantitative one, have been adopted in order to perform experiments in freely-moving, unstressed rats (see methodological details in Porro et al., 1991a,b; Porro and Cavazzuti, 1992). Experiments were initiated at different times (2, 30 or 60 min) after subcutaneous injection of a dilute formaldehyde solution into a forepaw or a hindpaw. The results of these studies will be briefly compared to those obtained in rat models of chronic pain.

Pain-related metabolic patterns in the spinal cord

During the first phase, and the beginning of the second phase of the behavioral response to s.c.

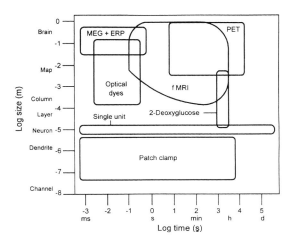

Fig. 2. Schematic diagram representing in a logarithmic scale the approximate ranges of spatial and temporal boundaries of application of techniques for studying brain function, relative to the size scale of neural structures. Lower limits in the vertical and horizontal axis correspond roughly to the best currently achievable spatial and temporal resolution, respectively. The spatial resolution may be defined as the minimal distance at which two different foci of activity can be separately identified. Functional imaging mapping techniques such as 2-DG, PET, and fMRI provide integrated values over a given time interval, corresponding to the temporal resolution. They are characterized by less spatial and temporal resolutions than single-unit and patch-clamp electrophysiological recording methods, but they allow the simultaneous investigation of functional activity levels of multiple neuronal assemblies. ERP and MEG have a better temporal, but a coarser spatial resolution. Within some approaches (e.g, PET or fMRI), the spatial and temporal resolutions may vary widely according to various factors, such as the measurement under investigation, the tracer (for PET), and the hardware and software characteristics of the instrumentation. Abbreviations: ERP: event-related potentials; MEG: magnetoencephalography; fMRI: functional magnetic resonance imaging; PET: positron-emission tomography. [Adapted from Churchland and Sejnowski (1988) *Science*, 242: 741–745; Belliveau et al. (1991) *J. Neuroimaging*, 1: 36–41].

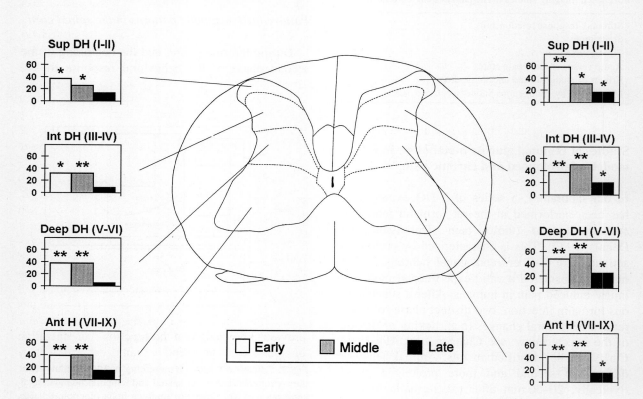

Fig. 3. Time profile of the mean percentage increases of the 2-DG uptake over control values in different gray matter regions of the cervical enlargement of the rat spinal cord, after the injection of a dilute formalin solution into the subcutaneous tissue of a forepaw. In control animals an equal volume of saline was injected. Mean metabolic activity values were obtained from the Superficial (Sup), Intermediate (Int), and Deep Dorsal Horns (DH) (corresponding approximately to laminae I-II, III-IV, and V-VI, respectively), and from the Anterior Horns (Ant H, laminae VII-IX), ipsilateral and contralateral to the injected paw. In the Early, Middle and Late groups, the 2-DG experiments were initiated 2, 30 and 60 min after s.c. formalin injection, respectively.*, ** significantly different from controls, $P < 0.05$ and $P < 0.01$, respectively.

formalin (2–30 min after the injection: 'early' group), widespread increments of the ^{14}C-2-DG uptake over uninjected or saline-injected controls were found in the cervical or lumbar enlargements of the spinal cord, following a forepaw (Porro et al., 1991a) or hindpaw injection (Aloisi et al., 1993), respectively. Bilateral metabolic increases were observed in anterior and posterior horns, as well as in the central gray region (lamina X) and in the ipsilateral dorso-lateral funiculus (DLF). No change was found in the contralateral DLF, nor in the dorsal columns. The superficial region (laminae I-II) of the ipsilateral dorsal horn displayed the highest percentage increase over control values (Fig. 3). The rostro-caudal extent of the activated region was greater in the deep dorsal horn, central gray matter and anterior horns, and more limited in the superficial region of the ipsilateral dorsal horn. In the dorsal horn contralateral to the side of injection a 'mirror image' of the distribution of the ^{14}C-2-DG uptake was observed.

The time profile of formalin-induced activation is different in various gray matter regions (Fig. 3). During the second "tonic" phase of the behavioral response ('middle' group: 2-DG experiments initiated 30 min after formalin injection) the metabolic increase of the ipsilateral superficial dorsal horn was much lower than that observed in the 'early' group. By contrast, percentage increases in the 2-DG uptake of the intermediate and deep dorsal horn were comparable to those found during the early phase. In experiments initiated 60 min after formalin injection ('late' phase), when the licking response had almost disappeared but flexion of the injected limb was still present, the overall activation was much lower. A significant enhancement over control values was still found, however, on the ipsilateral side of the cord, both in dorsal and ventral horns.

Price and colleagues (Price et al., 1991; Mao et al., 1992) investigated by the 2-DG method the metabolic activity pattern in the spinal cord of mononeuropathic rats, 10 days following a loose ligation of the sciatic nerve of one side (Bennett and Xie, 1988). At this time, animals displayed ongoing behaviors indicative of persistent spontaneous pain. In the absence of overt peripheral stimulation, the metabolic pattern observed at the level of the lumbar spinal cord in neuropathic rats bears several similarities to that found during the second phase of pain-related behavior after s.c. formalin injection. The common features in the two situations include: (a) a conspicuous bilateral increase in glucose utilization rates over controls, extending throughout the lumbar enlargement of the cord and the dorsal and ventral horns; (b) an overall greater intensity of metabolic activity on the side of the cord ipsilateral to the affected paw; (c) peak values of the 2-DG uptake in the deep ipsilateral dorsal horn (laminae V-VI); and (d) a somatotopic arrangement in both the superficial and deep regions of the dorsal horns, medial portions displaying higher metabolic rates than lateral ones over most of the rostro-caudal extent of the activated area.

Several observations suggest that these changes are predominantly due to the activity of pain-related neural circuits, rather than to the increased mechano- and proprioceptive sensory inflow induced by behavioral responses. Indeed, selective stimulation of limb muscles, mechanoceptive or proprioceptive fibers induces metabolic increases of lower intensities, which are restricted to the ipsilateral dorsal horn (mainly laminae III-IV), at least in normal rats (Berenberg, 1984; Santori et al., 1986; Juliano et al., 1987; Coghill et al., 1993). Furthermore, the metabolic pattern observed during neuropathic pain is not modified in animals paralyzed with a neuromuscular block (Mao et al., 1992), whereas functional activity changes after formalin injection are reduced but not abolished by pentobarbital anesthesia (Porro et al., 1991a). A bilateral enhancement of spinal cord metabolic rates has also been described after unilateral noxious heat stimulation in anesthetized cats (Abram and Kostreva, 1986) and spinal rats (Coghill et al., 1991), although contralateral activation was much lower in the latter study.

Acute prolonged somatic pain and chronic neuropathic pain appear to involve the activation of populations of spinal neurones located at the same sites. These are likely to participate both in spinal motor circuits, and in spino-supraspinal pathways (see below). Altogether, these studies extend our previous knowledge on the functional organization of spinal cord nociceptive units, derived mainly from electrophysiological recordings in anesthetized or spinal animals (Willis, 1985; Price, 1988; Willis and Coggeshall, 1991; Woolf, 1994). The intensity and spatial extent of metabolic activation of deep gray matter layers during the second phase of the response to s.c. formalin injection correspond to those found in neuropathic rats. Wide dynamic range neurons in the deep dorsal horn laminae have larger receptive fields, and they are more susceptible to excitability changes after repeated or prolonged noxious stimulation (Willis and Coggeshall, 1991). By contrast, neurons in the superficial laminae appear to be activated predominantly during the first phase of the response to acute tissue injury. It is conceivable that the time profile of activation of neurons in the superficial layers and of some of their projection sites (such as mesencephalic structures, see below) may be related to their different role in pain mechanisms (e.g., a more specific involvement in the "warning" signalling of acute intense injury).

Pain-related metabolic patterns in the brain

The results summarized in the preceeding section show that regions from which spino-supraspinal pathways originate (Willis and Coggeshall, 1991) display a time-dependent activation during acute prolonged pain. In studies performed with the conventional 2-DG technique, which lacks cellular resolution, it is impossible to discriminate between the activity of interneurons or projection neurons. Furthermore, it has been shown that increases in glucose utilization rates within a given area may be related to the activity of both excitatory and inhibitory synapses (see References in Roland, 1993). However, changes in the activity of projection neurons within the spinal cord may be indirectly assessed by examining metabolic levels of target regions in the brain.

Indeed, during the first and second phase of the behavioral response to s.c. formalin, regional increases in glucose utilization rates are found in several brain structures (Table 2) involved both in nociceptive and antinociceptive mechanisms (Porro et al., 1991b, unpublished; Porro and Cavazzuti, 1992). Significant bilateral increases have been detected in several portions of the medullary and pontine reticular formation and in parabrachial nuclei. Other brainstem regions involved in processing noxious information, such as the raphe nuclei, locus coeruleus, the periaqueductal gray matter, and the deep layers of the superior colliculus were also significantly activated. Several diencephalic and telencephalic sites, including medial, lateral and posterior thalamic nn., lateral habenula, and portions of the parietal, cingulate, anterior insular and frontal cortex also displayed bilaterally enhanced levels of metabolic activity. It is noteworthy that the increases were widespread but not totally non-specific. For instance, no change relative to controls was found in thalamic or cortical visual or auditory areas (lateral and medial geniculate bodies, occipital and temporal cortex).

The metabolic pattern observed in the brain during formalin-induced pain changes over time, and the time profile varies in different structures. For instance, in raphe nuclei and the deep layers of the superior colliculus pain-related increases were found only during the first phase of the response to s.c. formalin; on the other hand, in the lateral reticular nucleus and some medial thalamic regions significant enhancements of the 2-DG uptake persisted over the whole investigated period (Porro et

TABLE 2

Brain regions displaying increased rates of glucose utilization during prolonged formalin-induced pain in rats

Medulla
Medullary reticular formation, dorsal
Lateral reticular n.
Gigantocellular reticular n.
Paragigantocellular reticular n.
Parvocellular reticular n.
Raphe magnus n.

Pons
Pontine reticular n.
Lateral parabrachial region
Locus coeruleus

Midbrain
Periacqueductal gray
Deep Mesencephalic n.
Anterior pretectal n.
Deep layers of the superior colliculus

Diencephalon
Parafascicular n.
Posterior thalamic n.
Ventral medial thalamic n.
Ventral lateral thalamic n.
Central lateral-central-medial n.
Gelatinosus n.
Ventral posterolateral n.
Ventral posteromedial n.
Reticular thalamic n., post. part
Hypothalamic arcuate n.
Lateral habenula
Globus pallidus

Telencephalon
Caudate n.
Accumbens n.
Cingulate cortex
Somatosensory cortex, primary and secondary
Insular cortex
Anterior frontal cortex
Orbital cortex

al., 1991b; Porro and Cavazzuti, 1992). Frontal and orbital cortical regions, together with portions of the basal ganglia, were heavily activated during the second phase (Porro et al., unpublished). These time-related changes might in turn be related to involvement of specific structures in different aspects of pain (e.g., the arousal reaction or motor adjustments).

A direct comparison of the results of 2-DG studies performed during the second phase of formalin pain (Porro et al., unpublished) or neuropathic pain (Mao et al., 1993) can be found in Fig. 4. The spatial distribution of functional activation appears to be largely similar in the two painful conditions; most structures which are involved in pain induced by s.c. formalin also display increased functional activity levels during chronic neuropathic pain. The most obvious difference is the intensity of the activation, percentage increases in glucose utilization rates over control values being much larger, at least at supraspinal sites, during a chronic neuropathic state than during prolonged somatic (formalin-induced) pain. Moreover, some regions (e.g., the amygdala and retrosplenial cortex) were activated only in neuropathic pain. The very large increases in metabolic rates found in neuropathic rats indicate a profound enhancement of synaptic activity in the brain, an observation which parallels the altered pattern of spontaneous and stimulation-evoked neuronal discharges observed at thalamic and cortical sites in electrophysiological experiments during chronic pain in rodents. Accordingly, Juliano et al. (1987) reported that joint (ankle) rotation induces a much greater metabolic response at different levels of the neuraxis in lightly-anesthetized arthritic rats than in normal controls. Both the intensity and the spatial extent of 2-DG uptake appear to be increased in chronic pain models, suggesting that a larger number of neurons modify their activity.

Human studies of regional cerebral blood flow patterns related to pain

Several PET studies performed in healthy volunteers have shown increased regional cerebral blood flow (rCBF) in cortical and subcortical structures (Table 3) during brief noxious heat, as compared to nonnoxious thermal stimulation, or noxious chemical stimulation of the skin. Foci of activation were observed in regions corresponding to the thalamus, basal ganglia, anterior cingulate cortex, insula, and primary (SI) and secondary (SII) somatosensory cortical areas, as well as in the cerebellum, hypothalamus and periaqueductal gray matter (Jones et al., 1991; Talbot et al., 1991; Casey et al., 1994a; Coghill et al., 1994, 1995; Derbyshire et al., 1994; Hsieh et al., 1995a). Most rCBF changes were on the side of the brain contralateral to the stimulated arm, but ipsilateral activation has also been described in the thalamus, basal ganglia, and anterior insular, anterior cingulate, and parietal cortex (Casey et al., 1994a; Coghill et al., 1994; Derbyshire et al., 1994; Hsieh et al., 1995a). The activation of contralateral SI and SII during acute pain was not detectable in all the aforementioned PET studies (see Jones et al., 1991; Jones and Derbyshire, 1994); however, it was recently confirmed using high-resolution fMRI techniques following painful heat or electrical nerve stimulation (Wen et al., 1993; Gelnar et al., 1994; Davis et al., 1995). When more prolonged noxious thermal stimuli were employed, both rCBF decreases (Apkarian et al., 1992) and increases (Di Piero et al., 1994) in the primary sensory-motor region have been found by low-resolution SPECT studies, possibly depending on the intensity of stimulation. There is also evidence for involvement of the right prefrontal cortex (corresponding to Brodmann area 10) during acute pain; however, both rCBF increases (Derbyshire et al., 1994; Hsieh et al., 1995a) and decreases (Coghill et al., 1994) have been reported by different groups. Interestingly, the rCBF increases induced by prolonged noxious heat stimulation in several subcortical and cortical regions, such as the thalamus, primary sensory-motor cortex, SII, anterior cingulate cortex, rostral insula, and motor-related structures (basal ganglia, cerebellum) appear to be related to the intensity of perceived pain (Duncan et al., 1994) and the time elapsed from the beginning of stimulation (Casey et al., 1994b).

Some recent functional imaging studies have attempted to determine the neural correlates of

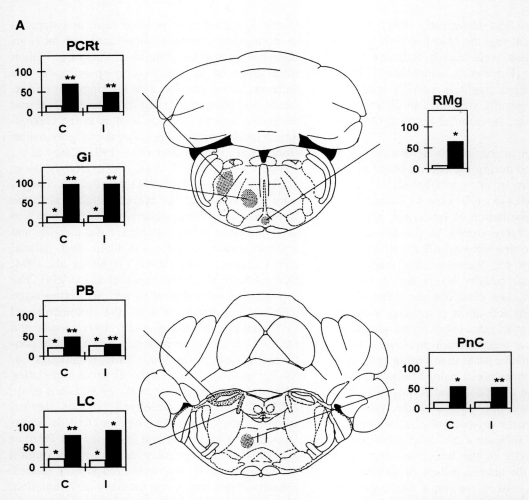

Fig. 4a,b. Schematic drawings of coronal sections of the rat brain (adapted from the Paxinos and Watson Atlas, Academic Press, 1986), depicting the intensity of mean functional activity increases in selected brain areas during acute prolonged formalin-induced pain (white columns), or chronic mononeuropathic pain (black columns). Histograms represent mean percentage changes of local glucose utilization rates over values of the respective control groups, on the side of the brain contralateral (C) or ipsilateral (I) to the affected limb. Formalin data were obtained in our laboratory (Porro et al., unpublished); data in neuropathic animals are from Mao et al., 1993. The fully quantitative 2-DG technique was employed in male Sprague-Dawley rats. Only structures investigated in both studies were included. *, **, *** significantly different from controls, $P < 0.05$, $P < 0.01$ and $P < 0.001$, respectively. Abbreviations: PCRt: parvocellular reticular nucleus. RMg: raphe magnus nucleus. Gi: gigantocellular reticular nucleus. PB: parabrachial nuclei. PnC: pontine reticular nucleus, caudal part. LC: locus coeruleus. RSG: retrosplenial granular cortex. Po: posterior thalamic nucleus. VPL: ventral posterolateral thalamic nucleus. VM: ventromedial thalamic nucleus. VL: ventrolateral thalamic nucleus. HL: hindlimb area of the cortex. Par 2: parietal cortex, area 2. FL: forelimb area of the cortex. Cg: cingulate cortex. CPu: caudate putamen.

chronic pain by comparing "resting" (namely, in the absence of overt peripheral stimulation) rCBF patterns obtained during pain to those of a pain-free state. In patients suffering from chronic neuralgia due to a peripheral mononeuropathy, Hsieh et al. (1995b) found significantly higher rCBF rates in PET scans obtained during the patients' habitual pain state than after pain alleviation (following a successful regional nerve block with lidocaine) in bilateral foci located in the anterior insula (area 14), inferior prefrontal cortex (areas 10/47), superior (area 7) and inferior (area 40) pa-

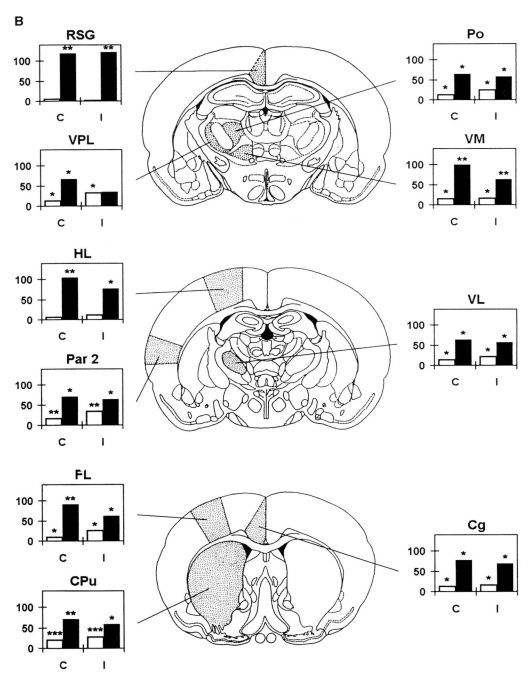

rietal regions, posterior cingulate/precuneate cortex (31/23/7), as well as in the cerebellar vermis. Moreover, a focus of activation was detected during the painful state in the posterior aspect of the right anterior cingulate cortex, irrespective of the affected side. However, in the posterior thalamic region rCBF rates were lower during pain than the remission state. No difference between the two conditions was found in regions corresponding to SI and SII. Asymmetry of thalamic blood flow during chronic unilateral pain, with relatively lower rCBF rates in the contralateral thalamus,

TABLE 3

Brain regions displaying focal cerebral blood flow changes during acute somatic pain in humans

---

Subcortical structures
Cerebellar vermis
Medial dorsal midbrain (periaqueductal gray)
Hypothalamus
Thalamus
Lentiform n./putamen

Cortical areas
Primary somatosensory area
Secondary somatosensory area
Supplementary motor area
Superior and inferior parietal areas
Anterior insula
Anterior cingulate
Prefrontal cortex

The table summarizes the main findings of PET or fMRI studies on the effects of acute noxious heat, chemical or electrical stimulation. It has to be underlined that in many studies only some of the reported areas showed significant rCBF changes relative to controls. Moreover, some structures, such as the primary somatosensory and prefrontal cortex, showed foci of enhanced or decreased rCBF in different studies. See references in the text.

was also detected in other PET studies (Di Piero et al., 1991; Iadarola et al., 1995). In a group of cancer patients with chronic pain, the asymmetry in thalamic rCBF values appeared to normalize after an anterolateral cordotomy providing pain relief (Di Piero et al., 1991). Iadarola et al. (1995) found rCBF decreases both in the anterior and posterior halves of the thalamus contralateral to the painful site, in a group of neuropathic pain patients showing pronounced mechano-allodynia. These rCBF changes were not accompanied by clear anatomical abnormalities, such as a reduction in the volume of thalamic tissue. Resting-state rCBF, as assessed by SPECT, was also bilaterally decreased in the thalamus and head of the caudate nucleus in women suffering from fibromyalgia, a disorder characterized by chronic musculoskeletal pain and fatigue (Mountz et al., 1995).

Significant differences between chronic pain patients and age-matched controls have also been described in their functional cerebral responses to brief noxious heat stimuli. In a group of six women suffering from atypical facial pain, the anterior cingulate cortex (area 24) displayed an enhanced rCBF response, whereas the orbitofrontal cortex (area 10), which was activated during pain in normal controls, and other frontal and parietal areas showed decreased rCBF rates during acute heat pain compared with non-painful heat (Derbyshire et al., 1994). Patients with rheumatoid arthritis had significantly lower rCBF values during heat pain in the thalamus, lentiform nucleus, anterior cingulate cortex, and prefrontal cortex (Jones et al., 1993). Similarly, noxious heat decreased the activity in the contralateral SI and SII in patients suffering from reflex sympathetic dystrophy with chronic pain; following a successful sympathetic block, noxious stimulation increased the activity in the same areas, as assessed by fMRI (Krauss et al., 1995). These observations show that, at least in some chronic pain states, the reactivity of the pain system to acute noxious input is altered.

Perspectives, interpretations and speculations

To date, most of the available knowledge on CNS mechanisms of pain derives from experimental work performed at the spinal cord level. These studies are of great value to deepen our understanding of nociceptive spinal circuits and to develop new therapeutic strategies. However, the spinal cord is but the first central station of the pain system, and pain perception is likely to depend upon activation of a distributed network of supraspinal neuronal populations (Melzack and Casey, 1968; Willis, 1985; Price, 1988). Therefore, we need more information on the activity of brain centers, and on spino-supraspinal relationships, during pain states. The advantage of functional imaging techniques lies in their potential ability to provide a three-dimensional map of functional activity levels of the entire CNS in animals, or large portions of the brain in humans, which may in turn be related to behavioral or perceptual correlates of pain. Since local rates of glucose metabolism are thought to reflect predominantly synaptic activity, the information gathered by the 2-DG method in experimental pain models are different from, and in several ways complementary to those obtained by single-unit electrophysiological recordings on the one hand, and immediate-early genes (IEGs) mapping studies (Chapter 18 by Zimmermann and Herdegen) on the other (see Porro and Cavazzuti, 1993).

Perspectives of functional imaging studies of pain in humans

Until recently, human PET studies have been hampered by a relatively poor spatial resolving power and by the amounts of radioactive tracers required, limiting the possibility of repeating measurements in the same subject. As a consequence, it was usually necessary to pool data from several subjects. However, intersubject averaging further decreases the spatial resolution and obscures interindividual regional variability of both anatomy and function in the human brain. If spatial resolution exceeds the size of the activated tissue (e.g., a small thalamic or cortical gray matter region), local metabolic or blood flow changes may be undetected, due to the so-called partial volume effect (Roland, 1993). Therefore, negative results must be interpreted cautiously. This may perhaps explain some of the discrepancies found in the literature, such as the inconstant activation of the thalamus during acute pain. Recent technical advances, however, has led to an improvement in the sensitivity of PET instrumentation, and therefore in the quality of studies (see Jones and Derbyshire, 1994). The radiation dose per scan may be reduced, and more scans may be performed in the same subjects. This allows individual-based analyses, which are likely to provide new important information in the near future. Functional MRI techniques, although still semi-quantitative in nature, have a better spatial and temporal resolution than PET, the latter being limited to the time course of hemodynamic changes which accompany neuronal activation (Cohen and Bookheimer, 1994; Kwong, 1995), and they are currently employed by several groups. In order to follow brain functional changes during pain on a more complete time frame, electrophysiological (Chen, 1993) approaches might be combined with PET or fMRI.

Spatial distribution of CNS functional activation during pain

Several similarities exist between the spatial distribution of CNS functional activation during acute prolonged formalin-induced pain and chronic neuropathic pain in the rat (Fig. 4). Beside providing a detailed picture of the spatial and temporal patterns of pain-related activity in the spinal cord, the 2-DG mapping studies summarized in the preceeding sections demonstrate the simultaneous activation of several subcortical regions, including brainstem, thalamic, basal ganglia, and hypothalamic areas, which are conceivably involved in sensory, motor, and vegetative responses related to pain. Despite different methodologies and the obvious caution when comparing the results of studies performed in such different species, some common features may also be detected between the spatial pattern of brain activation in rats and humans, at least during acute pain (see Tables 2 and 3). Indeed, functional imaging studies confirm and extend previous experimental and clinical observations, suggesting a role in pain mechanisms of cortical regions such as the primary and secondary somatosensory cortex, anterior cingulate, and portions of the frontal and orbital cortices (Willis, 1985; Price, 1988; Kenshalo and Willis, 1990; Vogt et al., 1993), and they provide evidence for pain-related activation of the anterior insula, for which electrophysiological evidence was lacking.

Altogether, the results of experimental and human studies performed during acute pain suggest a parallel activation of different spino-supraspinal pathways processing noxious information. In addition to the involvement of the "lateral" thalamo-cortical pathway, reaching the primary and secondary somatosensory cortices in the parietal lobe, the activation of structures related to limbic and motor circuits (Vogt et al., 1993), such as medial thalamus, cingulate and insular cortex, and portions of the basal ganglia, is likely to be related to both perceptual and behavioral reactions to actual or potential tissue damage. It is also worth mentioning that the activation shown in 2-DG studies of nuclei giving rise to diffuse ascending projection systems, such as the locus coeruleus, strengthens the hypothesis of extrathalamic modulation of cortical function by aminergic neurons, which may contribute to the affective-motivational dimension of pain (see Chapter 5 by Chapman).

In both formalin-induced and neuropathic pain models, injury at one side of the body induces a bilateral activation of neuronal populations of the spinal cord and brain in almost all of the investigated regions in unanesthetized rats. These findings have a behavioral counterpart in the presence of hyperalgesia contralateral to the injured side, described in different experimental and human models, and in the occasional occurrence of spontaneous 'mirror' pain-related behavior after s.c. formalin injection (see Attal et al., 1990; Aloisi et al., 1993). In humans, the picture is more complex. Most studies have shown a contralateral or bilateral activation of subcortical and cortical structures. However, rCBF changes lateralized to the right hemisphere, irrespective of the side of perceived pain, have been described in portions of the frontal, cingulate and inferior parietal cortices during acute or chronic pain (Derbyshire et al., 1994; Hsieh et al., 1995b). These findings are interesting in light of the involvement of parietal and prefrontal areas, more specifically on the right side of the brain, in emotional and attentional mechanisms (Ladavas et al., 1984; Shallice, 1988; Pardo et al., 1991; Posner and Rothbart, 1991), which are likely to be altered in chronic pain states. Additional studies will be required to assess the relative importance of several factors, such as the quality and duration of pain and the degree of coexistent depression, for right hemisphere activation.

An interesting issue concerns to what extent the pain-related CNS pattern of activation differs from that induced by nonnoxious somatic (e.g., mechanical) stimuli. In the spinal cord, the peak intensity and spatial extent of metabolic increases are much higher during acute noxious stimulation or prolonged pain than during intense nonnoxious stimulation; these findings are likely to reflect the recruitment of a larger number of spinal neurons (Coghill et al., 1993). Pain appears also to be associated with a more widespread activation in brainstem, diencephalic, and telencephalic structures than somatosensory stimulation or motor tasks, both in animal (see Sharp, 1984; Gonzalez and Sharp, 1985; Collins et al., 1986; Santori et al., 1986; Porrino et al., 1990; Porro and Cavazzuti, 1992; Mao et al., 1993) and human studies (e.g., Coghill et al., 1995; Davis et al., 1995). Thus, the overall CNS pattern of activation (including the spatial distribution and the intensity) seems to differentiate an acute painful from a non-painful state. Some brain areas, such as the anterior cingulate and anterior insular cortex, appear to display a consistent and pronounced activation in pain states, but different subregions may be preferentially involved in different motor, sensory or cognitive tasks (see Coghill et al., 1994; Hsieh et al., 1995b).

Comparison of metabolic and blood flow patterns in different pain states

Comparing the spatial and temporal profiles of brain activity in different experimental models and different kinds of pain in humans may help clarifying the functional role of individual regions, or better of different circuits, in pain states. For instance, pharmacologically-induced angina pectoris is associated with increased rCBF rates in some regions (hypothalamus, periaqueductal gray, left antero-inferior cingulate cortex, and the prefrontal cortex) and decreased rCBF in others (dorsal cingulate cortex, fusiform gyrus, and left posterior cingulate and parietal cortex). The decreased activity in the dorsal cingulate cortex, which is consistently activated during somatic and neuropathic pain, might be related to avoidance of movements which is commonly associated with cardiac pain (Rosen et al., 1994). As for somatic pain, primary and secondary somatosensory areas appear to be activated during phasic noxious stimulation, when the subject is more concerned with sensory-discriminative aspects of pain (e.g., localization and intensity coding), whereas during chronic pain cortical regions receiving information from medial thalamic structures, such as the anterior cingulate and prefrontal cortex, appear to be preferentially involved. Thus, different operational mechanisms may be involved in different pain states (Derbyshire et al., 1994; Hsieh et al., 1995b).

An intriguing difference between the results of animal and human studies concerns functional changes at the thalamic level. Metabolic activity levels in medial, lateral, and posterior thalamic nuclei are consistently increased during formalin-induced or neuropathic pain in rats (Porro and Cavazzuti, 1992; Mao et al., 1993; Porro et al., unpublished), suggesting enhanced synaptic activity. By contrast, regional blood flow in the thalamus appears to be *reduced* during chronic pain states in humans (Di Piero et al., 1991; Hsieh et al., 1995b; Iadarola et al., 1995; Mountz et al., 1995), whereas focal rCBF thalamic increases have been found during acute noxious stimulation (Jones et al., 1991; Casey et al., 1994a; Coghill et al., 1994; Hsieh et al., 1995a). At present, it can not be fully established whether these conflicting observations are related to different pathophysiological mechanisms or methodological issues. It is well known that, during acute intense nonnoxious sensory or motor stimulation, larger increases occur of rCBF (30–60%) than oxygen consumption and, probably, glucose utilization (Roland, 1993) in specific brain regions. This clearly facilitates the detection of phasic brain activation by blood-flow-sensitive methods. However, the relationships between regional blood flow, metabolic rates and neuronal function during prolonged changes of the activity of neuronal assemblies, and specifically during pain states, are far from being elucidated. With few exceptions, indices of blood flow rates (e.g., relative to whole-brain CBF), rather than absolute values, were indeed obtained in human studies on pain, and a recent quantitative PET study showed an unexpected global *reduction* of CBF during acute severe pain (Coghill et al., 1995). Additional quantitative studies with high-resolution imaging techniques for the study of cerebral metabolism and blood flow will help to unravel this issue.

Plasticity of neural systems in pain states

A fundamental issue in the neurobiology of pain concerns the plastic changes of the activity of central networks following tissue and nerve injury. These may be studied at several levels, from subcellular to behavioral, and by different methods. Functional imaging techniques appear well suited to investigate at a system level the time profile and circuitry of CNS activation during pain states. Our results in the formalin model suggest indeed that different channels processing noxious information may be differently modulated over time (Porro and Cavazzuti, 1993). The time-related changes in the activity of the pain system are likely to be related to different mechanisms, such as the pattern of peripheral sensory inflow, plastic changes of central nociceptive neurons, and modulation by endogenous antinociceptive systems (Dubner, 1991; Willis, 1994; Chapter 7 by Herz). To date, the bulk of our knowledge on the anatomy and function of endogenous antinociceptive systems concerns descending pathways and the spinal cord, whereas supraspinal mechanisms have received relatively little attention (Fields and Basbaum, 1994). The importance of endogenous antinociceptive mechanisms during acute prolonged pain in rats is highlighted by the results of a recent 2-DG study (Porro et al., 1993). Experiments were performed in formalin-injected rats in which the activity of the central beta-endorphinergic system was reduced by intracerebroventricular injection of specific antibodies. In these animals, metabolic rates were higher than the corresponding control group (treated with pre-immune gamma-globulins) in several diencephalic and telencephalic regions during the second peak of the response to s.c. formalin injection, in parallel with an increase of a pain-related behavior (licking the affected paw). Interestingly, formalin-induced metabolic activation at the spinal cord level was comparable in the two groups. This study shows that the metabolic pattern observed in the brain during tonic pain is indeed the net result of the balance between the activity of nociceptive and antinociceptive systems. Moreover, it suggests a direct action of endogenous opioids at supraspinal sites. It is worth mentioning that increases in opioid receptor binding in the orbitofrontal, cingulate and temporal cortices have been detected by PET in patients with rheumatoid arthritis, in association with a reduction of pain. These findings may be indicative of an increased activity of the central opioid system during chronic inflammatory pain (Jones et al., 1994). It would be interesting to investigate by similar approaches the involvement of central endogenous antinociceptive systems in other experimental pain models and different pain patients. Hopefully, these kinds of studies, including the investigation of the effects of analgesic drugs and techniques, will increase our understanding of the pathophysiology of prolonged pain states.

Acknowledgements

Supported by grants of Ministero Universita' Ricerca Scientifica Tecnologica and Consiglio Nazionale Ricerche, Italy.

References

Abram, S.E. and Kostreva, D.R. (1986) Spinal cord metabolic response to noxious radiant heat stimulation of the cat hind footpad. *Brain Res.*, 385: 143–147.
Aloisi, A.M., Porro, C.A., Cavazzuti, M., Baraldi, P. and Carli, G. (1993) 'Mirror pain' in the formalin test: behavioral and 2-deoxyglucose studies. *Pain,* 55: 267–273.
Apkarian, A.V., Stea, R.A., Manglos, S.H., Szeverenyi, N.M., King, R.B. and Thomas, F.D. (1992) Persistent pain inhibits contralateral somatosensory cortical activity in humans. *Neurosci. Lett.*, 140: 141–147.
Attal, N., Jazat, F., Kayser, V. and Guilbaud, G. (1990) Further evidence for 'pain-related' behaviors in a model of unilateral peripheral mononeuropathy. *Pain,* 41: 235–251.
Bennett, G.J. and Xie, Y.-K. (1988) A peripheral mononeuropathy in rat that produces disorders of pain sensation like those seen in man. *Pain,* 33: 87–107.
Berenberg, R.A. (1984) Recovery from partial deafferentation increases 2-deoxyglucose uptake in distant spinal segments. *Exp. Neurol.*, 84: 627–642.
Casey, K.L., Minoshima, S., Berger, K.L., Koeppe, R.A., Morrow, T.J. and Frey, K.A. (1994a) Positron Emission Tomographic analysis of cerebral structures specifically activated by repetitive noxious heat stimuli. *J. Neurophysiol.*, 71: 802–807.
Casey, K.L., Minoshima, S., Koeppe, R.A., Weeder, J. and Morrow, T.J. (1994b) Temporo-spatial dynamics of human forebrain activity during noxious heat stimulation. *Soc. Neurosci. Abstr.*, 20: 1573.
Chen, A.C.N. (1993) Human brain measures of clinical pain: a review. I. Topographic mappings, *Pain,* 54: 115–132.
Coderre, T.J., Katz, J., Vaccarino, A.L. and Melzack, R. (1993) Contribution of central neuroplasticity to pathological pain: review of clinical and experimental evidence. *Pain,* 52: 259–285.
Coghill, R.C., Price, D.D., Hayes, R.L. and Mayer, D.J. (1991) Spatial distribution of nociceptive processing in the rat spinal cord. *J. Neurophysiol.*, 65: 133–139.
Coghill, R.C., Mayer, D.J. and Price, D.D. (1993) The roles of spatial recruitment and discharge frequency in spinal cord coding of pain: a combined electrophysiological and imaging investigation. *Pain,* 53: 295–309.
Coghill, R.C, Talbot, J.D, Evans, A.C, Meyer, E., Gjedde, A., Bushnell, M.C. and Duncan, G.H. (1994) Distributed processing of pain and vibration by the human brain. *J. Neurosci.*, 14: 4095–4108.
Coghill, R.C, Sang, C.N., Gracely, R.H., Max, M.B., Berman, K.F., Bennett, G.J. and Iadarola, M.J. (1995) Regional and global cerebral blood flow during pain processing by the human brain. *Soc. Neurosci. Abstr.*, 21: 1636.
Cohen, M.S. and Bookheimer, S.Y. (1994) Localization of brain function using magnetic resonance imaging. *Trends Neurosci.*, 17: 268–277.
Collins, R.C., Santori, E.M., Der, T., Toga, A.W. and Lothman, E.W. (1986) Functional metabolic mapping during limb movement in rat. I. Stimulation of motor cortex. *J. Neurosci.*, 6: 448–462.
Davis, K.D., Wood, M.L., Crawley, A.P. and Mikulis, D.J. (1995) fMRI of human somatosensory and cingulate cortex during painful electrical nerve stimulation. *NeuroReport*, 7: 321–325.
Derbyshire, S.W.G., Jones, A.K.P., Devani, P., Friston, K.J., Feinmann, C., Harris, M., Pearce, S., Watson, J.D.G. and Frackowiak, R.S.J. (1994) Cerebral responses to pain in patients with atypical facial pain measured by positron emission tomography. *J. Neurol. Neurosurg. Psychiatry*, 57: 1166–1172.
Di Piero, V., Jones, A.K.P., Iannotti, F., Powell, M., Perani, D., Lenzi, G.L. and Frackowiak, R.S.J. (1991) Chronic pain: a PET study of the central effects of percutaneous high cervical cordotomy. *Pain,* 46: 9–12.
Di Piero, V., Ferracuti, S., Sabatini, U., Pantano, P., Cruccu, G. and Lenzi, G.L. (1994) A cerebral blood flow study on tonic pain activation in man. *Pain,* 56: 167–173.
Dubner, R. (1991) Neuronal plasticity and pain following peripheral tissue inflammation or nerve injury. In: M.R. Bond, J.E. Charlton and C.J. Woolf (Eds.), *Proc. VIth World Cong. Pain,* Pain Research and Clinical Management, Vol. 4. Elsevier, Amsterdam, pp. 263–276.
Dubuisson, D. and Dennis, S.G. (1977) The formalin test: a quantitative study of the analgesic effects of morphine, meperidine, and brain stem stimulation in rats and cats. *Pain,* 4: 161–174.
Duncan, G.E. (1992) High resolution autoradiographic imaging of brain activity patterns with 2-deoxyglucose: regional topographic and cellular analysis. In: F. Gonzalez-Lima, Th. Finkenstädt and H. Scheich (Eds.), *Advances in Metabolic Mapping Techniques for Brain Imaging of Behavioral and Learning Functions.* Kluwer, Dordrecht, pp. 151–172.
Duncan, G.H., Morin, C., Coghill, R.C., Evans, A., Worsley, K.J. and Bushnell, M.C. (1994) Using psychophysical rat-

ings to map the human brain regression of regional cerebral blood flow (rCBF) to tonic pain perception. *Soc. Neurosci. Abstr.*, 20: 1572.

Fields, H.L. and Basbaum, A.I. (1994) Central nervous system mechanisms of pain modulation. In: P.D. Wall and R. Melzack (Eds.), *Textbook of Pain*, 3rd edn., Churchill Livingstone, Edinburgh, pp. 243–257.

Gelnar, P.A., Szeverenyi, N.M. and Apkarian, A.V. (1994) SII has the most robust response of the multiple cortical areas activated during painful thermal stimuli in humans, using multi-slice functional MRI. *Soc. Neurosci. Abstr.*, 20: 1572.

Gonzalez, M.F. and Sharp, F.R. (1985) Vibrissae tactile stimulation: 14C-2-deoxyglucose uptake in rat brainstem, thalamus and cortex. *J. Comp. Neurol.*, 231: 457–472.

Hsieh, J.-C., Ståhle-Bäckdahl, M., Hägermark Ö., Stone-Elander, S., Rosenquvist G. and Ingvar, M. (1995a) Traumatic nociceptive pain activated the brain defense system investigated by PET. *Human Brain Mapping*, Suppl. 1: 182.

Hsieh, J.-C., Belfrage, M., Stone-Elander, S., Hansson, P. and Ingvar, M. (1995b) Central representation of chronic ongoing neuropathic pain studied by positron emission tomography. *Pain*, 63: 225–236.

Iadarola, M.J., Max, M.B., Berman, K.F., Byas-Smith, M.G., Coghill R.C., Gracely, R.H., Zeffiro, T. and Bennett, G.J. (1995) Unilateral decrease in thalamic activity observed with positron emission tomography in patients with chronic neuropathic pain. *Pain*, 63: 55–64.

Jones, A.K.P. and Derbyshire, S.W.G. (1994) Positron emission tomography as a tool for understanding the cerebral processing of pain. In: J. Boivie, P. Hansson and U. Lindblom (Eds.), *Touch, Temperature, and Pain in Health and Disease: Mechanisms and Assessments*, Progress in Pain Research and Management, Vol. 3. IASP Press, Seattle, WA, pp. 491–520.

Jones, A.K.P., Brown, W.D., Friston, K.J., Qi, L.Y. and Frackowiak, R.S.J. (1991) Cortical and subcortical localization of response to pain in man using positron emission tomography. *Proc. R. Soc. London B*, 244: 39–44.

Jones, A.K.P., Derbyshire, S.W.G. and Pearce, S. (1993) Reduced central responses to pain in patients with rheumatoid arthritis. *Br. J. Rheumatol.*, 32 (Suppl. 2): S20.

Jones, A.K.P., Cunningham, V.J., Ha-kawa, S., Fujiwara, T., Luthra, S.K., Silva, S., Derbyshire, S. and Jones, T. (1994) Changes in central opioid receptor binding in relation to inflammation and pain in patients with rheumatoid arthritis, *Br. J. Rheumatol.*, 33: 909–916.

Juliano, S.L., Bernard, J.-F., Peschanski, M., and Besson, J.-M. (1987) Altered metabolic activity patterns in arthritic rats evoked by somatic stimulation. In: J.-M. Besson, G. Guilbaud and M. Peschanski (Eds.), *Thalamus and Pain*, Excerpta Medica, Amsterdam, pp. 155–170.

Kenshalo, Jr., D.R. and Willis, Jr., W.D. (1990) The role of the cerebral cortex in pain sensation. In: A. Peters and E.G. Jones (Eds.), *Cerebral Cortex*, Vol. 9. Plenum, New York, pp. 153–212.

Krauss, B.R., Apkarian, A.V., Thomas, P.S. and Szeverenyi, N. (1995) Reciprocal reversal of frontal and parietal cortical pain activation by blocking chronic RSD pain: an fMRI study. *Human Brain Mapping*, Suppl. 1: 424.

Kwong, K.K. (1995) Functional magnetic resonance imaging with echo planar imaging. *Magn. Reson. Quart.*, 11: 1–13.

Kwong, K.K., Belliveau, J.W., Chesler, D.A., Goldberg, I.E., Weisskoff, R.M., Poncelet, B.P., Kennedy, D.N., Hoppel, B.E., Cohen, M.S., Turner, T., Cheng, H.-M., Brady, T.J. and Rosen, B.R. (1992) Dynamic magnetic resonance imaging of human brain activity during primary sensory stimulation. *Proc. Natl. Acad. Sci. USA*, 89: 5675–5679.

Ladavas, E., Nicoletti, R., Umiltà, C. and Rizzolatti, G. (1984) Right hemisphere interference during negative affect: a reaction time study. *Neuropsychologia*, 22: 479–485.

Mao, J., Price, D.D., Coghill, R.C., Mayer, D.J. and Hayes, R.L. (1992) Spatial patterns of spinal cord 14C-2-deoxyglucose metabolic activity in a rat model of painful peripheral mononeuropathy. *Pain,* 50 : 89–100.

Mao, J., Mayer, D.J. and Price, D.D. (1993) Patterns of increased brain activity indicative of pain in a rat model of peripheral mononeuropathy. *J. Neurosci.*, 13: 2689–2702.

Melzack, R. and Casey, K.L. (1968) Sensory, motivational, and central control determinants of pain. A new conceptual model. In: R. Kenshalo (Ed.), *The Skin Senses*. CC Thomas, Springfield, IL, pp. 423–443.

Mountz, J.M., Bradley, L.A., Modell, J.G., Alexander, R.W., Triana-Alexander, M., Aaron, L.A., Stewart, K.E., Alarcón, G.S. and Mountz, J.D. (1995) Fibromyalgia in women. Abnormalities of regional cerebral blood flow in the thalamus and the caudate nucleus are associated with low pain threshold levels. *Arthritis Rheumatism*, 38: 926–938.

Pardo, J.V., Fox, P.T. and Raichle, M.E. (1991) Localization of a human system for sustained attention by positron emission tomography. *Nature*, 349: 61–64.

Phelps, M.E., Huang, S.C., Hoffman, E.J., Selin, C., Sokoloff, L. and Kuhl, D.E. (1979) Tomographic measurement of local cerebral glucose metabolic rate in humans with (F-18)2-Fluoro-2-Deoxy-D-Glucose: validation of method. *Ann. Neurol.*, 6: 371–388.

Porrino, L.J., Huston-Lyons, D., Bain, G., Sokoloff, L. and Kornetsky, C. (1990) The distribution of changes in local cerebral energy metabolism associated with brain stimulation reward to the medial forebrain bundle of the rat. *Brain Res.*, 511: 1–6.

Porro, C.A. and Cavazzuti, M. (1992) Functional correlates of acute prolonged pain in the rat central nervous system: 2-DG studies. In: F. Gonzalez-Lima, Th. Finkenstädt and H. Scheich (Eds.), *Advances in Metabolic Mapping Techniques for Brain Imaging of Behavioral and Learning Functions*. Kluwer, Dordrecht, pp. 319–342.

Porro, C.A. and Cavazzuti, M. (1993) Spatial and temporal aspects of spinal cord and brainstem activation in the formalin pain model. *Prog. Neurobiol.*, 41: 565–607.

Porro, C.A., Cavazzuti, M., Galetti, A., Sassatelli, L. and Barbieri, G.C. (1991a) Functional activity mapping of the rat spinal cord during formalin-induced noxious stimulation. *Neuroscience*, 41: 655–665.

Porro, C.A., Cavazzuti, M., Galetti, A. and Sassatelli, L. (1991b) Functional activity mapping of the rat brainstem during formalin-induced noxious stimulation. *Neuroscience*, 41: 667–680.

Porro, C.A., Cavazzuti, M., Giuliani, D., Panerai, A. and Tedeschi, P. (1993) 2-deoxyglucose mapping of the rat CNS during tonic pain: effects of blockade of the central beta-endorphin system, *Abstracts Book VIIth World Congress on Pain,* IASP Publications, Seattle, WA, p. 501.

Posner, M.I. and Rothbart, M.K. (1991) Attentional mechanisms and conscious experience. In: A.D. Milner and M.D. Rugg (Eds.), *The Neuropsychology of Consciousness*. Academic Press, London, pp. 91–111.

Price, D.D. (1988) *Psychological and Neural Mechanisms of Pain*, Raven, New York, 241 pp.

Price, D.D., Mao, J., Coghill, R.C., d'Avella, D., Cicciarello, R., Fiori, M.G., Mayer, D.J. and Hayes, R.L. (1991) Regional changes in spinal cord glucose metabolism in a rat model of painful neuropathy. *Brain Res.*, 564: 314–318.

Raichle, M.E. (1987) Circulatory and metabolic correlates of brain function in normal humans. In: F. Plum (Ed.), *Handbook of Physiology, Section 1, The Nervous System*, Vol. 5, Higher Functions of the Brain. Oxford University Press, New York, pp. 643–674.

Roland, P.E. (1993) *Brain Activation*, Wiley, New York, 589 pp.

Rosen, S.D., Paulesu, E., Frith, C.D., Frackowiak, R.S.J., Davies, G.J., Jones, T. and Camici, P.G. (1994) Central nervous pathways mediating angina pectoris. *Lancet*, 344: 147–150.

Santori, E.M., Der, T. and Collins, R.C. (1986) Functional metabolic mapping during forelimb movement in rat. II. Stimulation of forelimb muscles, *J. Neurosci.*, 6: 463–474.

Shallice, T. (1988) *From Neuropsychology to Mental Structure*, Cambridge University Press, Cambridge, UK.

Sharp, F.R. (1984) Regional (14C)2-deoxyglucose uptake during forelimb movements evoked by rat motor cortex stimulation: cortex, diencephalon, midbrain. *J. Comp. Neurol.*, 224: 259–285.

Sokoloff, L. (1983) Measurement of local glucose utilization in the central nervous system and its relationship to local functional activity. In: A. Lajtha (Ed.), *Handbook of Neurochemistry*, Vol. 3. Plenum, New York, pp. 225–257.

Sokoloff, L., Reivich, M., Kennedy, C., Des Rosiers, M.H., Patlak, C.S., Pettigrew, K.D., Sakurada, O. and Shinohara, M. (1977) The 14C-deoxyglucose method for the measurement of local cerebral glucose utilization: theory, procedure, and normal values in the conscious and anesthetized albino rat. *J. Neurochem.*, 28: 897–931.

Talbot, J.D., Marrett, S., Evans, A.C., Meyer, E., Bushnell, C.M. and Duncan, G.H. (1991) Multiple representations of pain in human cerebral cortex. *Science*, 251, 1355–1358.

Vogt, B.A., Sikes, R.W. and Vogt, L.J. (1993) Anterior cingulate cortex and the medial pain system, In: B.A. Vogt and M. Gabriel (Eds.), *Neurobiology of Cingulate Cortex and Limbic Thalamus*. Birkhäuser, Boston, MA, pp. 313–344.

Wen, H., Wolff, R., Balaban, R., Turner, R., Kenshalo, D., Berman, K.F. and Iadarola, M.J. (1993) Imaging pain in humans with high resolution functional magnetic resonance imaging (MRI) at 4 Tesla (4T). *Soc. Neurosci. Abstr.*, 19: 1074.

Willis, W.D. (1985) *The Pain System*, Karger, Basel, 346 pp.

Willis, W.D. (1994) Central plastic responses to pain. In: G.F. Gebhart, D.L. Hammond and T.S Jensen (Eds.), *Proc. 7th World Cong. Pain,* Progress in Pain Research and Management, Vol. 2. IASP Press, Seattle, WA, pp. 301–324.

Willis, W.D. and Coggeshall, R.E. (1991) *Sensory Mechanisms of the Spinal Cord*, 2nd edn. Plenum, New York, 575 pp.

Woolf, C.J. (1994) The dorsal horn: state-dependent sensory processing and the generation of pain. In: P.D. Wall and R. Melzack (Eds.), *Textbook of Pain*, 3rd edn. Churchill Livingstone, Edinburgh, pp. 101–112.

CHAPTER 5

Limbic processes and the affective dimension of pain

C. Richard Chapman

Departments of Anesthesiology, Psychiatry and Behavioral Sciences, University of Washington School of Medicine, Seattle, WA 98195, USA and Pain and Toxicity Research Program, Fred Hutchinson Cancer Research Center, Seattle, WA, USA

Introduction

Most of us think of pain as an unpleasant sensation that originates in traumatized tissues, but pain also has emotional qualities. Ancient philosophers by and large considered pain an emotion. Aristotle, for example, called it a passion of the soul. A contemporary writer described pain's qualities as including extreme aversiveness, an ability to annihilate complex thoughts and other feelings, an ability to destroy language, and a strong resistance to objectification (Scarry, 1985). Her perspective resonates with the lessons of everyday life: while pain has sensory features and lends itself to sensory description, it is above all else a powerful and demanding feeling state. Put more simply, pain is in part an emotion.

The International Association for the Study of Pain has acknowledged the central role of emotion in its keystone definition:

"Pain [is] an unpleasant sensory and *emotional* experience associated with actual or potential tissue damage, or described in terms of such damage" (Merskey, 1979, italics added).

The definition clearly emphasizes the role of affect as an intrinsic component of pain. Emotion is not simply a consequence of pain sensation that occurs after a noxious sensory message arrives at somatosensory cortex. Rather, it is a fundamental part of the pain experience. Until recently, this conspicuous and oft-quoted definition merited only lip service. Researchers concerned with basic mechanisms of pain addressed sensory processing almost exclusively, simply ignoring the question of why pain disturbs us and compels us to seek relief. Fortunately, recent work on imaging of pain processes has begun to shed light on the complexity of human pain and its emotional character.

This chapter calls attention to the affective characteristics of pain in an attempt to bridge knowledge gained in laboratory research to clinical pain phenomena. Literature on the neurophysiology of emotion and the neuroendocrinology of stress, both rich sources of literature, suggest approaches for studying the affective component of pain. Moreover, psychological research and theory in these areas also provides a valuable resource and a fresh perspective. In this chapter I briefly review these areas, derive from them hypotheses about the affective character of pain, and call attention to the congruity of recent data from brain imaging studies with these hypotheses.

I offer a limited model to describe the central mechanism of the affective dimension of pain. Models are, by definition, intentional oversimplifications, and I put this one forth as an example of how one might study pain as an emotion rather than as a definitive approximation of truth. The brain and human consciousness are more complex by far than the model allows, but we must begin somewhere if we are to progress. In brief, I propose that tissue trauma: (1) excites spinoreticular as well as spinothalamic pathways; (2) generates concomitant affective and sensory processes that subserve complementary adaptive functions; (3) activates predominantly noradrenergic limbic structures to produce the affective dimension of pain; and (4) the hypothalamically-mediated stress response plays a role in pain chronicity. Before addressing pain directly, I briefly review the field of emotion research.

Emotion, its functions and its expressions

The field of emotion research suffers from a lack of consensus on basic definitions. There are many theories of emotion drawn from many disciplines, and the individual theories have spawned their own definitions. Review of the many theories exceeds the scope of this chapter. However, I find strong agreement among mainstream emotion researchers on several fundamental points, and these serve to define the field. Good consensus exists on the following:

(1) Emotional phenomena evolved to foster survival of the individual, and the species and emotional responses to stimuli and emotional expression foster biological adaptation;
(2) Emotions impute positive or negative hedonic qualities to a stimulus in accordance with the biological importance and meaning of that stimulus;
(3) The central neuroanatomy for emotion corresponds to the limbic brain;
(4) Emotions activate - they produce impulses to act or to express one's self;
(5) Emotions communicate, and the negative emotional expression of one individual will tend to produce negative emotion in another; and
(6) Human cognitions and emotions function interdependently. What we think determines what we feel, and the reverse holds true also.

These points of agreement help to clarify what science currently means by emotion, but a conclusive, consensual and generic definition for emotion still eludes us.

In approaching the emotional dimension of pain, I favor a sociobiological (evolutionary) framework that interprets feeling states, related physiology and behavior in terms of adaptation and survival. Nature has equipped us with the capability for negative emotion for a purpose; bad feelings are not simply accidents of human consciousness. By understanding the emotional dimension of pain from this perspective, we may gain some insight about how to prevent or control emotions that foster suffering. Implementation of this approach as a world-view of pain requires that we dispense with conventional language habits that involve describing pain as a transient sensory event. Instead, I argue that we construe pain as a state of the individual which has as its primary defining feature awareness of, and homeostatic adjustment to, tissue trauma.

Adaptive functions of emotion

Emotions and the emotional dimension of pain characterize mammals exclusively and appear to foster mammalian adaptation. MacLean (1990) contended that emotions "impart subjective information that is instrumental in guiding behavior required for self-preservation and preservation of the species. The subjective awareness that is an affect consists of a sense of bodily pervasiveness or by feelings localized to certain parts of the body." As emotion evolved to facilitate adaptation and survival, and negative emotion plays an important defensive role. The ability to impute threat to certain types of environmental events protects against life-threatening injury.

Within consciousness threat manifests as a feeling state, and in humans threatening events that are not immediately present can exist as emotionally-colored somatosensory images. We can react emotionally to the mental image of a painful event before it happens (e.g., venipuncture), or for that matter we can respond emotionally to the sight of another person's tissue trauma. The emotional intensity of such a feeling marks the adaptive significance of the event that produced the experience. The threat of a minor injury normally provokes less feeling than one that incurs a high risk of death. The emotional magnitude of a pain, therefore, is the internal representation of the threat associated with the event that produced the pain. The key point is that the strength of emotional arousal indicates, and expresses, perceived threat to the biological integrity of the individual.

Emotions and behavior

Emotions compel action and also expression through vocalization, posture, variations in facial musculature patterns and alterations of activity. This enhances communication and social support, thus contributing to survival. Darwin (1872), observing animals, noted that emotions enable communication through vocalization, startle, posture, facial expression and specific behaviors. Contemporary investigators who study emotions and human or animal social behavior emphasize that communication is a fundamental adaptive function of emotion (Ploog, 1986). Social mammals, including humans, use one another or their social group as resources for adaptation and survival. The emotional expression of pain in the presence of supportive persons is socially powerful; it draws upon a fundamental sociobiological imperative, communicating threat and summoning assistance.

Central neuroanatomy of emotion

The limbic brain represents an anatomical common denominator across mammalian species (MacLean, 1990), and this suggests that emotion represents a common feature in consciousness across mammals. Early investigators focused on the role of olfaction in limbic function. Papez (1937) linked the limbic brain to emotion, stating that: "It is proposed that the hypothalamus, the anterior thalamic nuclei, the gyrus cinguli, the hippocampus and their interconnections constitute a harmonious mechanism which may elaborate the functions of central emotion, as well as participate in emotional expression." Emotion may have evolutionary roots in olfactory perception.

MacLean (1952) introduced the term "limbic system" four decades ago and characterized its functions. Currently, he identifies three main subdivisions of the limbic brain: amygdala, septum and thalamocingulate (MacLean, 1990). Fig. 1 illustrates three main subdivisions of the limbic brain. These represent sources of afferents to parts of limbic cortex (MacLean, 1990). He also postulated that the limbic brain responds to two basic types of input: interoceptive and exteroceptive. These refer to sensory information from internal and external environments, respectively.

Pain research has yet to address the links between nociception and limbic processing definitively. However, one can find anecdotal evidence that implicates limbic structures in the distress of pain. Radical frontal lobotomies, once performed upon patients for psychosurgical purposes, typically interrupted pathways projecting from hypothalamus to cingulate cortex and putatively relieved the suffering of intractable pain without destroying sensory awareness (Fulton, 1951). Such neurosurgical records help clarify recent positron emission tomographic observations of human subjects undergoing painful cutaneous heat stimulation: noxious stimulation activates contralateral anterior cingulate and several other limbic areas (vide infra).

Emotion in learning and memory

Organisms that can learn readily from experience have adaptive advantages over those that cannot. That which promotes learning promotes survival. The affective component of pain contributes to both operant (instrumental) learning and classical conditioning (learning by association). Operant learning requires reinforcers, and reinforcers are events accompanied by emotions. Classical conditioning represents the formation of an association between a normally neutral event and the negative emotion associated with the onset of pain. Memory of past events, like learning, depends heavily upon emotion, and memories of past experience tend to shape expectations for the present and future.

Operant learning can occur in any setting where patients are active and reinforcing events take place. A reinforcer is an event that alters the future likelihood of a behavior recurring when it follows an instance of that behavior (Fordyce, 1990). Events that create pleasant feelings function as rewards (positive reinforcers); events that produce negative feelings are punishments (i.e., they suppress behaviors). The positive or negative nature of reinforcers and their personal significance occur in conscious awareness as feelings (Rolls, 1986).

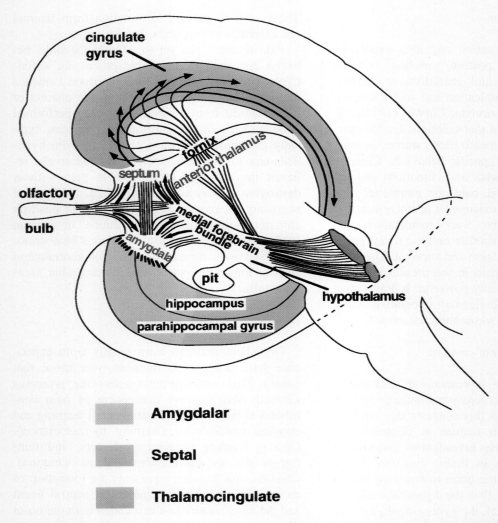

Fig. 1. Three subdivisions of the limbic brain. MacLean (1990) proposed a three-part grouping of limbic structures and functions: amygdalar, septal and thalamocingulate subdivisions. The groupings appear as color shadings. This figure, modified from MacLean's illustration (MacLean, 1990), portrays the hippocampus as an upright arch joining the septum at one end and the amygdala at the other. MacLean noted connections between thalamocingulate structures and those within the internal capsule.

Put another way, events that shape behavior are those that are emotionally prominent. Emotion-free events have no reinforcing properties and therefore cannot contribute to adaptive learning.

Fear accompanying pain can become associated with non-noxious stimuli through classical conditioning. In fear conditioning, the repeated pairing of a neutral stimulus with a noxious one can condition the perceiver so that the neutral stimulus, occurring alone, acts as a trigger to elicit fear. Many people develop fear or frank phobia in dentist's offices through classical conditioning.

Biologically, fear conditioning supports survival by fostering avoidance of potentially dangerous situations. Through conditioning, ordinarily neutral stimuli become warning cues for danger (Staddon, 1983). It also helps the individual martial a flight or fight response to a challenge after pre-exposure to it. Osborne et al. (1975) found that MHPG (3-methoxy-4-hydroxyphenylglycol), an indicator of norepinephrine turnover in brain, provided a marker of fear conditioning; exposure to a painful event increased MHPG in a manner that tracked the conditioning process.

Conditioned emotional responses are essentially sensory-affective associations. The amygdala appears to be the key structure in the linking of sensory experience to emotional arousal and in the conditioning of negative emotional associations (Gray, 1982; LeDoux et al., 1990). It probably contributes to the emotional evaluation of cognitive events (via corticofugal pathways) as well as sensory events that reach it via the dorsal noradrenergic bundle or sensory thalamus. Aggleton and Mishkin (1986) described the amygdala as a gateway to the emotions for stimuli (simple or complex) in all sensory modalities, both conditioned and unconditioned.

Sensory processing (in the case of pain, spinothalamic processing) can elicit complex, negative emotional processes through Pavlovian conditioning. This is not a "post-sensory" cortical association but rather a by-product of thalamic processing. LeDoux et al. (1988, 1990) working with auditory stimuli, determined that projections from acoustic thalamus to the amygdala allow the classical conditioning of emotional responses to normally neutral auditory stimuli in experimental animals. To condition animal subjects, they paired tones with footshock, evaluating autonomic responses and emotional behaviors. Their lesion work implicates separate efferent projections from the amygdala in conditioning of autonomic and behavioral responses. Moreover, their work indicating that emotional memories established by conditioning of subcortical systems strongly resist extinction (LeDoux, 1993). These and other observations suggest that emotion is a complex process sustained by several mechanisms. Under controlled circumstances individual mechanisms can be independently conditioned.

Fear conditioning almost certainly occurs in patients who undergo repeated painful diagnostic or treatment procedures. Fear conditioning can exacerbate the affective dimension of pain in cases where minor pain and intense affective arousal have been paired. Moreover, it can form associations between the environment surrounding a painful event and affective processing of that event so that the environment alone could elicit some elements of the affective dimension of pain.

Emotion associated with pain probably influences memory. Memory researchers surmise that both limbic and nonlimbic mechanisms contribute to memory processes (Gabriel et al., 1986). Emotional significance controls at least some and perhaps much memory formation: evidence exists that the brain preferentially stores information that has strong emotional loadings (Bower, 1981; Tucker et al., 1990). Heath (1986) proposed that learning and memory are "rooted in feeling and emotion" and identified hippocampus, cortical medial amygdala and cingulate gyrus as key areas involved in negative emotions.

In sum, the emotional component of pain appears to support adaptation and survival by facilitating learning, memory and related cognitive processes. It provides a bridge by which pain can influence the psychological status of the individual and his/her behavioral tendencies. Inadvertent conditioning can cause anticipatory anxiety or exacerbate the emotional distress of a painful event.

Emotion and cognition

Negative emotions appear to be much more than reactions to undesirable events; in nature they help an individual to determine which things benefit and which things threaten survival, and they compel behavior consistent with such evaluations. Moreover, emotional expression allows the individual communicate this judgment to others and thus set up group approach or avoidance behaviors. As noted above MacLean (1990) described emotion as a process that imparts subjective information. In these respects, emotion approximates a crude intelligence. If emotion is a proto-intelligence, then evolutionarily newer structures, namely the later stages of cortical development, should have demonstrable links with limbic structures and functions.

Such interconnections exist. Parts of frontal lobe (the dorsal trend) appear to have developed from rudimentary hippocampal formation while other parts (the paleocortical trend) originated in olfactory cortex. While these two areas interconnect anatomically, the former analyzes sensory information while the latter contributes emotional tone to that sensory information (Pandya et al., 1987). Pribram (1980) noting that limbic function involves frontal and temporal cortex, offered a bottom-up concept for how cognition relates to feelings: that is, emotion determines cognition. However, the multimodal neocortical association areas project corticofugally to limbic structures (Turner et al., 1980) and this suggests that cognitions may drive emotions.

Central mechanisms for the emotional dimension of pain

Nociception and central noradrenergic processing

Central sensory and affective central pain processes share common sensory mechanisms in the periphery: A-delta and C fibers serve as tissue trauma transducers (nociceptors) for both, the chemical products of inflammation sensitize these nociceptors, and peripheral neuropathic mechanisms such as ectopic firing excite both processes. Differentiation of sensory and affective processing begins at the dorsal horn of the spinal cord. Sensory transmission follows spinothalamic pathways and transmission destined for affective processing takes place in spinoreticular pathways. As others have described the sensory processing of nociception well, it need not be reviewed here (see Peschanski and Weil-Fugacza, 1987).

Nociceptive centripetal transmission engages both spinoreticular and spinothalamic pathways (Villanueva et al., 1989). The spinoreticular tract contains somatosensory and viscerosensory afferent pathways that arrive at different levels of the brain stem. Spinoreticular axons possess receptive fields that resemble those of spinothalamic tract neurons projecting to medial thalamus, and, like their spinothalamic counterparts, they transmit tissue injury information (Villanueva et al., 1990). Most spinoreticular neurons carry nociceptive signals and many of them respond preferentially to noxious input (Bowsher, 1976; Bing et al., 1990).

Processing of nociceptive signals to produce affect commences in reticulocortical pathways. Four extrathalamic afferent pathways project to neocortex: the dorsal noradrenergic bundle (DNB) originating in the locus coeruleus (LC); the serotonergic fibers that arise in the dorsal and median raphe nuclei; the dopaminergic pathways of the ventral tegmental tract that arise from substantia nigra; and the acetylcholinergic neurons that arise principally from the nucleus basalis of the substantia innominata (Foote and Morrison, 1987). Of these, the noradrenergic pathway links most closely to negative emotional states (Gray, 1982, 1987). The set of structures receiving projections from this complex and extensive network corresponds to classic definition of the limbic brain (Papez, 1937; Isaacson, 1982; Gray, 1987; MacLean, 1990).

Although other processes governed predominantly by other neurotransmitters almost certainly play important roles in the complex experience of emotion during pain, I emphasize the role of central noradrenergic processing here. This limited perspective offers the advantage of simplicity and permits the model to tell a well focused story. This processing involves two central noradrenergic pathways: the dorsal and ventral noradrenergic bundles.

Locus coeruleus and the dorsal noradrenergic bundle

The pontine nucleus, locus coeruleus (LC) resides bilaterally near the wall of the fourth ventricle as Fig. 2 shows. The locus has three major projections: ascending, descending and cerebellar. The ascending projection, the dorsal noradrenergic bundle (DNB), is the most extensive and important (Fillenz, 1990). Fig. 3 illustrates the DNB schematically. Projecting from the LC throughout limbic brain and to all of neocortex, the DBN accounts for about 70% of all brain norepinephrine (Svensson, 1987). The LC gives rise to most central noradrenergic fibers in spinal cord, hypothalamus, thalamus, hippocampus (Aston-Jones et al., 1985) in addition to its projections to limbic cortex and neocortex. Consequently this visually

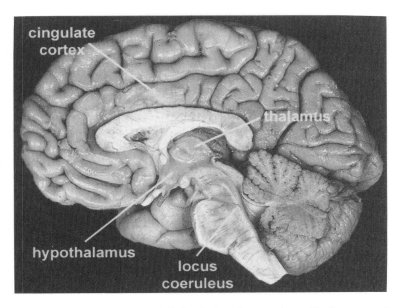

Fig. 2. A saggital view of the right hemisphere in a specimen. The locus coeruleus, hypothalamus, thalamus and cingulate cortex provide landmarks that help link this picture to the schematic in Fig. 3.

inauspicious nucleus exerts almost an enormous influence on brain activity.

The LC reacts to signaling from sensory stimuli that potentially threaten the biological integrity of the individual or signal damage to that integrity. Nociception inevitably and reliably increases activity in neurons of the LC, and LC excitation appears to be a consistent response to nociception (Korf et al., 1974; Stone, 1975; Morilak et al., 1987; Svensson, 1987). Notably, this does not require cognitively-mediated attentional control since it occurs in anesthetized animals. Foote et al. (1983) reported that slow, tonic spontaneous activity at the locus in rats changed under anesthesia in response to noxious stimulation. Experimentally induced phasic LC activation produces alarm and apparent fear in primates (Redmond and Huang, 1979), and lesions of the LC eliminate normal heart rate increases to threatening stimuli (Redmond, 1977).

The LC reacts consistently, but it does not respond exclusively, to noxious sensory input. LC activity increases following nonpainful threatening occurrences such as strong cardiovascular stimulation (Elam et al., 1985; Morilak et al., 1987) and certain visceral events such as distention of the bladder, stomach, colon or rectum (Elam et al., 1986b; Svensson, 1987). Thus, while it reacts to nociception consistently, the LC is not a nociceptive-specific nucleus. Rather, it responds to biologically threatening events, of which tissue injury is a significant subset. One can describe the LC as a central analog of the sympathetic ganglia (Amaral and Sinnamon, 1977).

Invasive studies confirm the linkage between LC activity and threat. Direct activation of the DNB and associated limbic structures in laboratory animals produces sympathetic nervous system response and elicits emotional behaviors such as defensive threat, fright, enhanced startle, freezing and vocalization (McNaughton and Mason, 1980). This indicates that enhanced activity in these pathways corresponds to negative emotional arousal and behaviors appropriate to perceived threat.

Normally, activity in the LC increases alertness; tonically enhanced LC and DNB discharge corresponds to hypervigilance and emotionality (Foote et al., 1983; Butler et al., 1990). The DNB is the mechanism for vigilance and orientation to affec-

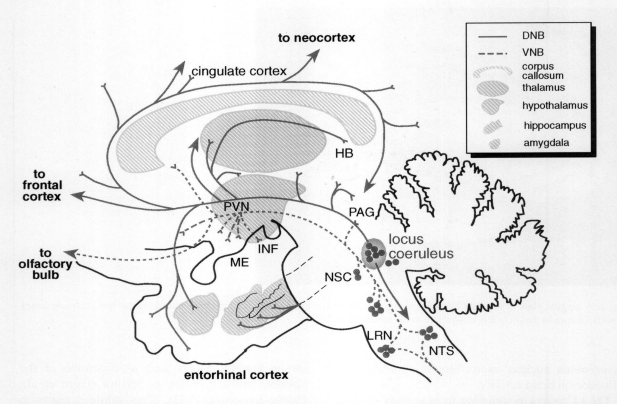

Fig. 3. Central corticopetal noradrenergic transmission in a primate brain (parasagittal view). The cell bodies of neurons that produce norepinephrine appear as circles. The major projections of these cell bodies are the dorsal noradrenergic bundle (DNB) shown in solid red and the ventral noradrenergic bundle (VNB) shown in dashed red. Tissue trauma signals from spinoreticular pathways excite the primarily noradrenergic locus coeruleus, activating the DNB which extends throughout the limbic brain and to neocortex. ME: median eminence, PAG: periaqueductal gray, HB: habenula, NSC: nucleus subcoeruleus, LRN: lateral reticular nucleus, NTS: nucleus tractus solitarius, INF: infundibulum, PVN: paraventricular nucleus of the hypothalamus.

tively relevant and novel stimuli. It also regulates attentional processes and facilitates motor responses (Elam et al., 1986a; Foote and Morrison, 1987; Gray, 1987; Svensson, 1987). In this sense, the LC influences the stream of consciousness on an ongoing basis and readies the individual to respond quickly and effectively to threat when it occurs.

LC and DNB support biological survival by making global vigilance for threatening and harmful stimuli possible. Siegel and Rogawski (1988) hypothesized a link between the LC noradrenergic system and vigilance, focusing on rapid eye movement (REM) sleep. They noted that LC noradrenergic neurons maintain continuous activity in both normal waking state and non-REM sleep, but during REM sleep these neurons virtually cease discharge activity. Moreover, an increase in REM sleep ensues after either lesions of the DNB or following administration of clonidine, an alpha-2 adrenoceptor agonist. Because LC inactivation during REM sleep permits rebuilding of noradrenergic stores, REM sleep may be necessary preparation for sustained periods of high alertness during subsequent waking. Siegel and Rogawski contended that "...a principal function of NE in the CNS is to facilitate the excitability of target neurons to specific high priority signals" (Siegel and Rogawski, 1988). Conversely, reduced LC activity periods (REM sleep) allow time for a suppression of sympathetic tone.

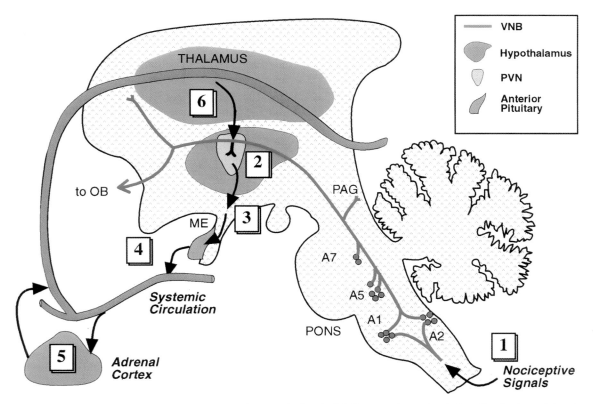

Fig. 4. Response of the hypothalamo-pituitary-adrenocortical axis (HPA) to noxious stimulation. Feedback-modulated response involves six steps. In the first, signals of tissue injury excite the ventral noradrenergic bundle (VNB), including several medullary and pontine nuclei (designated A1, A2, A5, and A7). When these signals reach the hypothalamus, they stimulate the paraventricular nucleus (PVN); this is Step 2. The PVN produces corticotropin releasing hormone (CRH). CRH-producing neurons extend from the PVN to the median eminence (ME) where they release CRH into the portal circulation, Step 3. At this point the tissue injury signals become neurohumoral rather than neuronal. The anterior pituitary responds to CRH by releasing adrenocorticotropin (ACTH) into the systemic circulation (Step 4). The adrenocortex responds to ACTH by releasing corticosteroids into the systemic circulation (Step 5). In addition to their extensive metabolic effects, the corticosteroids bind to receptors at the PVN (Step 6), thus closing the feedback loop. This mechanism contributes to the physiological arousal associated with the affective component of pain. (This figure is adapted with permission from Chapman, 1994).

These considerations, viewed collectively, suggest that the emotional dimension of pain shares central mechanisms with vigilance. This biologically important process, intensified by injury signals from within the organism, distressing environmental events from without the organism, or a combination of these, can generate a state that progresses to hypervigilance and beyond to panic. As a subjective experience, the emotional quality of pain seems, therefore, lends itself to description as awareness of immediate biological threat.

The ventral noradrenergic bundle and the hypothalamo-pituitary-adrenocortical (HPA) axis

The ventral noradrenergic bundle (VNB) is an ascending noradrenergic system that enters the medial forebrain bundle (see Fig. 3). Neurons in the medullary reticular formation project to hypothalamus via the VNB (Sumal et al., 1983). Sawchenko and Swanson (1982) identified two VNB-linked noradrenergic and adrenergic pathways to paraventricular hypothalamus in the rat:

the A1 region of the ventral medulla (lateral reticular nucleus, LRN), and the A2 region of the dorsal vagal complex (the nucleus tractus solitarius, NTS) which receives visceral afferents. These medullary neuronal complexes supply 90% of catecholaminergic innervation to the paraventricular hypothalamus via the VNB (Assenmacher et al., 1987). Regions A5 and A7 make comparatively minor contributions to the VNB.

Because it innervates the hypothalamus, the VNB is important for pain research. The noradrenergic axons in the VNB respond to noxious stimulation (Svensson, 1987) as does the hypothalamus (Kanosue et al., 1984). Moreover, nociception-transmitting neurons at all segmental levels of the spinal cord project to medial and lateral hypothalamus and several telencephalic regions (Burstein et al., 1988). These projections provide the major necessary neurophysiologic link between tissue injury and the hypothalamic response. Hormonal messengers may also play a part in some circumstances.

The hypothalamic paraventricular nucleus (PVN) serves as the coordinating center for the HPA axis. Neurons of the PVN receive afferent information from several reticular areas including ventrolateral medulla, dorsal raphe nucleus, nucleus raphe magnus, LC, dorsomedial nucleus, and the nucleus tractus solitarius (Sawchenko and Swanson, 1982; Peschanski and Weil-Fugacza, 1987; Lopez et al., 1991). Still other afferents project to the PVN from the hippocampus and amygdala. Nearly all hypothalamic and preoptic nuclei send projections to PVN.

In responding to potentially or frankly injurious stimuli the PVN initiates a complex series of events regulated by feed back mechanisms (see Fig. 4). These processes ready the organism for extraordinary behaviors that will maximize its chances to cope with the threat at hand (Selye, 1978). Analgesia during stress helps people and animals to cope with threat without the distraction of pain. Laboratory studies with rodents indicate that animals placed in restraint or subjected to cold water develop analgesia (Amir and Amit, 1979; Bodner et al., 1979; Kelly et al., 1993). Lesioning the PVN attenuates such stress induced analgesia (Truesdell and Bodner, 1987). Contemporary writers such as Henry (1986) and LeDoux (1988) hold that neuroendocrine arousal mechanisms are not limited to emergency situations even though most research emphasizes that such situations elicit them. In complex social contexts, submission, dominance and other transactions can elicit neuroendocrine and autonomic responses, modified perhaps by learning and memory. This suggests that neuroendocrine processes accompany all sorts of emotion-eliciting situations.

Strong links exist between hypothalamus and autonomic nervous system reactivity (Panksepp, 1986). Psychophysiologists hold that diffuse sympathetic arousal reflects, albeit imperfectly, negative emotional arousal. The PVN invokes autonomic arousal through neural as well as hormonal pathways. It sends direct projections to the sympathetic intermediolateral cell column in the thoracolumbar spinal cord and the parasympathetic vagal complex, sources of preganglionic autonomic outflow (Krukoff, 1990). In addition, it signals release of epinephrine and norepinephrine from the adrenal medulla. ACTH (adrenocorticotrophic hormone) release, while not instantaneous, is quite rapid: it occurs within about 15 s (Sapolsky, 1992). These considerations implicate the HPA axis in the neuroendocrinologic and autonomic manifestations of emotion during pain states.

In addition to controlling neuroendocrine and autonomic nervous system reactivity, the HPA axis coordinates emotional arousal with behavior (Panksepp, 1986). Direct stimulation of hypothalamus can elicit well-organized action patterns, including defensive threat behaviors, accompanied by autonomic manifestations (Jänig, 1985). The existence of demonstrable behavioral subroutines in animals suggests that the hypothalamus plays a key role in matching behavioral reactions and bodily adjustments to challenging circumstances or biologically relevant stimuli. Moreover, at high levels stress hormones, especially glucocorticoids, may affect central emotional arousal, lowering startle thresholds and influencing cognition (Sapolsky, 1992). Saphier (1987) observed that cortisol altered the firing rate of neurons in limbic forebrain. Put simply, the HPA axis takes executive responsibility for coordinating behavioral readiness with physiological capability, awareness, and cognitive function.

Pain, stress and chronicity

Tissue trauma stimulates the HPA axis and thus produces a complex adaptive stress response involving neural and endocrinologic changes. This process lends itself to physiological investigation of stress. Concomitantly, the perception of tissue trauma (pain) produces parallel changes in consciousness and behavior. These changes constitute psychological stress and invite study from a psychological perspective. In contrast to the field of emotion, where little consensus exists, physiologists and psychologists agree substantially on the nature of stress, and their theories complement one another. Indeed, these areas overlap. They meet on a common, albeit controversial, scientific ground: emotion research.

Physiologists investigating stress focus on the primarily endocrinologic, feed back dependent, HPA axis. Sapolsky (1992), addressing the impacts of acute and chronic stress response on the process of aging, defined the primary concepts of stress research in these terms: "A *stressor* can be defined in a narrow, physiological sense as any perturbation in the outside world that disrupts homeostasis, and the *stress-response* is the set of neural and endocrine adaptations that help reestablish homeostasis". Psychologists, by and large, accept this perspective, but they emphasize psychological rather than physiological reactions to personal threats and injuries that originate in the individual's environment. Both physiological and behavioral perspectives are essential for a comprehensive description of stress.

These considerations generate a concept of stress as a *state* of the individual that has both physiological and psychological manifestations. Accordingly, *a stressor is any event or circumstance, acute or chronic, that threatens the biological or psychological integrity of the individual.* In nature tissue trauma threatens the biological viability of the individual by definition, and continuing signals of tissue trauma (nociception) constitute a persisting stressor. Physiological stress responses therefore accompany the neurological signaling of tissue trauma like its shadow, and apart from the first few seconds following injury onset when the body is mobilizing such responses, they coexist with sensory pain.

Physiological stress responses interact in complex ways with the sensory qualities of pain and with psychological coping. For example, glucocorticoids released by the HPA axis during the stress response diminish inflammation and block the sensitization of nociceptors in injured tissue. This response minimizes peripheral sensitization and attenuates noxious signaling. At the same time, HPA arousal releases ACTH and other proopiomelanocortin derived peptides including beta-endorphin into the blood stream (see Fig. 4). Conclusive information eludes us, but sufficient evidence exists to entertain the hypothesis that the stress response may concomitantly increase beta endorphinergic activity at the hypothalamic infundibular nucleus and thus centrally modulate the sensory aspect of pain. These changes facilitate fight or flight and energize psychological coping.

It follows that when pain persists over weeks or months, some form of the stress response persists as well. Biologically, this response exists to equip the individual for the emergency situation of immediate threat. When stress persists over days, weeks or months, its consequences may become counter-adaptive: disrupted circadian rhythms (e.g., sleep disorder, fatigue, appetite disturbance).

Supporting findings from cerebral blood flow studies

Recent studies involving positron emission tomography (PET) of regional cerebral blood flow (CBF) in volunteers experiencing pain have yielded strong support for the hypothesis that noxious stimulation activates limbic structures. Changes in CBF index neuronal activity in specific brain regions. The emerging findings noted below have led some investigators to postulate a "medial pain system" that involves emotional components and may play a dominant role in pain chronicity (Derbyshire et al., 1994; Jones et al., 1994). They

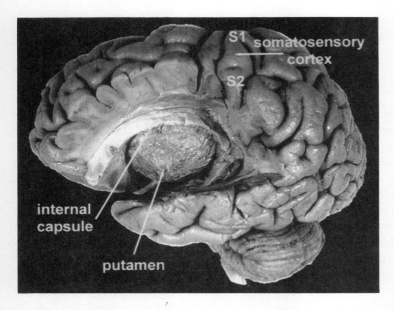

Fig. 5. This specimen of the left hemisphere demonstrates the location of the internal capsule and putamen. The globus pallidus is medial to the putamen and hidden from view. Interestingly, Chudler et al. (1993) demonstrated in rats that the globus pallidus receives nociceptive information. The specimen also reveals the location of primary (S1) and secondary (S2) somatosensory cortex.

contrast this to the lateral pain system which dominates in acute experimental pain and relates to somatosensory cortex.

Figs. 5–8 help clarify the outcomes by illustrating from specimens and magnetic resonance imaging the limbic brain structures associated with the pain experience. I encourage the reader to examine these figures in light of the findings summarized below and to refer to Figs. 1–3.

Pioneering studies examined both normals and patients. Jones et al. (1991) used a Peltier thermode to apply heat to the hands of normal volunteers, contrasting the CBF findings across three stimulus intensities ranging from noxious to innocuous. Pain related changes in CBF appeared in contralateral thalamus, lenticular nucleus and cingulate cortex. The same team studied CBF in 5 cancer patients with pain before and after percutaneous, ventrolateral cervical cordotomy. They compared patients before pain with normals and then compared patients with themselves before and after neurosurgical intervention. The comparison of patients with normals revealed significantly less blood flow in three out of four of the individual quadrants of the hemithalamus contralateral to the side of pain in the cancer patients. Cordotomy abolished the differences. Cordotomized patients demonstrated decreased CBF in the dorsal anterior quadrant of the thalamus contralateral to the side of pain, but no changes were evident in either primary somatosensory cortex or prefrontal cortex.

The lenticular nucleus, or lentiform nucleus, lies lateral to the thalamus and within the internal capsule, as shown in Fig. 5. It divides into two parts, the larger putamen and (medial to it and hidden from view) the smaller globus pallidus. The posterior limb of the internal capsule separates the globus pallidus from the thalamus. Fig. 5 also illustrates the locations of primary (S1) and secondary (S2) somatosensory cortex, areas associated with the sensory aspect of pain.

Talbot et al. (1991) stimulated the forearms of six normal volunteers with noxious heat from a contact thermode. Pain related CBF changes appeared in contralateral cingulate gyrus and in primary and secondary somatosensory cortex. Coghill et al. (1994) follow this with a PET study comparing CBF changes in normal volunteers during painful heat stimulation and vibrotactile stimulation. With painful stimulation subjects demonstrated CBF changes in contralateral thalamus, primary and secondary somatosensory cortices, anterior cingulate cortex, insula, and frontal cortex. With vibrotactile stimulation, changes appeared contralaterally in primary somatosensory cortex and bilaterally in secondary somatosensory cortex and insula. Both types of stimuli activated primary and secondary somatosensory cortical areas, but painful stimuli had a significantly greater effect on insula and produced in general a more widely dispersed effect.

Fig. 6 demonstrates the location of the insula in a specimen. Note that this is the same view as Fig. 5 but cut at a lesser depth. Figs. 7 and 8 indentify insula, cingulate gyrus and straight gyrus in axial and coronal magnetic resonance images of the brain. In these pictures one can see the lateral proximity of the insula to the lentiform nucleus. Maclean (1990) noted that the insula, or claustrocortex, is the only limbic cortex in which cells respond directly to somatic stimulation and that it lies in close proximity to primary and secondary somatosensory areas of the neocortex.

Casey et al. (1994) delivered noxious and innocuous heat pulses to the forearms of normal volunteers during PET analysis of CBF. Significant CBF increases occurred contralaterally with painful stimulation in thalamus, cingulate cortex, primary and secondary somatosensory cortex, and insula. Ipsilaterally, secondary somatosensory cortex, thalamus, medial dorsal midbrain and cerebellar vermis also showed increases in CBF.

Jones et al. (1994) examined rheumatoid arthritis patients with chronic inflammatory pain in a test of the hypothesis that such pain alters endogenous opioid binding at receptors in the brain. High concentration of such receptors exists in periaqueductal gray, medial thalamus, lentiform nucleus, anterior cingulate cortex and insular cortex. If chronic pain is associated with increased production of endogenous opioids and increased binding at receptors, then an exogenously introduced opioid substance should find fewer binding sites in these areas. The investigators used PET scanning to tracer quantities of ^{11}C diprenorphine following its intravenous injection in four patients, in pain and after pain relief. The observed significant changes in superior and inferior frontal cortex, straight gyrus, anterior and posterior cingulate, superior and midtemporal cortices. See Figs. 7 and 8 for localization of straight gyrus and cingulate.

Derbyshire et al. (1994) studied CBF in six patients with atypical facial pain, applying noxious and innocuous heat stimuli to the back of their hands contrasting their regional blood flow patterns to those of normal controls. Both patients and controls showed marked CBF differences between painful and nonpainful conditions in thalamus, anterior cingulate cortex, lentiform nucleus, insula, and prefrontal cortex. The patient group showed increased blood flow in anterior cingulate cortex but decreased blood flow in prefrontal cortex.

Do chronic and acute pain produce different patterns in regional CBF? Mountz and colleagues studied women diagnosed with fibromyalgia syndrome, a disorder characterized by widespread chronic musculoskeletal pain and fatigue (Mountz et al., 1995). They examined resting state CBF in ten patients and compared their data with those of

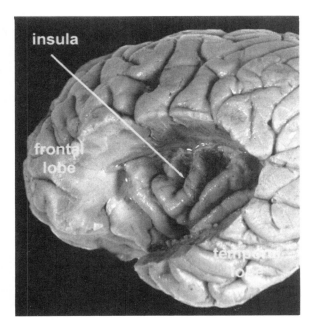

Fig. 6. In this specimen of the left hemisphere, removal of the opercula has exposed the insula (or central lobe), which lies deep within the lateral sulcus.

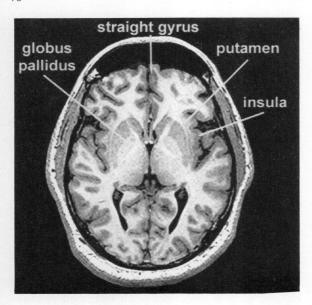

Fig. 7. This axial plane magnetic resonance image of the brain displays the relative locations of several of the structures associated with the limbic aspects of pain. The putamen and globus pallidus reside within the internal capsule, the lateral bordering on the thalamus. Adjacent to them, the insula lies within the lateral sulcus. The straight gyrus, only partially visible here, extends along the base of each frontal lobe medial and adjacent to the olfactory bulb and olfactory tract.

7 normal women. Resting regional bilateral blood flow was significantly lower in the fibromyalgia patients than in normals at thalamus, at the head of the caudate nucleus and in cortex. The observation of lower rather than higher CBF levels in the chronic pain patients led the authors to speculate that chronic pain may eventually reduce blood flow in certain brain areas. They postulated that a release of C fiber neuropeptides in response to chronic noxious stimulation together with diminished CBF altered central nervous system sensitivity to normally mildly noxious stimulation in fibromyalgia patients. These findings open new hypotheses about central differences in acute and chronic pain.

Collectively, these studies support the hypothesis that there is a limbic component of pain. They demonstrate that painful experience is associated with activation of the limbic brain as well as the better defined somatosensory pathways. Thalamus, cingulate cortex, insula, and frontal cortex emerge with high consistency across studies of both normal volunteers and patients with pain. Fig. 1 suggests that the response pattern associated with pain corresponds to MacLean's thalamocingulate division. Fig. 3 models the DNB and demonstrates that the results of the various PET studies are broadly consistent with the hypothesis that noradrenergic activation plays a major role in the affective component of pain.

Perspectives, interpretation and speculation

What role might the proposed affective dimension of pain play in chronic pain? Broadly, chronic pain breaks down into three types: (a) pain associated with enduring or slowly progressing tissue trauma (e.g., arthritis, neoplasia); (b) pain of neuropathic origin (e.g., lumbar disc herniation; post-mastectomy arm pain due to intercostal nerve injury; phantom limb pain); and (c) pain persisting in

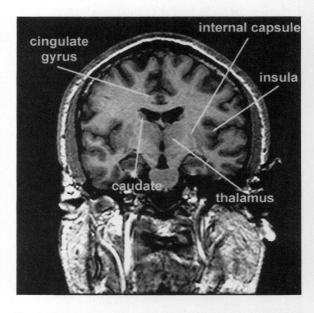

Fig. 8. This coronal plane magnetic resonance brain image demonstrates the location of cingulate gyrus immediately rostral to the corpus callosum. It also clarifies the relative positions of the thalamus, internal capsule, insula and the caudate nucleus. An elongated "C" shaped structure, the caudate nucleus arches along the dorsolateral border of the thalamus. Rostrally, the head of the caudate and the putamen are continuous.

the absence of adequate evidence of tissue trauma that may or may not have a historical link to injury or disease. While the mechanisms for the affective dimension of pain serve an adaptive biological function during or immediately following acute tissue trauma, they serve no biological purpose in any of the three types of chronic pain. A general rule of thumb in contrasting acute and chronic pain is that fear and anxiety are associated acute pain while depression more often accompanies chronic pain. This provides a clue to the role of affective processes in chronicity.

In cancer patients with diffuse bone metastasis, pain tends to emerge at certain metastatic foci but not others, to persist for variable durations of time, and to quiet down again. Overall, patients tend to have a constant bout of pain that has variable, often multiple locations. Such pain occurs in the context of high perceived threat and vigilance directed at subjective markers of disease progression. Thus, pain contributes to and sustains existing negative emotions such as fear. It also fosters depression by disturbing the HPA axis, degrading sleep and disrupting other biorhythms essential for personal comfort. Emerging studies indicate that depression in cancer patients with poorly controlled pain often depends directly upon the pain. Relief of the pain via skillful use of opioid or anti-inflammatory medications commonly alleviates the depression as well. This suggests that the affective component of pain is central to the suffering of patients with pain who are afflicted with life threatening disease. Accordingly, it points to a need for the development of interventions that directly target the affective component of pain.

Patients with neuropathic pain present with problems that have distinctive sensory qualities. The sensory features of patient complaint can distract care providers from the affective dimension of such pain syndromes. Because neuropathic pain responds poorly and sometimes not at all to treatment, patients sustain a perpetual low grade or moderate grade stress response. Poor sleep, fatigue, poor concentration, diminished sexual interest and general lethargy are consequences of HPA axis dysrythmia. The central noradrenergic components of the pain contribute hypervigilance and diffuse anxiety to the picture of dysphoria. Causalgia patients provide a striking example of pain induced hypervigilance.

Patients in the third group with the so-called benign chronic pain syndrome collectively defy any singular explanation. Fibromyalgia syndrome is only one example of this type of chronic pain. Often, however, affective disturbance is conspicuous, and this is the case with fibromyalgia. In some cases (and fibromyalgia is not necessarily among them), affective disturbance does not result from pain as it does for (a) and (b). Rather, emotional disorder rooted in domestic conflict, vocational maladjustment, or other psychosocial problems finds expression in the language and behaviors of pain. As I described above, emotion compels communication and communication requires a language. While some persons suffering emotional tension and distress can express their discomfort verbally, others find it more natural and fulfilling to express dysphoria in somatic complaints, abnormal posture or gait, invalid life style, and/or diminished activity levels. In such cases the expressed pain is not about tissue trauma; rather tissue trauma is a metaphor for threat to the self as a psychological entity. In short, pain is a language and not a symptom for some chronic pain patients.

Unfortunately, the third type of patient has received the most attention to date in the pain clinician's literature. Because the emotional and cognitive abnormalities of presentation are often conspicuous and perplexing in such patients, the literature is replete with material on the behavioral aspects of chronic pain. This material is valuable for the assessment and management of the "benign" chronic pain patient, but it has fostered the inappropriate generalization that emotional expression on the part of any patient with pain reflects psychological problems unrelated to tissue trauma. For patients in categories (a) and (b) and probably some in (c), this generalization simply does not apply. The pain has an affective dimension, and this dimension contributes directly to patient suffering. These considerations point to the need for research that clarifies the affective aspects of chronic pain related to tissue trauma and neuropathy.

Summary

Emotion is a fundamental characteristic of mammals and therefore of humans. The capability for emotion evolved to serve adaptation and promote survival. The intimate interdependence of emotional expression and pain behavior suggests that emotion is more than a reaction to the sensory awareness of pain. Closer examination shows that it is an integral part of the experience of pain caused by tissue trauma. Neurophysiological evidence indicates that signals indicating tissue trauma excite both spinothalamic pathways that convey them to somatosensory cortex and spinoreticular pathways that lead to limbic areas via extensive noradrenergic projections. In part, the affective aspect of pain involves excitation of central limbic structures involved in vigilance, fear and panic. In addition, it probably involves excitation of the hypothalamo-pituitary-adrenocortical axis and the feedback dependent stress response. During pain, sensory and affective processing take place simultaneously and in parallel.

While one can usefully describe the sensory aspects of pain in terms of linear, step-wise processes, limbic mechanisms and the stress response are inherently nonlinear, recursive and highly complex. These responses involve multiple interconnected structures and hormonal as well as neural signaling. And unlike sensory processes, one cannot specify a single point of termination that corresponds by analogy to primary somatosensory cortex. Limbic structures are complexly interconnected. Moreover, emotional processing of noxious signaling may involve several central phenomena and can greatly outlast the sensory processing. The emotional aspect of pain thus differs from the sensory aspect in several fundamental respects. It lends itself more readily to description as a state of the individual than as a sensation. I have argued that we construe pain as a state of the individual which has as its primary defining feature awareness of, and homeostatic adjustment to, tissue trauma.

These considerations suggest that the emotional aspect of pain may be more important for the clinical presentation of pain and its control than is the sensory aspect. Patients do not suffer with pain because of its sensory intensity but rather because of pain's negative emotional quality. In our efforts to prevent and control pain, we need to acknowledge the primacy of its emotional character and to make the emotional well being of the patient a high priority goal of pain control intervention.

Acknowledgements

Professor John Sundsten and the Digital Anatomist CGI Package, Department of Biological Structure, University of Washington, Seattle, WA graciously provided the images that appear in Figs. 5–8. The Digital Anatomist Program is a World Wide Web server located at http://www1.biostr.washington.edu/DigitalAnatomist.html.

References

Aggleton, J.P. and Mishkin, M. (1986) The amygdala: sensory gateway to the emotions. In: R. Plutchik and H. Kellerman (Eds.), *Emotion: Theory, Research and Experience*, Vol. 3, Academic Press, Orlando, FL, pp. 281–299.

Amaral, D.B. and M., S.H. (1977) The locus coeruleus: neurobiology of a central noradrenergic nucleus. *Prog. Neurobiol.*, 9: 147–196.

Amir, S. and Amit, Z. (1979) The pituitary gland mediates acute and chornic pain responsiveness in stressed and nonstressed rats. *Life Sci.*, 24: 439–448.

Assenmacher, I., Szafarczyk, A., Alonso, G., Ixart, G. and Barbanel, G. (1987) Physiology of neuropathways affecting CRH secretion. In: W.F. Ganong, M.F. Dallman and J.L. Roberts (Eds.), *The Hypothalamic-pituitary-adrenal Axis Revisited*, Vol. 512, pp. 149–161.

Aston-Jones, G., Foote, S.L. and Segal, M. (1985) Impulse conduction properties of noradrenergic locus coeruleus axons projecting to monkey cerebrocortex. *Neuroscience*, 15: 765–777.

Bing, Z., Villanueva, L. and Le Bars, D. (1990) Ascending pathways in the spinal cord involved in the activation of subnucleus reticularis dorsalis neurons in the medulla of the rat. *J. Neurophysiol.*, 63: 424–438.

Bodner, R.J., Glusman, M., Brutus, M., Spiaggia, A. and Kelly, D. (1979) Analgesia induced by cold-water stress: attenuation following hypophysectomy. *Physiol. Behav.*, 23: 53–62.

Bower, G.H. (1981) Mood and memory. *Am. Psychol.*, 36: 129–148.

Bowsher, D. (1976) Role of the reticular formation in responses to noxious stimulation. *Pain*, 2: 361–378.

Burstein, R., Cliffer, K.D. and Giesler, G.J. (1988) The spinohypothalamic and spinotelecephalic tracts: direct nociceptive projections from the spinal cord to the hypothalamus and telencephalon. In: R. Dubner, G.F. Gebhart and M.R. Bond (Eds.), *Proc. 5th World Congress on Pain*, Elsevier, New York, pp. 548–554.

Butler, P.D., Weiss, J.M., Stout, J.C. and Nemeroff, C.B. (1990) Corticotropin-releasing factor produces fear-enhancing and behavioral activating effects following infusion into the locus coeruleus. *J. Neurosci.*, 10: 176–183.

Casey, K.L., Minoshima, S., Berger, K.L., Koeppe, R.A., Morrow, T.J. and Frey, K.A. (1994) Positron emission tomographic analysis of cerebral structures activated specifically by repetitive noxious heat stimuli. *J. Neurophysiol.*, 71: 802–807.

Chapman, C.R. (1994) The affective dimension of pain: psychobiology and clinical implications. In: W.S. Nimmo, D.J. Bowbotham and G. Smith (Eds) *Anaesthesia*. 2nd edn. Blackwell Scientific, Oxford, UK, pp. 1557–1569.

Chudler, E.H., Sugiyama, K. and Dong, W.K. (1993) Nociceptive responses in the neostriatum and globus pallidus of the anesthetized rat. *J. Neurophysiol.*, 69: 1890–1903.

Coghill, R.C., Talbot, J.D., Evans, A.C., Meyer, E., Gjedde, A., Bushnell, M.C. and Duncan, G.H. (1994) Distributed processing of pain and vibration by the human brain. *J. Neurosci.*, 14: 4095–4108.

Darwin, C. (1872) *The Expression of the Emotions in Man and Animals*, John Murray, London.

Derbyshire, S.W., Jones, A.K., Devani, P., Friston, K.J., Feinmann, C., Harris, M., Pearce, S., Watson, J.D. and Frackowiak, R.S. (1994) Cerebral responses to pain in patients with atypical facial pain measured by positron emission tomography. *J. Neurol. Neurosurg. Psychiatry*, 57: 1166–72.

Elam, M., Svensson, T.H. and Thoren, P. (1985) Differentiated cardiovascular afferent regulation of locus coeruleus neurons and sympathetic nerves. *Brain Res.*, 358: 77–84.

Elam, M., Svensson, T.H. and Thoren, P. (1986a) Locus coeruleus neurons and sympathetic nerves: activation by cutaneous sensory afferents. *Brain Res.*, 366: 254–261.

Elam, M., Svensson, T.H. and Thoren, P. (1986b) Locus coeruleus neurons and sympathetic nerves: activation by visceral afferents. *Brain Res.*, 375: 117–125.

Fillenz, M. (1990) *Noradrenergic Neurons*, Cambridge University Press, Cambridge, UK.

Foote, S.L. and Morrison, J.H. (1987) Extrathalamic modulation of corticofunction. *Annu. Rev. Neurosci.*, 10: 67–95.

Foote, S.L., Bloom, F.E. and Aston-Jones, G. (1983) Nucleus locus ceruleus: new evidence of anatomical and physiological specificity. *Physiology Rev.*, 63: 844–914.

Fordyce, W.E. (1990) Contingency management. In: J.J. Bonica (Ed.), *The Management of Pain*, Vol. II, Lea & Febiger, Philadelphia, PA, pp. 1702–1710.

Fulton, J.E. (1951) *Frontal Lobotomy and Affective Behavior*. WW Norton, New York.

Gabriel, M., Sparenborg, S.P. and Stolar, N. (1986) The neurobiology of memory. In: J.E. LeDoux and W. Hirst (Eds.), *Mind and Brain: Dialogues in Cognitive Neuroscience*, Cambridge University Press, Cambridge, UK, pp. 215–254.

Gray, J.A. (1982) *The Neuropsychology of Anxiety: an Enquiry into the Functions of the Septo-hippocampal System*, Oxford University Press, New York.

Gray, J.A. (1987) *The Psychology of Fear and Stress*, Cambridge University Press, Cambridge, UK.

Heath, R.G. (1986) The neural substrate for emotion. In: R. Plutchik and H. Kellerman (Eds.), *Emotion: Theory, Research, and Experience*, Vol. 3, Academic Press, New York, pp. 3–35.

Henry, J.P. (1986) Neuroendocrine patterns of emotional response. In: R. Plutchik and H. Kellerman (Eds.), *Emotion: Theory, Research and Practice*, Vol. 3, Academic Press, Orlando, FL, pp. 37–60.

Isaacson, R.L. (1982) *The Limbic System*, Plenum Press, New York.

Jänig, W. (1985b) Systemic and specific autonomic reactions in pain: efferent, afferent and endocrine components. *Eur. J. Anaesth.*, 2: 319–346.

Jones, A.K., Brown, W.D., Friston, K.J., Qi, L.Y. and Frackowiak, R.S. (1991a) Cortical and subcortical localization of response to pain in man using positron emission tomography. *Proc. R. Soc. London B, Biol. Sci.*, 244: 39–44.

Jones, A.K., Cunningham, V.J., Ha-Kawa, S., Fujiwara, T., Luthra, S.K., Silva, S., Derbyshire, S. and Jones, T. (1994) Changes in central opioid receptor binding in relation to inflammation and pain in patients with rheumatoid arthritis. *Br. J. Rheumatol.*, 33: 909–16.

Kanosue, K., Nakayama, T., Ishikawa, Y. and Imai-Matsumura, K. (1984) Responses of hypothalamic and thalamic neurons to noxious and scrotal thermal stimulation in rats. *J. Thermobiol.*, 9: 11–13.

Kelly, D.D., Silverman, A.-J., Glusman, M. and Bodner, R.J. (1993) Characterization of pituitary mediation of stress-induced antinociception in rats. *Physiol. Behav.*, 53: 769–775.

Korf, J., Bunney, B.S. and Aghajanian, G.K. (1974) Noradrenergic neurons: morphine inhibition of spontaneous activity. *Eur. J. Pharmacol.*, 25: 165–169.

Krukoff, T.L. (1990) Neuropeptide regulation of autonomic outflow at the sympathetic preganglionic neuron: anatomical and neurochemical specificity. *Ann. N. Y. Acad. Sci.*, 579: 162–167.

LeDoux, J.E. (1993) Emotional memory: in search of systems and synapses. In: F.M. Crinella and J. Yu (Eds.), *Brain Mechanisms. Ann. N. Y. Acad. Sci.*, 702: 149–157.

LeDoux, J.E., Iwata, J., Cicchetti, P. and Reis, D.J. (1988) Different projections of the central amygdaloid nucleus mediate autonomic and behavioral correlates of conditioned fear. *J. Neurosci.*, 8: 2517–2529.

LeDoux, J.E., Farb, C. and Ruggiero, D.A. (1990) Topographic organization of neurons in the acoustic thalamus that project to the amygdala. *J. Neurosci.*, 10: 1043–1054.

Lopez, J.F., Young, E.A., Herman, J.P., Akil, H. and Watson, S.J. (1991) Regulatory biology of the HPA axis: an integrative approach. In: S.C. Risch (Ed.), *Central Nervous System Peptide Mechanisms in Stress and Depression*, American Psychiatric Press, Washington, DC, pp. 1–52.

MacLean, P.D. (1952) Some psychiatric implications of physiological studies on frontotemoral portion of limbic system (visceral brain). *Electroencehalogr. clin. Neurophysiol.*, 4: 407–418.

MacLean, P.D. (1990) *The Triune Brain in Evolution: Role in Paleocerebral Functions*, Plenum Press, New York.

McNaughton, N. and Mason, S.T. (1980) The neuropsychology and neuropharmacology of the dorsal ascending noradrenergic bundle - a review. *Prog. Neurobiol.*, 14: 157–219.

Merskey, H. (1979) Pain terms: a list with definitions and a note on usage. Recommended by the International Association for the Study of Pain (IASP) Subcommittee on Taxonomy. *Pain*, 6: 249–252.

Morilak, D.A., Fornal, C.A. and Jacobs, B.L. (1987) Effects of physiological manipulations on locus coeruleus neuronal activity in freely moving cats. II. Cardiovascular challenge. *Brain Res.*, 422: 24–31.

Mountz, J.M., Bradley, L.A., Modell, J.G., Alexander, R.W., Triana-Alexander, M., Aaron, L.A., Stewart, K.E., Alarcon, G.S. and Mountz, J.D. (1995) Fibromyalgia in women: abnormalities of regional cerebral blood flow in the thalamus and the caudate nucleus are associated with low pain threshold levels. *Arthritis Rheumatism*, 38: 926–938.

Osborne, F.H., Mattingley, B.A., Redmon, W.K. and Osborne, J.S. (1975) Factors affecting the measurement of classically conditioned fear in rats following exposure to escapable versus inescapable signaled shock. *J. Exp. Psychol.*, 1: 364–373.

Pandya, D.N., Barnes, C.L. and Panksepp, J. (1987) Architecture and connections of the frontal lobe. In: E. Perecman (Ed.), *The Frontal Lobes Revisited*, Lawrence Erlbaum Associates, Hillsdale, NJ, pp. 41–72.

Panksepp, J. (1986) The anatomy of emotions. In: R. Plutchik and H. Kellerman (Eds.), *Emotion: Theory, Research and Experience*, Vol. 3, Academic Press, Orlando, FL, pp. 91–124.

Papez, J.W. (1937) A proposed mechanism of emotion. *Arch. Neurolog. Psychol.*, 38: 725–743.

Peschanski, M. and Weil-Fugacza, J. (1987) Aminergic and cholinergic afferents to the thalamus: experimental data with reference to pain pathways. In: J.M. Besson, G. Guilbaud and M. Paschanski (Eds.), *Thalamus and Pain*, Excerpta Medica, Amsterdam, pp. 127–154.

Ploog, D. (1986) Biological foundations of the vocal expressions of emotions. In: R. Plutchik and H. Kellerman (Eds.), *Emotion: Theory, Research, and Experience*, Vol. 3, Academic Press, New York, pp. 173–198.

Pribram, K.H. (1980) The biology of emotions and other feelings. In: R. Plutchik and H. Kellerman (Eds.), *Emotion: Theory, Research, and Experience*, Vol. 1, Academic Press, New York, pp. 245–269.

Redmond, D.E.J. (1977) Alteration in the functions of the nucleus locus coeruleus: a possible model for studies of anxiety. In: I. Hannin and E. Usdin (Eds.), *Animal Models in Psychiatry and Neurology*, Pergamon Press, New York, pp. 293–306.

Redmond, D.E.J. and Huang, Y.G. (1979) Current concepts. II. New evidence for a locus coeruleus-norepinephrine connection with anxiety. *Life Sci.*, 25: 2149–2162.

Rolls, E.T. (1986) Neural systems involved in emotion in primates. In: R. Plutchik and H. Kellerman (Eds.), *Emotion: Theory, Research, and Experience*, Vol. 3, Academic Press, New York, pp. 125–144.

Saphier, D. (1987) Cortisol alters firing rate and synaptic responses of limbic forebrain units. *Brain Res. Bull.*, 19: 519–524.

Sapolsky, R.M. (1992) *Stress, the Aging Brain, and the Mechanisms of Neuron Death*, The MIT Press, Cambridge, MA.

Sawchenko, P.E. and Swanson, L.W. (1982) The organization of noradrenergic pathways from the brain stem to the paraventricular and supraoptic neuclei in the rat. *Brain Res. Rev.*, 4: 275.

Scarry, E. (1985) *The Body in Pain: the Making and Unmaking of the World*, Oxford University Press, New York.

Selye, H. (1978) *The Stress of Life*, McGraw-Hill, New York.

Siegel, J.M. and Rogawski, M.A. (1988) A function for REM sleep: regulation of noradrenergic receptor sensitivity. *Brain Res. Rev.*, 13: 213–233.

Staddon, J.E.R. (1983) *Adaptive Behavior and Learning*, Cambridge University Press, London.

Stone, E.A. (1975) Stress and catecholamines. In: A.J. Friedhoff (Ed.), *Catecholamines and Behavior*, Vol. 2, Plenum Press, New York, pp. 31–72.

Sumal, K.K., Blessing, W.W., Joh, T.H., Reis, D.J. and Pickel, V.M. (1983) Synaptic interaction of vagal afference and catecholaminergic neurons in the rat nucleus tractus solitarius. *J. Brain Res.*, 277: 31–40.

Svensson, T.H. (1987) Peripheral, autonomic regulation of locus coeruleus noradrenergic neurons in brain: putative implications for psychiatry and psychopharmacology. *Psychopharmacology*, 92: 1–7.

Talbot, J.D., Marrett, S., Evans, A.C., Meyer, E., Bushnell, M.C. and Duncan, G.H. (1991) Multiple representations of pain in human cerebral cortex [see comments]. *Science*, 251: 1355–8.

Truesdell, L.S. and Bodner, R.J. (1987) Reduction in coldwater swim analgesia following hypothalamic paraventricular nucleus lesions. *Physiol. Behav.*, 39: 727–731.

Tucker, D.M., Vannatta, K. and Rothlind, J. (1990) Arousal and activation systems and primitive adaptive controls on cognitive priming. In: N.L. Stein, D. Leventhal and T. Trabasso (Eds.), *Psychological and Biological Approaches to Emotion*, Lawrence Erlbaum Associates, Hillsdale, NJ, pp. 145–166.

Turner, B.H., Mishkin, M. and Knapp, M. (1980) Organization of the amygdalopedal projections from modality-specific cortical association areas in the monkey. *J. Comp. Neurol.*, 19: 515–543.

Villanueva, L., Bing, Z., Bouhassira, D. and Le Bars, D. (1989) Encoding of electrical, thermal, and mechanical noxious stimuli by subnucleus reticularis dorsalis neurons in the rat medulla. *J. Neurophysiol.*, 61: 391–402.

Villanueva, L., Cliffer, K.D., Sorkin, L.S., Le Bars, D. and Willis, W.D.J. (1990) Convergence of heterotopic nociceptive information onto neurons of caudal medullary reticular formation in monkey (Macaca fascicularis). *J. Neurophysiol.*, 63: 1118–1127.

Section II

Neuropeptides, Inflammation and Neuropathic Injuries

CHAPTER 6

Neurogenic mechanisms and neuropeptides in chronic pain

A. Dray

Sandoz Institute for Medical Research, 5 Gower Place, London WC1E 5BN, UK

Introduction: neuropeptide containing afferent neurons

All tissues are innervated by fine afferent fibres but the properties and physiological function of these fibres may differ depending on whether they are somatic or visceral afferents. A large subgroup of primary afferent fibres are nociceptive but this is less clear in visceral systems under normal physiological conditions. However a significant portion of afferent fibres become responsive once sensitised by irritants or inflammatory mediators (Habler et al., 1990; Schmelz et al., 1994). Most afferents also contain one or more peptides but the pattern of peptide coexistence and regulation is not well understood. Neuropeptide-mediated cell signalling is complex, and it is not yet clear what the entire range of functions is. Changes in neuropeptide synthesis and release are related to the pain symptoms which follow chronic inflammation and neuropathic injuries. These specific features of neuropeptide containing afferent neurones are discussed in this article but readers are also directed to several recent publications (Holzer, 1988; Willis and Coggeshall, 1991; Scott, 1992; Levine et al., 1993; Rang et al., 1994) that have described some the properties of sensory neurones in considerably greater detail.

Neuropeptides and the orchestration of inflammation

Various products of tissue damage and inflammation stimulate afferent fibres to induce pain, hyperalgesia and to release neuropeptides which produce a variety of effects in the periphery. Some of the mediators of inflammation whose actions are best understood are summarised in Fig. 1. These include peptidic growth factors and cytokines released from target tissues and immune cells as well as neuropeptides released from primary afferents (substance P) and from sympathetic neurones (neuropeptide Y).

The most comprehensively investigated of the sensory neuropeptides are substance P, neurokinin-A (NK-A) and calcitonin gene-related peptide (CGRP) which play a critical role in the responses elicited by sensory nerves. These are important for orchestrating a number of events that occur in inflammation (Dray, 1994) (Fig. 2). For example substance P causes vasodilatation in part via the release of NO from vascular endothelium. In addition contraction of endothelial cells in venules allows the extravasation of plasma, immune cells, and other active substances (bradykinin, ATP, 5-HT, histamine) which thus gain access to the site of tissue injury. However vasodilatation and extravasation do not always occur together when fine afferent nerves are stimulated (Janig and Lisney, 1989) supporting the functional heterogeneity of afferent fibres and the release of different peptides. Indeed CGRP produces vasodilatation of arterioles with little direct effect on vascular permeability. The increases blood flow into venules results in synergistic actions with substance P in causing plasma extravasation (Brain and Williams, 1985; Gamse and Saria, 1985; Green et al., 1992). Mast cell degranulation, by substance P, also releases other inflammatory mediators including histamine and 5HT as well as proteolytic enzymes which

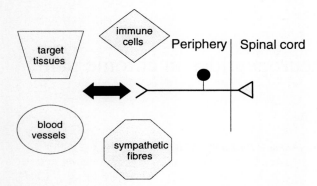

Fig. 1. Interactions between afferent nerve terminals and other cells and structures in peripheral tissues. The mediators responsible for most of the interactions are mainly peptides. Thus growth factors and cytokines are released from target tissues and immune cells while neuropeptides such as substance P, and neuropeptide -Y are released from peripheral afferent fibres and sympathetic fibres respectively.

catalyse the production of kinins. The latter may also evoke pain and induce a number of pro-inflammatory effects (Dray and Perkins, 1994; Rang et al., 1994). Substance P, NK-A and CGRP, when injected intradermally, are also capable of causing hyperalgesia (Nakamura-Craig and Gill, 1991). Presently, it is not clear whether this can be attributed to direct stimulation of sensory fibres or to a number of indirect mechanisms involving sympathetic nerve fibres or the vasculature. It is useful to remember that the effects of neurokinins may be complemented or modified by the concomitant release of other sensory neuropeptides. For example, galanin, somatostatin and neuropeptide Y may have inhibitory actions on afferent excitability and thereby reduce substance P release from sensory fibres during neurogenic inflammation (Gazelius et al., 1981; Green et al., 1992). This mechanism may be important in some pathological situations, e.g. rheumatoid arthritis, in which substance P plays a critic role (Levine et al., 1993) and where the expression of sensory neuropeptides may be altered by inflammatory processes.

Lymphatic tissues are innervated by sensory fibres containing substance P and CGRP (Weihe et al., 1991) and it is likely that these neuropeptides regulate lymphoid tissues. Indeed the immune response (secretion of antigen-specific antibody) to antigen challenge is reduced following pretreatment of rats with capsaicin, to selectively damage sensory fibres and deplete neuropeptides. The reduced antibody response can be restored by exogenously administered neurokinins (Helme et al., 1987; Eglezos et al., 1991). Stimulation and recruitment of inflammatory cells is another important role of substance P in inflammation (Payan et al., 1983). For example, stimulation of cytokine production from monocytes (Lotz et al., 1988) leads to the expression and activation of adhesion molecules necessary for the attraction and movement of leukocytes along the vascular endothelium. Effects such as these offer an explanation for the observed close proximity of peptide-containing fibres to immune cells (Weihe et al., 1991). In addition, the persistence of macrophages in neuroma tissue following nerve cuts or ligations suggest that interactions between nerves and immune cells may be important for chronic pain symptoms as well as neuronal regeneration and remodelling (Friesen et al., 1993). The release of neurokinins has also been suggested to be involved directly in receptor mediated regenerative processes particularly in the control of NGF release, the regrowth of connective tissue and the revascularization that follows an injury (Nilsson et al., 1985; White et al., 1987; Ziche et al., 1990; Fan et al., 1993).

Ultimately, the effects of neurokinins are medi-

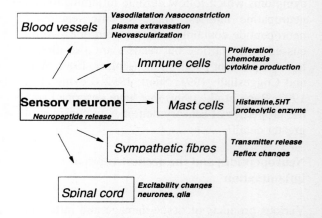

Fig. 2. Neuropeptides orchestrate peripheral inflammation and alter spinal excitability.

ated through the activation of one or other of the specific neurokinin (NK) receptors, NK1, NK2 or NK3 to which substance P, NKA and NKB respectively are the preferred naturally occurring ligands (Maggi et al., 1993). So far, actions mediated by NK1 and NK2 receptors have been implicated in the effects of neurokinins outlined above. This characterisation has been facilitated by the availability of several selective NK1 (CP99994, RP67580, SR140333) and NK2 receptor antagonists (MEN 10376, FK888, SR48968) (Maggi et al., 1992). Little work has been done on the role of NK3 receptors, due to the present lack of specific pharmacological antagonists. However increased expression of NK3 receptors in the spinal cord has been demonstrated only after peripheral inflammation (McCarson and Krause, 1994). Interestingly the mismatch between neuropeptide release and synaptic receptors in the spinal cord suggests that effects via non-synaptic elements may also be of importance during inflammation (Liu et al., 1994; Valtschanoff et al., 1995). Indeed, NK receptor activation may induce the release of neuroactive substances (ref) including glutamate from glial cells made reactive by inflammatory processes.

A specific pathophysiological role for neuropeptide-containing afferents, in the trigeminovascular system, has been suggested in the etiology of migraine (Moskowitz and Macfarlane, 1993). Thus activation of perivascular trigeminal neurones leads to release of a number of peptides (substance P, CGRP) producing vasodilatation and extravasation in the dura mater. Experimentally, the extravasation induced by capsaicin or trigeminal nerve stimulation can be suppressed by a number of drugs (sumatriptan, ergot alkaloids) whose activity correlates with their potency against clinical migraine. Vasoconstriction or effects on afferent excitability have been suggested to explain the mechanism of action of these drugs (Humphrey and Feniuk, 1991; Moskowitz and McFarlane, 1993). Recently, a role for neurokinins in plasma extravasation in the dura has been further supported by the fact that NK1 receptor antagonists (Moussaoui et al., 1993; Shepheard et al., 1993) potently reduce extravasation in experimental models.

Changes in peptidergic fibre functions imposed by tissue injury and inflammation

During inflammation the events described above are likely to be amplified and the contribution of peptidergic afferent fibres increased. This is partly because of increased neural activity and peptide release; but more importantly because increased peptide synthesis and nerve sprouting occur as a result of increased trophin or growth factor activity (Noguchi et al., 1988; Donnerer et al., 1994). During inflammation or nerve injury the neuropeptide content of sensory nerves may be initially reduced (Gillardon et al., 1991) due to nerve activation and exhaustion of releasable stores of peptide. Following this, a number of changes occur due to the effects of NGF and other neurotrophins which are secreted in greater amounts (Donnerer et al., 1994) by a variety of cell types (fibroblasts, keratinocytes, Schwann cells) upon stimulation by inflammatory mediators such as the cytokines interleukin 1-β (IL1-β) and tumor necrosis factor -α (TNF-α). Increased NGF has been measured in several inflammatory conditions including pleurisy, rheumatoid arthritis as well as in blister fluid and in the skin following experimental inflammation (Woolf et al., 1994). NGF binds with a specific tyrosine kinase receptor (trk A) on small sensory neurones and is transported to the cell body where it stimulates increased mRNA production coding for neurokinin precursor peptides. This involves increasing gene transcription by stimulating transcription activators (e.g. Oct-2) or gene promoters (e.g. calgcat) (Watson et al., 1995). Tracer experiments suggest that increased amounts of neuropeptide may be secreted or leak from the terminals of afferent fibres both in the spinal cord (Valtschanoff et al., 1992, 1995) as well as in the periphery. NGF-induced sprouting of sensory fibres may further amplify these events (Diamond et al., 1992). The effects of NGF or other neurotrophins such as ciliary neurotrophic factor (CNTF) or neurotrophin-3 (NT3) are not exerted uniformly on sensory neurones, since their corresponding trk receptors trkA, trkB and trkC are heterogeneously distributed on different populations of neurones. However, it is likely that some neurotrophins affect sensory neurones by interaction which do not involve the known trk receptors (McMahon et al., 1994). NGF also produces a number of other indirect effects on afferent fibres via the release of proinflammatory mediators such as histamine and prostanoid derivatives such as leukotriene C4 (LTC_4) from mast cells and leukocytes.

As a result of these effects, NGF induces a prolonged increase in afferent fibre sensitivity and synaptic efficacy resulting in hyperalgesia particularly to mechanical stimuli. Consistent with this are studies showing that inflammation, or the injection of NGF increase tissue sensitivity and responsiveness to noxious stimuli. Treatment with anti-NGF antibodies reduce these effects (Lewin and Mendell, 1993; Woolf et al., 1994) and also prevent the increased substance P and CGRP mRNA expressed in sensory neurones (Woolf et al., 1994). In addition there may be secondary contributions to NGF induced hyperalgesia through opioid and kinin mechanisms (Apfel et al., 1993; Rueff and Mendell, 1994).

Changes in neuropeptide nerve function induced by nerve lesions

Peripheral nerve lesions or nerve ligations, which remove or prevent the transport of the target tissue sources of NGF and possibly other neurotrophins, reduce neurokinin and CGRP synthesis in small sensory nerves. Deficits in substance P and CGRP are also seen in small afferent fibres following experimental diabetes induced by streptozotocin. This is partly due to the lack of NGF in skin and muscle as well as other metabolic and neurotrophic insufficiencies affecting fine afferent fibres. The peptide deficit can be corrected following insulin treatment (Fernyhough et al., 1994). The hyperalgesia associated with peripheral nerve injuries produced by nerve ligation can also be *reduced* by NGF infusions (Thomas et al., 1993) possibly due to an increase in the survival of damaged afferents. On the other hand NGF has been shown to be an important mediator of the hyperalgesia caused by inflammatory stimuli (Woolf et al., 1994) and treatment of normal animals with NGF induces hyperalgesia to thermal and mechanical stimuli (Lewin and Mendell, 1993). Several mechanisms may account for this including overexpression and release of peptidergic mediators of nociception as mentioned earlier as well as the proliferation of sensory and sympathetic nerve fibres.

While nerve lesions reduce the expression of substance P and CGRP, similar nerve injuries increase the expression of other peptides such as galanin, neuropeptide Y (NPY), vasoactive intestinal peptide (VIP) and their receptors which are not normally detectable in sensory neurones (Villar et al., 1991; Wakisaka et al., 1992; Nahin et al., 1994). This may be explained in part by the absence of target-derived inhibitory factors which normally suppress the expression of some neuropeptides. It is also conceivable that injured tissues induce hitherto unidentified trophins which are able to alter the phenotype of specific sensory neurones. Large myelinated neurones undergo a number of spectacular changes; they begin to express neurokinins (Marchand et al., 1994), they become abnormally excitable and discharge spontaneously, and they may sprout abnormally to innervate areas of the spinal cord which normally transmit specific pain signals (Woolf and Doubell, 1994). These changes are likely to contribute to post injury pain and allodynia following stimulation of low threshold A-fibres, but no particular correlation has been established so far between peptide release and A-fibre induced pain.

Although it is not clear what function the newly expressed peptides serve; neurokinins are likely to be facilitatory, while other neuropeptides induce a net inhibition in the spinal cord to compensate for the increased excitability seen after peripheral inflammation or injury. For example, NPY which is normally present in sympathetic fibres, may be released both from the sprouting sympathetic fibres which innervate sensory neurones after peripheral nerve injury (McLachlan et al., 1993) and from large DRG neurones which express NPY after injury (Wakisuka et al., 1992; Itogawa et al., 1993). Inhibition may occur through NPY receptors which are upregulated on small DRG neurones (Mantyh et al., 1994; Zhang et al., 1994) and

NPY may cause antinociceptive by blocking calcium conductance and transmitter release (Colmers and Bleakman, 1994). On the other hand NPY has also been associated with producing hyperalgesia following neuropathy but this appears to be via activation of Y2 receptors located on sympathetic fibres (Tracey et al., 1995).

Interestingly, while peripheral nerve transection produces a decreased synthesis of neurokinins and CGRP, dorsal rhizotomy induces an *increase* in the synthesis of these peptides in sensory neurones. This would suggest that neuronal phenotype may also be regulated by centrally derived neurotrophins. So far however the identity of these substances has not been determined (Villar et al., 1991; Inaishi et al., 1992).

Sensory neurones and growth regulators

Although much attention has been focused on neurotrophins and their role in sensory neurone regulation, the function of these cells may also be dramatically influenced by other cellular growth regulators. For example acidic fibroblast growth factor (aFGF) has been identified in sensory neurones (Elde et al., 1991) from which it may be released to affect nearby cells. Furthermore depletion of acidic FGF upon nerve injury may play a role in the pathophysiology of neuropathic pain as treatment with acidic FGF accelerates the regeneration and functional recovery of sensory fibres after nerve injury (Laird et al., 1995). The release of acidic FGF from sensory nerve terminals may also affect the growth of nearby tissues or act as a growth regulator for sensory neurones themselves, as acidic FGF receptors have been identified on sensory cells. In addition transforming growth factor α (TGFα) which is synthesised by skin keratinocytes, also facilitates sensory nerve regeneration as it enhances the survival of dorsal root ganglion cells grown in culture (Chalazonitis et al., 1992) and regulates the production of target tissue derived NGF (Buchman et al., 1994). Presently it is not known whether growth regulators exert specific effects on sensory neurone excitability but they have been shown to regulate early gene expression (c-Fos, c-Jun) (Gold et al., 1993; Lo and Cruz, 1995) and thus are capable of producing phenotypic changes in sensory neurones. Whether this directly affects the expression of sensory neuropeptides is not known at present. Since constant regulatory interactions are likely to occur between sensory neurones and surrounding tissues, changes in sensory neurone function may develop because of some dysfunction in the neurochemistry of target tissue following injury. It is conceivable that such abnormalities also contribute to chronic pain.

Sympathetic neurons, neuropeptides and afferent fibres

Sympathetic nerves are important in the generation of certain types of chronic pain, but the reasons for this are poorly understood (see review by McMahon, 1991). Interestingly sympathetic neurones, which contain catecholamine transmitters, may also normally make neuropeptides (e.g. NPY) but in addition are able to express other peptides when stimulated by inflammatory products such as cytokines. Curiously NGF is required for the survival of sympathetic neurones but does not normally cause the expression of neurokinins in these cells. However, another cellular regulator, leukaemia inhibitory factor (LIF) also generated during inflammation has recently been shown to induce the expression of substance P (Jonakait, 1994). In addition, a number of interactions between sympathetic and afferent neurones have been described. Thus, neurokinins, released from activated sensory neurones, may stimulate post ganglionic sympathetic fibres to alter vascular calibre, induce changes in local blood flow, and indirectly affect plasma extravasation. In keeping with this, sympathectomy reduces the plasma extravasation induced by either noxious stimulation or by the administration of inflammatory mediators. However, in the knee joint, the sympathetic transmitters noradrenaline and NPY reduced plasma extravasation (Levine et al., 1993); probably due to an inhibition of calcium permeability necessary for neuropeptide release (Colmers and Bleakman, 1994). Direct interactions of sympathetic nerves or sympathetic transmitters with afferent fibres have not been easy to demonstrate (McMahon, 1991;

Treede et al., 1992) except after peripheral nerve damage or inflammation. Thus, afferent fibres can be sensitized during inflammation to induce hyperalgesia, partly by the release of prostanoids from sympathetic fibres (Levine et al., 1993). In addition, sympathetic nerve stimulation, or the direct administration of noradrenaline, excited some C-fibre afferents after a partially injury to a sensory nerve trunk (Sato and Perl, 1991) or after sciatic nerve transection (Devor et al., 1994). These effects were attenuated by phentolamine and block of α- adrenergic receptors which are presumed to be expressed on C-fibres (Sato and Perl, 1992) as well as on large A fibre afferents (Devor et al., 1994). Clearly, further characterization of the specific adrenoceptors expressed on afferent fibres is necessary.

Sensory neuropeptides in the spinal cord

To date there is little evidence that neurokinins directly alter the excitability of the central terminals from which they are released. However, direct depolarization of DRG neurones by substance P has been described (Dray and Pinnock, 1994) and NK1 receptors on primary afferents have been postulated (Malcangio and Bowery, 1994). These observations may explain why spinal administration of substance P and the depolarization of primary afferent nerve terminals enhanced plasma protein extravasation in the skin; though stimulation of other supraspinal pathways or preganglionic sympathetic neurones may also have contributed to this (Kerouac et al., 1987). In addition, the inhibition of afferent nerve excitability that was observed following substance P administration into the spinal dorsal horn may have been due to depolarization block of afferent nerve terminals (Davies and Dray, 1980). It remains an intriguing possibility that presynaptic actions of NKs may be important for regulating peripheral nerve excitability as well as the excitability of afferent terminals within the spinal cord. Other neuropeptides (e.g. somatostatin, opioid peptides) can inhibit afferent terminal excitability and thereby control transmitter release in the spinal cord. As in peripheral terminals, these substances act upon receptors, often coupled with G-protein which regulate ionic conductances and thereby calcium coupled transmitter release (Levine et al., 1993; Rang et al., 1994).

Most small (capsaicin-sensitive) primary sensory neurones terminate in the superficial dorsal horn but some fibres also project to the contralateral dorsal horn and a significant portion of these contain substance P (Ogawa et al., 1985). It is therefore very likely that ipsilateral stimulation of fine afferents induces contralateral excitability changes in the spinal cord, especially in inflammatory conditions where their signaling capacity can be increased by nerve sprouting and by increased peptide synthesis and release. Indeed, following the establishment of an inflammatory injury, the release of neuropeptides (substance P and neurokinin A) which is normally too small to measure, can be readily detected in the spinal dorsal horn but can also be detected at some distance from the site of release (Schaible et al., 1990). These data suggest that neurokinins are normally efficiently removed following neurosecretion and they remain restricted to the dorsal horn. However, during inflammation their sphere of activity may be significantly extended so that they can participate more extensively in increasing spinal excitability.

In the spinal cord, post-synaptic receptors for neurokinins have been extensively characterised. Neurokinins play an essential role, together with glutamate which is also released from fine afferent nerve terminals, in enhancing the excitability of dorsal horn neurones (Urban et al., 1994). Normally neurokinins released by acute stimulation of nociceptors has little affect on the excitability of dorsal horn cells and any excitation is little affected by NK receptor antagonists. Repetitive stimulation of C-fibres and peripheral inflammation, which induces a prolonged activation of nociceptors, increases the appearance of NK1 receptors (McCarson and Krause, 1994) and enhances the sensitivity of spinal neurones to NK1 antagonists (Thompson et al., 1994; Urban et al., 1994) (Fig. 3). Indeed, during inflammation, spinal excitability is increased by specific NK1/NMDA receptor interactions on wide dynamic range, dorsal

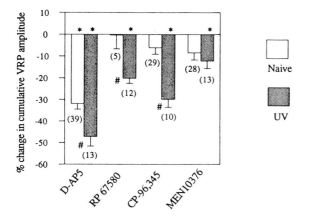

Fig. 3. This summarises the effects of the NMDA receptor antagonist D-AP5 and the changes in effectiveness of neurokinin receptor antagonists on spinal excitability in naïve rats and in animals that had received a peripheral injury with ultraviolet irradiation which induced hyperalgesia of the skin. D-AP5 (20 µM) reduced the VRP in both naïve and UV treated animals whereas RP67580 and CP-96345 (NK1 antagonists) and MEN10376 (NK2 antagonists) only reduced responses following the induction of hyperalgesia. (modified from Thompson et al., 1994).

horn cells and by the release of glutamate and several inflammatory mediators from glial cells by NK1 receptor activation (Urban et al., 1994). In keeping with this, the NK1 antagonists RP67580 and CP96345 are more efficacious in producing analgesia in conditions of chronic rather than acute pain (Birch et al., 1992).

Several reports indicate that different modalities of noxious stimulation determine the type of peptide released in the spinal cord (Wiesenfeld-Hallin, 1986). For example, somatostatin release occurred only during heat or chemical stimulation whereas substance P was released by mechanical stimulation (Kuraishi et al., 1985). Such data further support the possible functional heterogeneity of nociceptive afferents but so far similar studies have not been performed during chronic pain to indicate what the pattern of neuropeptide release might be. This is an important issue which requires further investigation.

Summary and perspectives

Fine afferent fibres contain a multiplicity of neuropeptides whose functions, for the most part, are unknown. This discussion has focused on the neurogenic effects of the neurokinins which are localised in many afferents and which have been most extensively studies. The effects of substance P, exemplify the diversity of actions which have been described for sensory neuropeptides. In the spinal dorsal horn, substance P induces changes in excitability of afferent nerve terminals, spinal neurones and glial cells; events associated with the propagation of nociceptive signals and the modulation of acute and chronic pain responses. Following peripheral inflammation, neuropeptide synthesis and released are increased and there some indications that release may also occur without afferent nerve stimulation though the significance of this is unclear (Valtschanoff et al., 1992). On the other hand, reduced synthesis and release of neurokinins occur after peripheral nerve injury when target-tissue derived trophic factors are removed. In addition, there is evidence for the expression of several neuropeptides in small and *large* sensory neurones which are not normally found there. The significance of this for chronic pain or for the processes of regeneration after injury are unknown. In the periphery, a multiplicity of effects produced by the release of neuropeptides suggests that they orchestrate several components of inflammation. Indeed, sensory neuropeptides provide an important chemical interface for neuroimmune regulation.

Neuropeptides also appear to have an emerging role in the regeneration and remodelling of peripheral nerves and tissues after injury, but further studies are necessary here. It is clear that all features of afferent nerve function can be tempered by their chemical environment and involves dynamic interactions between afferent (and sympathetic) fibres and surrounding tissue. This affects neural excitability, but perhaps more significantly over the long term there are changes in the peptide composition, concentration and release. The plastic features of afferents make difficulties for precise neurochemical and functional classsification of different groups of fibres. However, these changes are likely to play an important role in producing the diversity of symptoms observed in chronic pain.

References

Apfel, S.C., Newell, M.E. and Kessler, J.A. (1993) Opioid antagonists block NGF induced thermal hyperalgesia. *Soc. Neurosci. Abstr.*, 19: 665.

Birch, P.J., Harrison, S.M., Hayes, A.G., Rogers, H. and Tyers, M.B. (1992) The non-peptide NK1 receptor antagonist, (±)-CP-96345, produces antinociception and anti-oedema effects in the rat. *Br. J. Pharmacol.*, 105: 508–510.

Brain, S.D. and Williams, T.J. (1985) Inflammatory oedema induced by synergism between calcitonin gene related peptide and mediators of increased vascular permeability. *Br. J. Pharmacol.*, 86: 855–860.

Buchman, V.L., Sporn, M. and Davies, A.M. (1994) Role of transforming growth factor-beta isoforms in regulating the expression of nerve growth factor and neurotrophin-3 mRNA levels in embryonic cutaneous cells at different stages of development. *Development*, 120: 1621–1629.

Chalazonitis, A., Kessler, J.A., Twardzik, D.R. and Morrison, R.S. (1992) Transforming growth factor α, but not epidermal growth factor, promotes the survival of sensory neurons in vitro. *J. Neurosci.*, 12: 583–594.

Colmers, W.F. and Bleakman, D. (1994) Effects of neuropeptide Y on the electrical properties of neurons. *Trends Neurosci.*, 17: 373–379.

Davies, J. and Dray, A. (1980) Depression and facilitation of synaptic responses in cat dorsal horn by substance P administered into substantia gelatinosa. *Life Sci.*, 27: 2037–2042.

Devor, M., Janig, W. and Michaelis, M. (1994) Modulation of activity in dorsal root ganglion neurons by sympathetic activation in nerve-injured rats. *J. Neurophysiol.*, 71: 38–47.

Diamond, J., Holmes, M. and Coughlin, M, (1992) Endogenous NGF and nerve impulses regulate the collateral sprouting of sensory axons in the skin of the adult rat. *J. Neurosci.*, 12: 1454–1466.

Donnerer, J., Schuligoi, R. and Stein, C. (1994) Increased content and transport of substance P and calcitonin gene-related peptide in sensory nerves innervating inflamed tissue: evidence for a regulatory function of nerve growth factor in vivo. *Neuroscience*, 49: 693–698.

Dray, A. (1994) Tasting the inflammatory soup: the role of peripheral neurones. *Pain Rev.*, 1: 153–171.

Dray, A. and Pinnock, R.D. (1982) Effects of substance P on rat sensory ganglion neurones in vitro. *Neurosci. Lett.*, 33: 61–66.

Dray, A. and Perkins, M. (1993) Bradykinin and inflammatory pain. *Trends Neurosci.*, 16: 99–104.

Eglezos, A., Andrews, P.V., Boyd, R.L. and Helme, R.D. (1991) Tachykinin-mediated modulation of the primary antibody response in rats: evidence for mediation by NK2 receptor. *J. Neuroimmunol.*, 32: 11–18.

Elde, R., Cao, V., Cintra, A., Brelje, T.C., Pelto-Huikko, M., Junttila, T., Fuxe, K., Pettersson, R.F. and Hokfelt, T. (1991) Prominent expression of acidic fibroblast growth factor in motor and sensory neurons. *Neuron*, 7: 349–364.

Fan, T.-P.D., Hu, D.-E., Guard, S., Gresham, G.A. and Watling, K.J. (1993) Stimulation of angiogenesis by substance P and interleukin-1 in the rat and its inhibition by NK1 or interleukin-1 receptor antagonists. *Br J. Pharmacol.*, 110: 43–49.

Fernyhough, P., Diemel, L.T., Brewster, W.J. and Tomlinson, D.R. (1994) Deficits in sciatic nerve neuropeptide content coincide with a reduction in target tissue nerve growth factor messenger RNA in streptozotocin-diabetic rats: effects of insulin treatment. *Neuroscience*, 62: 337–344.

Friesen, J., Risling, M. and Fried, K. (1993) Distribution and axonal relations of macrophages in a neuroma. *Neuroscience*, 55: 1003–1013.

Gamse, R. and Saria, A. (1985) Potentiation of tachykinin-induced plasma protein extravasation by calcitonin gene-related peptide. *Eur. J. Pharmacol.*, 114: 61–66.

Gazelius, B., Brodin, E., Olgart, L. and Panopoulos, P. (1981) Evidence that substance P is a mediator of antidromic vasodilatation using somatostatin as a release inhibitor. *Acta Physiol. Scand.*, 113: 155–159.

Gillardon, F., Morano, I. and Zimmermann, M. (1991) Ultraviolet irradiation of the skin attenuates calcitonin gene-related peptide mRNA expression in rat dorsal root ganglion cells. *Neurosci. Lett.*, 124: 144–147.

Gold, B.G., Storm-Dickerson, T. and Austin, D.R. (1993) Regulation of the transcription factor c-JUN by nerve growth factor in adult sensory neurons. *Neurosci. Lett.*, 154: 129–133.

Green, P.G., Basbaum, A.I. and Levine, J.D. (1992) Sensory neuropeptide interactions in the production of plasma extravasation in the rat. *Neuroscience*, 50: 745–749.

Habler, H.J., Janig, W. and Koltzenburg, M. (1990) Activation of unmyelinated afferent fibres by mechanical stimuli and inflammation of the urinary bladder in the cat. *J. Physiol.*, 425: 545–562.

Helme, R.D., Eglezos, A., Dandie, G.W., Andrews, P.V. and Boyd, R.L. (1987) The effect of substance P on the regional lymph node antibody response to antigenic stimulation in capsaicin-treated rats. *J. Immunol.*, 139: 3470–3473.

Holzer, P. (1988) Local effector functions of capsaicin-sensitive sensory nerve endings: involvement of tachykinins, calcitonin gene-related peptide and other neuropeptides. *Neuroscience*, 24: 739–768.

Humphrey, P.P.A. and Feniuk, W. (1991) Mode of action of the anti-migraine drug sumatriptan. *Trends Pharmacol. Sci.*, 12: 444–446.

Inaishi, Y., Kashihara, Y., Sakaguchi, M., Nawa, H. and Kuno, M. (1992) Cooperative regulation of calcitonin gene-related peptide levels in rat sensory neurons via their central and peripheral processes. *J. Neurosci.*, 12: 518–524.

Itogawa, T., Yamanaka, H., Wakisaka, S., Sasaki, Y., Kato, J., Kurisu, K. and Tsuchitani, Y. (1993) Appearance of neuropeptide-Y like immunoreactivity in the rat trigeminal ganglion following dental injuries. *Arch. Oral Biol.*, 38: 725–728.

Janig, W. and Lisney, J.W. (1989) Small diameter myelinated

afferents produce vasodilatation but not plasma extravasation in rat skin. *J. Physiol.*, 415: 477–486.

Jonakait, G.M. (1993) Neural-immune interactions in sympathetic ganglia. *Trends Neurosci.*, 10: 419–423.

Kerouac, R., Jacques, L. and Couture, R. (1987) Intrathecal administration of substance P enhances cutaneous plasma protein extravasation in pentobarbital anaesthetized rat. *Eur. J. Pharmacol.*, 133: 129–136.

Kuraishi, Y., Hirota, N., Sato, Y., Hino, Y., Satoh, M. and Takagi, H. (1985) Evidence that substance P and somatostatin transmit separate information related to pain in the spinal dorsal horn. *Brain Res.*, 325: 294–298.

Laird, J.M.A., Mason, G.S., Thomas, K.A., Hargreaves, R.J. and Hill, R.G. (1995) Acidic fibroblast growth factor stimulates motor and sensory axon regeneration after sciatic nerve crush in the rat. *Neuroscience*, 65: 209–216.

Levine, J.D., Fields, H.L. and Basbaum, A.I. (1993) Peptides and the primary afferent nociceptor. *J. Neurosci.*, 13: 2273–2286.

Lewin, G.R. and Mendell, L.M. (1993) Nerve growth factor and nociception. *Trends Neurosci.*, 16: 353–358.

Liu, H., Brown, J.L., Jasmin, L. Maggio, J.E., Vigna, S.R., Mantyh, P.W. and Basbaum, A.L. (1994) Synaptic relationship between substance P and the substance P receptor: light and electron microscopic characterization of the mismatch betwen neuropeptides and their receptors. *Proc. Natl. Acad. Sci. USA*, 91: 1009–1013.

Lo, Y.Y. and Cruz, T.F. (1995) Involvement of reactive oxygen species in cytokine and growth factor induced c-fos expression in chondrocytes. *J. Biol Chem.*, 270: 11727–11730.

Lotz, M., Vaughan, J.H. and Carson, D.A. (1988) Effect of neuropeptides on production of inflammatory cytokines by human monocytes. *Science*, 241: 1218–1221.

Maggi, C.A., Patacchini, R., Rovero, P. and Giachetti, A. (1992) Tachykinin receptors and tachykinin receptor antagonists. *J. Auton. Pharmacol.*, 13: 23–93.

Malcangio, M. and Bowery, N.G. (1994) Effect of tachykinin NK1 receptor antagonist, RP 67580 and SR 140333, on electrically-evoked substance P release from rat spinal cord. *Br. J. Pharmacol.*, 113: 635–641.

Mantyh, P.W., Allen, C.J., Rogers, S., DeMaster, E., Ghilard, J.R., Mosconi, T., Kruger, L., Mannon, P.J., Taylor, I.L. and Vigna, S.R. (1994) Some sensory neurones express neuropeptide Y receptors: potential paracrine inhibition of primary afferent nociceptors following peripheral nerve injury. *J. Neurosci.*, 14: 3958–3968.

Marchand, J.E., Wurm, W.H., Kato, T. and Kream, R.M. (1994) Altered tachykinin expression by dorsal root ganglion neurons in a rat model of neuropathic pain. *Pain*, 58: 219–231.

McCarson, K.E. and Krause, J.E. (1994) NK-1 and NK-3 type tachykinin receptor mRNA expression in the rat spinal cord dorsal horn is increased during adjuvant or formalin induced nociception. *J. Neurosci.*, 14: 712–720.

McLachlan, E.M., Janig, W., Devor, M. and Michaelis, M. (1993) Peripheral nerve injuries triggers noradrenergic sprouting within dorsal root ganglia. *Nature*, 363: 543–546.

McMahon, S.B. (1991) Mechanisms of sympathetic pain. *Br. Med. Bull.*, 47: 584–600.

McMahon, S.B. and Gibson, S. (1987) Peptide expression is altered when afferent nerves reinnervate inappropriate tissue. *Neurosci. Lett.*, 73: 9–15.

McMahon, S.B., Armanini, M.P., Ling, L.H. and Phillips, H.S. (1994) Expression and coexpression of trk receptors in subpopulation of adult primary afferent sensory neurons projecting to identified peripheral targets. *Neuron*, 12: 1161–1171.

Moskowitz, M.A. and Macfarlane, R. (1993) Neurovascular and molecular mechanisms in migraine headache. *Cerebrovasc. Brain Metab. Rev.*, 5: 159–177.

Moussaoui, S.M., Phillipe, L., Le Prado, N. and Garret, C. (1993) Inhibition of neurogenic inflammation in the meninges by a non-peptide NK-1 receptor antagonists RP 67580. *Eur. J. Pharmacol.*, 238: 421–424.

Nahin, R.L., Ren, K., de Leon, M. and Ruda, M. (1994) Primary sensory neurons exhibit altered gene expression in rat model of neuropathic pain. *Pain*, 38: 95–108.

Nakamura-Craig, M. and Gill, B.K. (1991) Effect of neurokinin A, substance P and calcitonin gene related peptide in peripheral hyperalgesia in the rat paw. *Neurosci. Lett.*, 124: 49–51.

Nilsson, J., von Euler, A.M. and Dalsgaard, C.-J. (1985) Stimulation of connective tissue cell growth by substance P and substance K. *Nature*, 315: 61–63.

Noguchi, K., Morita, Y., Kiyama, H., Ono, K. and Tohyama, M. (1988) A noxious stimulus induces the protachykinin-a gene expression in rat dorsal root ganglion: a quantitative study using in situ hybridization histochemistry. *Mol. Brain Res.*, 4: 31–35.

Ogawa, T., Kanazawa, I. and Kimura, S. (1985) Regional distribution of substance P, neurokinin α and neurokinin β in rat spinal cord, nerve roots and dorsal root ganglia and effects of dorsal root section or spinal transection. *Brain Res.*, 359: 152–157.

Payan, D.G., Brewster, D.R. and Goetzl, E.J. (1983) Specific stimulation of human T-lymphocytes by substance P. *J. Immunol.*, 131: 1613–1615.

Rang, H.P., Bevan, S.J. and Dray, A. (1994) Nociceptive peripheral neurones: cellular properties. In: P.D. Wall and R. Melzack (Eds.), *Textbook of Pain*, Churchill Livingstone: Edinburgh, pp. 57–78.

Rueff, A. and Mendell, L.M. (1994) NGF-induced thermal hyperalgesia in adult rats involves the activation of bradykinin B1-receptors. *Soc. Neurosci. Abstr.*, 20: 671.

Sato, J. and Perl, E.R. (1991) Adrenergic excitation of cutaneous pain receptors induced by peripheral nerve injury. *Science*, 251: 1608–1610.

Schaible, H.-G., Jarrott, B., Hope, P.J. and Duggan, A.W. (1990) Release of immunoreactive substance P in the spinal cord during development of acute arthritis in the knee joint of the cat: a study with antibody microprobes. *Brain Res.*, 529: 214–223.

Schmelz, M., Schmidt, R., Ringkamp, M., Handwerker, H.O. and Torebjork, H.E. (1994) Sensitization of insensitive branches of C-nociceptors in human skin. *J. Physiol.*, 480: 389–394.

Scott, S.A. (1992) *Sensory Neurons; Diversity, Development and Plasticity*, Oxford University Press, New York.

Shepheard, S.L., Williamson, D.J., Hill, R.G. and Hargreaves, R.J. (1993) The non-peptide neurokinin-1 antagonist RP67580 blocks neurogenic plasma extravasation in the dura mater of rats. *Br. J. Pharmacol.*, 108: 11–12.

Thomas, D.A., Ren, K. and Dubner, R. (1993) Nerve growth factor alleviates a painful peripheral neuropathy in rats. *Soc. Neurosci. Abstr.*, 19: 969.

Thompson, S.W.N., Dray, A. and Urban, L. (1994) Injury-induced plasticity of spinal reflex activity: NK1 neurokinin receptor activation and enhanced A- and C-fiber mediated responses in the rat spinal cord in vitro. *J. Neurosci.*, 14: 3672–3687.

Tracey, D.J., Romm, M.A. and Yao, N.N. (1995) Peripheral hyperalgesia in experimental neuropathy: exacerbation by neuropeptide Y. *Brain Res.*, 669: 245–254.

Treede, R.-D., Meyer, R.A., Raja, S.N. and Campbell, J.N. (1992) Peripheral and central mechanisms of cutaneous hyperalgesia. *Prog. Neurobiol.*, 38: 397–421.

Urban, L., Thompson, S.W.N. and Dray, A. (1994) Modulation of spinal excitability: cooperation between neurokinin and excitatory amino acid transmission. *Trends Neurosci.*, 17: 432–438.

Valtschanoff, J.G., Weinberg, R.J. and Rustioni, A. (1992) Peripheral injury and anterograde transport of wheat germ agglutinin-horse radish peroxidase to the spinal cord. *Neuroscience*, 50: 685–696.

Valtschanoff, J.G., Weinberg, R.J. and Rustioni, A. (1995) Central release of tracer after noxious stimulation of the skin suggests non-synaptic signalling by unmyelinated fibres. *Neuroscience*, 64: 851–854.

Villar, M.J., Wiesenfelt-Hallin, Z., Xu, X.-J., Theodorsson, E., Emson, P.C. and Hokfelt, T. (1991) Further studies on galanin-, substance P- and CGRP-like immunoreactivities in primary sensory neurons and spinal cord: effects of dorsal rhizotomies and sciatic nerve lesions. *Exp. Neurol.*, 112: 29–39.

Wakisaka, S., Kajander, K.C. and Bennett, G.J. (1992) Effects of peripheral nerve injuries and tissue inflammation on the levels of neuropeptide Y-like immunoreactivity in rat primary afferent neurons. *Brain Res.*, 598: 349–352.

Watson, A., Ensor, E., Symes, A., Winter, J., Kendall, G. and Latchman, D. (1995) A minimal CGRP gene promoter is inducible by nerve growth factor in adult rat dorsal root ganglion neurons but not in PC12 phaeochromocytoma cells. *Eur. J. Neurosci.*, 7: 394–400.

Weihe, E., Nohr, D. Muller, S., Buchler, M., Friess, H. and Zentel, H.-J. (1991) The tachykinin neuroimmune connection in inflammatory pain. *Ann. N. Y. Acad. Sci.*, 632: 283–295.

White, D.M., Ehard, P., Hardung, M., Meyer, D.K., Zimmerman, M. and Otten, U. (1987) Substance p modulates the release of locally synthesized nerve growth factor from rat saphenous nerve neuroma. *Arch. Pharmacol.*, 336: 587–590.

Wiesenfeld-Hallin, Z. (1986) Substance P and somatostatin modulate spinal cord excitability via physiologically different sensory pathways. *Brain Res.*, 372: 172–175.

Willis, W. and Coggeshall, R. (1991) *Sensory Mechanisms of the Spinal Cord*. Plenum Publishing, New York.

Woolf, C.J. and Doubell, T.P. (1994) The pathophysiology of chronic pain - increased sensitivity to low threshold Aβ-fibre inputs. *Curr. Biol.*, 4: 525–534.

Woolf, C.J., Safieh-Garabedian, B., Ma, Q.-P., Crilly, P. and Winter, J. (1994) Nerve Growth factor contributes to the generation of inflammatory sensory hypersensitivity. *Neuroscience*, 62: 327–331.

Zhang, X., Bao, L., Xu, Z.-Q. Kopp, J., Arvidsson, U., Elde, R. and Hokfelt, T. (1994) Localization of neuropeptide Y Y1 receptors in the rat nervous system with special reference to somatic receptors on small dorsal root ganglion neurons. *Proc. Natl. Acad. Sci. USA*, 91: 11738–11742.

Ziche, M., Morbidelli, L., Pacini, M., Alessandri, G. and Maggi, C.A. (1990) Substance P stimulates neovascularization in vivo and proliferation of cultured endothelial cells. *Microvasc. Res.*, 40: 264–278.

CHAPTER 7

Peripheral opioid analgesia - facts and mechanisms

Albert Herz

Max-Planck-Institute for Psychiatry, Department of Neuropharmacology, Am Klopferspitz 18a, D-82151 Martinsried, Germany

Introduction

Opioids are considered to exert antinociceptive (analgesic) effects through actions within the central nervous system. There is increasing evidence, however, that, under certain circumstances, peripheral sites of opioid action may also become involved in pain modulation. A prerequisite for the manifestation of such peripheral effects seems to be inflammation, accompanied by hyperalgesia of the tissue from which the nociceptive impulses arise. This fact is exemplified by experiments in which a variety of inflammatory agents such as prostaglandins, carrageenan, formalin, bradykinin, Freund's adjuvant etc. were injected locally (Ferreira and Nakamura, 1979; Joris et al., 1987; Russell et al., 1987; Levine and Taiwo, 1989; Taiwo and Levine, 1991). Very little, however, is known about the underlying mechanisms, e.g. the significance of the inflammatory process, the participation of various opioid receptor types, the role of endogenous opioid peptides and the intrinsic mechanisms involved in peripheral opioid analgesia (for review see Barber and Gottschlich, 1992; Junien and Wettstein, 1992; Herz, 1995). These questions are discussed here, in the context of experiments performed in this laboratory over the last couple of years.

Exogenously applied opioids in monoarthritic rats

Monoarthritis, induced by injection of Freund's complete adjuvant into the plantar site of the rat paw served as the model of inflammatory pain and the paw pressure test was used to measure nociception (Millan et al., 1988). Systemic application of low doses of morphine or fentanyl (μ-opioid receptor ligands) or U 50,488 (κ-opioid receptor ligand) increased the nociceptive threshold in the inflamed, but not in the non-inflamed paw (Fig. 1). These effects were antagonized by intraplantar injection of the universal (non-selective) opioid receptor antagonist naloxone, confirming the presumption of a local opioid action. This notion was finally proven by the finding that intraplantar injection of low doses of μ-, δ- or κ-opioid receptor agonists increased the nociceptive threshold in the inflamed, but not into the non-inflamed paw. These effects were blocked by antagonists selective for the respective opioid receptor types (Fig. 2). These findings indicate that μ-, δ- and κ-opioid receptors are involved in peripheral antinociception, and that all three opioid systems may independently modulate pain under inflammatory conditions (Stein et al., 1988a,b, 1989).

Stress-induced activation of endogenous opioids

There is much evidence that endogenous opioid peptides are released by different kinds of stress and the so-called "stress-induced analgesia" may, at least partly, be attributable to activation of intrinsic opioid mechanisms (Przewłocki, 1993). In the present experiments, cold water swim (CWS) was used as a stress model. The short-lasting (10–20 min) increase in paw pressure threshold, observed after swimming in ice-water for 1 min was confined to the inflamed paw and was antagonized

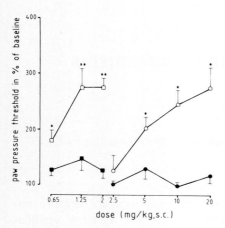

Fig. 1. Antinociceptive effects elicited by subcutaneous injections of either morphine or U 50,488 in monoarthritic rats (paw pressure test). (U 50,488 is a κ-opioid receptor selective bencenacetamide derivative). Local inflammation of one hind paw was induced by intraplantar injection of Freund's adjuvant 4 days before testing. The antinociceptive effects of low doses of each opiate were restricted to the inflamed paw (adapted from Stein et al., 1988). Values are means ± SE. $*P = < 0.05$; $**P = < 0.02$. ●, Morphine in inflamed paw; ○, U 50,488 in inflamed paw; ■, morphine in non-inflamed paw; ●, U 50,488 in non-inflamed paw.

by intraplantar injection of naloxone (Fig. 3). Systemic injection of quaternary naltrexone, which has very limited access to the brain, also antagonized this effect. These results indicate that this stress-induced analgesia is of peripheral origin, thus providing an appropriate model for further examination of the underlying mechanisms. The unilateral increase in nociceptive threshold was antagonizable by intraplantar injection of low doses of the μ-receptor antagonist CTOP and the δ-receptor antagonist ICI 174,864, but not by the κ-receptor antagonist nor-binaltorphimine. Also, intraplantar injection of antibodies directed against β-endorphin (β-EP), but not of antibodies directed against dynorphin, inhibited the stress-induced increase in nociceptive threshold. Thus, these experiments strongly suggest that the local antinociception results from the release of endogenous opioid peptides interacting with μ- and δ-, but not κ-receptors. β-EP seems to be a most likely peptide candidate (Parsons et al., 1990; Stein et al., 1990a,b).

Origin of the endogenous opioids

The question arises as to the sources of the opioid peptides involved in local, stress-induced analgesia. Extraction of peptides from inflamed paws revealed an almost 10-fold increase in the content of immunoreactive β-EP with respect to controls; the content of immunoreactive methionine-enkephalin was also, although somewhat less increased, while dynorphin could not be detected in neither inflamed nor non-inflamed paw-tissue. Immunocytochemistry revealed dense β-EP, and (somewhat less) methionine-enkephalin staining in the immune cells infiltrating the inflamed tissue, while dynorphin immunoreactivity was seen in a small number only of faintly stained cells. In non-inflamed tissue, opioid immunoreactivity was almost completely absent (Stein et al., 1990b). The identity of the cells containing β-EP was determined by means of double staining with immune-cell markers and β-EP antibodies: T- and B-

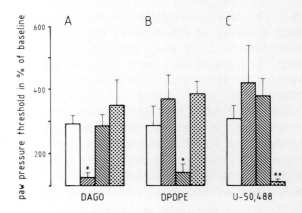

Fig. 2. Antinociceptive effects of selective opioid receptor agonists given by intraplantar injection into the inflamed rat paw, and antagonism by specific opioid receptor antagonists. The effect of the μ-receptor agonist DAGO (1 μg) was antagonized by the μ-receptor antagonist CTAP (1 μg), but by neither the δ-receptor antagonist ICI 174,864 (5 μg) nor the κ-receptor antagonist nor-BNI (50 μg) (A). Similarly, the antinociceptive effects of δ-receptor agonist DPDPE (40 μg) and the κ-receptor agonist U 50,488 (50 μg) were only antagonized by the respective receptor antagonists (B,C) (adapted from Stein et al., 1989). Values are means ± SE. $*P = < 0.05$; $**P = < 0.01$. White bar, NaCl; left diagonal bar, ICI; right diagonal bar, CTAP; dotted bar, nor-BNI.

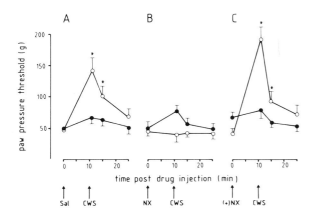

Fig. 3. Cold-water-swim (CWS) stress-induced antinociception. A rapid and short-lasting increase in nociceptive threshold was observed in the inflamed, but not in the non-inflamed paw (A). Naloxone (NX) (18 μg) intraperitoneally injected before swimming, antagonized the antinociceptive effect (B), while the pharmacologically inactive (+) enantiomer of naloxone was ineffective (C) (adapted from Stein et al., 1990a). Values are means ± SE. *$P = <0.05$; ○, inflamed paw; ●, non-inflamed paw.

lymphocytes, as well as monocytes and macrophages in the inflamed tissue displayed β-EP immunoreactivity. Finally, in-situ hybridization experiments showed that mRNAs coding for pro-opiomelanocortin and pro-enkephalin (but not for prodynorphin) are abundant in cells of the inflamed tissue (and almost completely absent in non-inflamed tissue), providing evidence that opioid peptides of the β-EP and enkephalin families are synthetized and processed within the various types of immune cells at the site of inflammation (Przewłocki et al., 1992) (Fig. 4). Autoradiographic studies with iodinated β-EP showed that these immune cells also bear opioid receptors (Hassan et al., 1993). The observation that pretreatment of rats with the immunosuppressant cyclosporin A as well as whole-body x-ray-irradiation suppressed CWS-induced antinociception provided a link between the histochemical and behavioural data (Stein et al., 1990a). Besides β-EP-like peptides also enkephalins (or enkephalin congeners) seem to participate in stress-induced antinociception (Stein et al., 1990a). In experiments in which the breakdown of such peptides was prevented by either local or systemic application of enkephalinase inhibitors the increase in nociceptive threshold in the inflamed paw after CWS was significantly prolonged, but remained unchanged in the uninflamed paw (Parsons and Herz, 1990) (Fig. 5). Taken together, these data indicate that opioid peptides present in the immune cells of inflamed tissue are involved in the manifestation of local CWS-induced antinociception.

The foregoing experiments point to a critical role of immune-cells associated β-EP and enkephalin-like peptides in the manifestation of stress-induced peripheral antinociception. The mechanism(s) triggering the release of these peptides, however, remains unclear. A possible role of cytokines in the opioid release mechanism results from experiments which revealed that cytokines, such as interleukin-6 and tumor necrosis factor, induce short-lasting antinociception when injected directly into the inflamed paw, but not when injected into the non-inflamed paw. These effects are antagonizable by local injection of naloxone or the μ-opioid receptor antagonist CTOP. Pretreatment of the rats with the immuno-suppressant cyclosporin A abolished the antinociceptive effect of these cytokines (Członkowski et al., 1993). Recent experiments in which corticotropin-releasing factor (CRF) was locally injected revealed antinociceptive effects similar to those produced by the cytokines. Using cell suspensions from inflamed lymph nodes an acute release of immunoreactive β-EP by interleukin-1 and CRF was demonstrated (Schäfer et al., 1994). This latter findings are of particular interest in view of data showing that the stress hormone CRF is present within and can be released from inflamed rat paw (Hargreaves et al., 1989).

Peripheral opioid receptors mediating antinociception

It is likely that the opioid receptors mediating the peripheral antinociceptive effects of externally applied and endogenously released opioids are located at the terminal region of fine afferent nerve fibers (Barthó et al., 1990; Andreev et al., 1994).

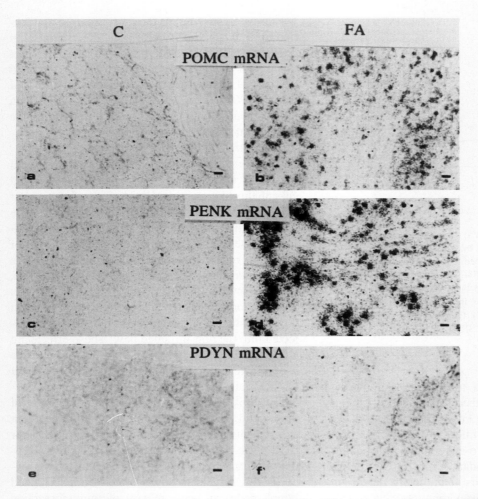

Fig. 4. In-situ hybridization of proopiomelanocortin- (POMC) mRNA (a,b), proenkephalin- (PENK) mRNA (c,d) and prodynorphin- (PDYN) mRNA (e,f) in subcutaneous tissue of non-inflamed (a,c,e) and inflamed (b,d,f) rat paws (4 days after induction of inflammation with Freund's adjuvant). The oligonucleotide probes were labelled with ^{35}S, the sections exposed for 6 weeks before developing and counterstaining with cresyl violet (adapted from Przewłocki et al., 1992). C, non-inflamed controls; FA, inflamed paw-tissues; Scale 5 μm.

By means of histochemistry and using anti-idiotypic opioid receptor antibodies with high affinity for μ- and δ-opioid receptors the presence of such receptors on small cutaneous nerve fibers of rat paw tissue was demonstrated (Stein et al., 1990b; Honda et al., 1994). Interestingly, immunostaining was also seen in non-inflamed paws, although to a lower extent. In a recent study the effect of inflammation on the axonal transport of opioid receptors in the sciatic nerve and the presence of such receptors in the rat paw was studied autoradiographically, using iodine labelled β-EP as ligand. In the absence of inflammation some opioid binding sites were already visible in non-inflamed paw tissue (Hassan et al., 1993) (Fig. 6). Three to four days after induction of inflammation of the paw there was massive increased β-EP binding on both sites of the ligated sciatic nerve and at the fine nerve fibers of the paw (as well as at the immune cells infiltrating the surrounding tissue). The ^{125}I-labelled β-EP as used in this study only labels μ- and δ-receptors. Other data, how-

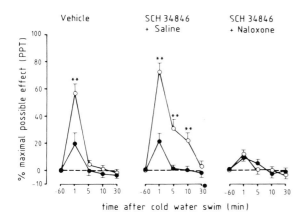

Fig. 5. Prolongation of cold-water-swim stress-induced antinociception in the inflamed paw of rats pretreated with the enkephalinase inhibitor SCH 34826 and inhibition of the antinociceptive effect by naloxone. 20 mg/kg SCH 34826 was applied i.p. 60 min prior to 1 min cold water swim. Naloxone was injected s.c. 5 min before the swim test (adapted from Parsons and Herz, 1990). PPT, paw-pressure threshold; Values are means ± SE. *$P = < 0.05$; ** $P = < 0.02$; ○, inflamed paw; ●, non-inflamed paw.

ever, indicate that κ-receptors may be also involved, possibly depending on the inflammatory agent or the time after induction of inflammation (Haley et al., 1990).

Clinical aspects of peripheral opioid analgesia

Sites of action of opioids outside the CNS in mediating analgesia offer interesting clinical implications. Quaternary opioids with alkaloid structure or highly hydrophilic opioid peptides may be expected not to permeate the blood-brain barrier, and thus to be devoid of undesirable central side-effects such as respiratory depression, nausea, sedation or dependence liability. Such compounds, synthesized and tested in appropriate animal models indeed showed antinociceptive activity attributable to peripheral sites of opioid action in the presence of inflammation (Abbott, 1988; Follenfant et al., 1988; Schiller et al., 1990; Rogers et al., 1992). There ist, however, little experience so far with such compounds in humans (Posner et al., 1990).

Several clinical trials have been performed in which opioids were applied directly to the peripheral site from which pain originates. In a double-blind study, a low dose of morphine (1 mg; ineffective upon systemic application) was injected after arthroscopic knee surgery directly into the knee-joint. Significantly lower pain scores were observed over several hours in the patients who received morphine in comparison to controls, and naloxone dose-dependently antagonized the opioid effects (Stein et al., 1991). Similar results have also been obtained in some recent studies (Joshi et al., 1993; Dalsgaard et al., 1994), whereas only marginal (Heard et al., 1992) or no (Raja et al., 1992) analgesic effects of morphine were seen in other studies, even when higher doses of morphine were injected into the knee joint. In contrast, intraarticular injection of naloxone in patients undergoing arthroscopic knee surgery resulted in increased pain scores; tissue samples of the synovia of these patients exhibited synovitis with inflammatory cells containing immunoreactive opioid peptides. This points to a continuous release of opioid peptides from the inflamed synovia (Stein et al., 1993). Thus, the local application of opioids offers new promising therapeutic perspectives - although the particular conditions for successful clinical application of this approach, e.g. the significance of the presence of inflammatory processes, has still to be elaborated.

Interpretations and perspectives

The mechanisms involved in peripheral opioid analgesia will be discussed according to the schematic drawing (Fig. 7) which illustrates the events at the terminal region of a nociceptive nerve fibre at various developmental stages of the inflammatory process. Planel "A" represents the "normal" condition in absence of inflammation or tissue injury. There is general agreement that in this case peripheral sites do not play a significant role in the antinociceptive effects of opioids. This is remarkable in view of the fact that opioid receptors (binding sites) can be demonstrated at the primary afferents of somatosensory fibres of the non-inflamed tissue (see above). It has to be concluded that these receptors are non-functional under such conditions.

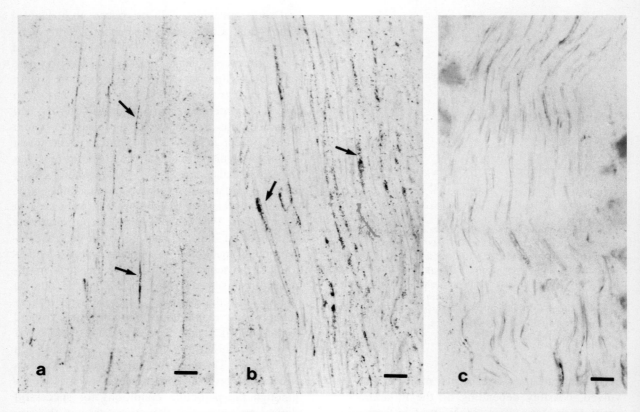

Fig. 6. Light microscopic autoradiography of ^{125}I-β-EP binding (arrows) in the cutaneous nerve fibers in paw tissue of non-inflamed rats (a) and four days after the induction of inflammation (b); (c) shows non-specific binding in the presence of unlabelled β-endorphin (adapted from Hassan et al., 1993). Scale bars = 15 μm.

The situation at the early stages of inflammation is represented in Panel B. In this case μ-, δ- as well as κ-opioid ligands exert antinociceptive effects (Fig. 2). This effect becomes manifest 6–12 h after instillation of Freund's adjuvant into the paw (Antonijevic et al., 1995). The swelling of the paw and increase of its temperature develop within hours, reaching maximal levels 6 h after injection of the agent. Thus it seems that the time courses of the development of antinociception and early stages of inflammation are similar. Such a parallelism can also be seen when inflammatory agents (such as formalin) are used, which have more rapid onsets of the inflammation, (Haley et al., 1990). It is interesting to note that different types of opioid receptors may be involved, depending on the type of inflammatory agent employed. Thus in the case of prostaglandins μ-opioid receptor ligands are effective (Levine and Taiwo, 1989), whereas κ-opioid receptor ligands are effective when formalin is applied (Haley et al., 1990).

Several points deserve discussion with regard to the question as to when opioids become involved during the process of inflammation. Inflammatory agents result in the release of mediators such as substance P, CGRP, bradykinin, prostaglandins etc. and it is suggested that these endogenouos substances may be causal in the sensitization of nociceptors (hyperalgesia) accompanying inflammation (Schmidt et al., 1994). There is considerable data which indicates that opioids prevent the release of these compounds (Lembeck and Donnerer, 1985; Yakskh, 1988; Barber, 1993), but it is nevertheless improbable that this represents the mechanism underlying the peripheral antinociceptive effect of opioids. At least under the

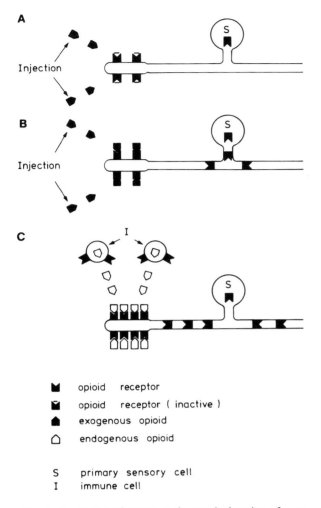

Fig. 7. Illustration of events at the terminal region of a nociceptive nerve fibre at various stages of the development of the inflammatory process (see text). Panel A, non-inflamed tissue; Panel B, early stages of inflammation; Panel C, later stages of inflammation.

Recently it has been suggested that a "perineurial defect" induced by the inflammatory process is a crucial determinant for peripheral opioid analgesia (Antonijevic et al., 1995). This implies the existence of a perineurial barrier preventing the permeation of endogenous or exogenous opioids to reach the opioid receptors; these receptors are most probably localized in the nerve terminal region, which is devoid of such a permeation barrier (Heppelmann et al., 1990). Also the fact that opioids with high lipid solubility (e.g. fentanyl) which easily permeate biological membranes behave similar to hydrophilic compounds as morphine and opioid peptides does not support such a view (Stein et al., 1988a). Thus, it appears most probable that changes at the receptor sites themselves may render opioids efficacious during inflammation.

It is most likely that the relevant opioid receptors are located in the terminal region of the somatosensory afferents. Unfortunately, the histochemical and autoradiographic methods used to visualize these receptors do not allow their precise anatomical localization and electron microscopic studies have obviously not yet been performed. However, indirect hints for the localization of these receptors in the terminal region come from the experiments with capsaicin: desensitization of capsaicin receptors, presumably localized at the (capsaicin-sensitive) somatosensory nerve endings, not only prevent inflammation-induced hyperalgesia, but also the increase of sensitivity to morphine (Barthó et al, 1990).

It has been shown recently that inflammation greatly enhances orthodromic and antidromic axonal flow of opioid receptors, resulting in an increase of the density of these binding sites in the terminal region on the somatosensory fibres in the paw (Hassan et al., 1993). This process has a slow time-course and 2–3 days are required before the number of receptors in the terminal region is significantly increased. Thus, this "up-regulation" of opioid receptors becomes manifest only in the later stages of inflammation (see below), whereas sensitivity for peripheral opioids starts much earlier.

Electrophysiological data obtained from rat saphenous nerve-hind paw skin preparations, in conditions of the present experiments the peripheral antinociceptive effect of the opioids is rather short-lasted and (other) signs of inflammation are obviously not significantly affected. In addition, the opioids not only overcome hyperalgesia, but increase the nociceptive threshold far above baseline levels. These observations indicate that the anti-inflammatory and antinociceptive effects can be separated.

which the polymodal nociceptor activity was tested are in line with the conclusions from behavioural experiments (Andreev et al., 1994): The spontaneouos activity of these nociceptors, developing after ultraviolet irradiation, was blocked by local application of μ- and κ-receptor opioids in a naloxone-reversible manner. This indicates that the nerve terminals become hyperpolarized and discharge activity is blocked by an opioid receptor-mediated change in ion conductance. Increase of potassium ion conductance and/or decrease of calcium ion conductance has been repeatedly documented as mechanisms of opioid-induced neural inhibition (North, 1986).

A remaining question concerns the mechanism(s) that underlie the "activation" of the opioid receptors during the development of inflammation. The pathology of inflammation is complex, and a variety of physicochemical and chemical changes (including pH) take place. A dominant role of acid pH in inflammatory excitation and sensitization of nociceptors in rat skin is well documented (Steen et al., 1995). Changes in the tertiary structure of opioid receptors and the coupling to G proteins, affecting the signal transduction, may be considered as possible mechanisms underlying such an activation. Recently it has been shown that low pH increases opioid agonist efficacy. It is suggested that this is due to decreased inactivation of G proteins augmenting the interaction of opioid receptors with effector systems (Childers, 1993; Selley et al., 1993).

A number of other points are relevant with regard to the mechanisms involved in opioid-induced inhibition of nociceptors during advanced stages of inflammation (Panel C). It is tentatively proposed that the enhanced transport of opioid receptors by axonal flow (see above), resulting in their accumulation at the nerve terminals, may enhance opioid analgesia in fully developed inflammation. As stated above, however, such an "up-regulation" which needs time does not seem to be an indispensable condition for the manifestation of peripheral opioid analgesia, although it may potentiate this effect.

Interesting aspects arise from the observation of an accumulation of inflammatory, immuno-competent cells, containing opioid peptides, in the inflamed tissue and the release of such peptides under conditions of stress. The importance of this mechanism results from the fact that immunosuppressive procedures, which inhibit the accumulation of such cells, abolish stress-induced analgesia (Stein et al., 1990b). Although cytokines (and CRF which also has cytokine-like functions) may be involved in the release of opioid peptides (Członkowski et al., 1993; Schäfer et al., 1994) their role remains unclear. One may speculate on the significance of similar opioid peptide release under clinical conditions. The finding that local application of naloxone enhances pain scores in patients undergoing arthroscopic knee surgery indicates that opioid peptides may be continually released from immuno-competent cells of the inflamed tissue (Stein et al., 1993). Further investigations are required to prove whether such an activation of opioidergic mechanisms during inflammation represents a counter-regulatory process in pain states of major significance.

In summary, the present view of peripheral opioid analgesia offers a variety of different facets; the mechanisms underlying them are presently only partly understood; their further elucidation is indispensible for successful introduction of these new aspects into clinical practice. Further investigations along these lines should provide new insights into the intrinsic mechanisms of pain modulation.

Acknowledgements

The author would like to thank Dr. Osborne Almeida for English corrections and Mrs. Kahleis for typing the manuscript.

References

Abbott, F. (1988) Peripheral and central antinociceptive actions of ethylketocyclazocine in the formalin test. *Eur. J. Pharmacol.*, 152: 93–100.

Andreev, N., Urban, L. and Dray, A. (1994) Opioids suppress spontaneous activity of polymodal nociceptors in rat paw skin induced by ultraviolet irradiation. *Neuroscience*, 58: 793–798.

Antonijevic I., Mousa, S.A., Schäfer, M. and Stein, C. (1995)

Perineurial defect and peripheral opioid analgesia in inflammation. *J. Neurosci.,* 15: 165–172.

Barber, A. (1993) μ- and κ-opioid receptor agonists produce peripheral inhibition of neurogenic plasma extravasation in rat skin. *Eur. J. Pharmacol.,* 236: 113–120.

Barber, A. and Gottschlich, R. (1992) Opioid agonists and antagonists: an evaluation of their peripheral actions in inflammation. *Med. Res. Rev.,* 12: 525–562.

Barthó, L., Stein, C. and Herz, A. (1990) Involvement of capsaicin-sensitive neurones in hyperalgesia and enhanced opioid antinociception in inflammation. *Naunyn Schmiedeberg's Arch. Pharmacol.,* 342: 666–670.

Childers, S.R. (1993) Opioid receptor-coupled second messenger systems. In: A. Herz (Ed.), *Handbook of Experimental Pharmacology, Vol. Opioids I,* Springer, Heidelberg, pp. 189–216.

Członkowski, A., Stein, C. and Herz, A. (1993) Peripheral mechanisms of opioid antinociception in inflammation: involvement of cytokines. *Eur. J. Pharmacol.,* 242: 229–235.

Dalsgaard, J., Felsby, S., Juelsgaard, P. and Froekjaer, J. (1994) Low-dose intra-articular morphine analgesia in day case knee arthroscopy: a randomized double-blinded prospective study. *Pain,* 56: 151–154.

Ferreira, S.H. and Nakamura, M. (1979) Prostaglandin hyperalgesia: the peripheral analgesic activity of morphine, enkephalins and opioid antagonists. *Prostaglandins,* 18: 191–200.

Follenfant, R.L., Hardy, G.W., Lowe, L.A., Schneider, C. and Smith, T.W. (1988) Antinociceptive effects of the novel opioid peptide BW443C compared with classical opiates; peripheral versus central actions. *Br. J. Pharmacol,* 93: 85–92.

Haley, J., Ketchum, S. and Dickenson, A. (1990) Peripheral κ-opioid modulation of the formalin response: an electrophysiological study in the rat. *Eur. J. Pharmacol.,* 191: 437–446.

Hargreaves, K.M., Costello, A.H. and Joris, J.L. (1989) Release from inflamed tissue of a substance with properties similar to corticotropin-releasing factor. *Neuroendocrinology,* 49: 470–482.

Hassan, A.H.S., Ableitner, A., Stein, C. and Herz, A. (1993) Inflammation of the rat paw enhances axonal transport of opioid receptors in the sciatic nerve and increases their density in the inflamed tissue. *Neuroscience,* 55: 185–195.

Heard, S.D., Edwards, W., Ferrari, D., Hanna, D., Wong, P.C., Liland, A. and Willock, M. (1992) Analgesic effect of intraarticular bupivacain or morphine after arthroscopic knee surgery: a randomized, protective double-blind study. *Anesth. Analg.* 74: 822–826.

Heppelmann,B., Messlinger, K., Neiss, W.F. and Schmidt, R.F. (1990). Ultrastructural three-dimensional reconstruction of group III and group IV sensory nerve endings ("free nerve endings") in the knee joint capsule of the cat: evidence for multiple receptive sites. *J. Comp. Neurol.,* 292: 103–116.

Herz, A. (1995) Opioid peptides, opioid receptors and peripheral analgesia. In: L.F. Tseng (Ed.), *The Pharmacology of Opioid Peptides.* Haarwood Academic, Amsterdam, pp. 287–301.

Honda, C.N., Dado, R.J., Riedl, M., Lee, J.H., Arvidsson, U., Wessendorf, M.W., Loh, H.H., Law, P.Y. and Elde, R. (1994) Distribution of delta- and mu-opioid receptors in small sensory neurons and axons. *Soc. Neurosci. Abstr.,* 20, 1728.

Joris, J.L., Dubner, R. and Hargreaves, K.M. (1987) Opioid analgesia at peripheral sites: a target for opioids released during stress and inflammation? *Anesth. Analg.,* 66: 1277–1281.

Joshi, G.P., McCarroll, S.M., O'Brien, T.M. and Lenane, P. (1993) Intraarticular analgesia following knee arthroscopy. *Anesth. Analg.,* 76: 333–336.

Junien, J.L. and Wettstein, J.G. (1992) Role of opioids in peripheral analgesia. *Life Sci.,* 51: 2009–2018.

Lembeck, F. and Donnerer, L. (1985) Opoid control of the function of primary afferent substance P fibres. *Eur. J. Pharmacol.,* 114: 241–246.

Levine, J.D. and Taiwo, Y.O. (1989) Involvement of the μ-opiate receptor in peripheral analgesia. *Neuroscience,* 32: 571–575.

Millan, M.J. Członkowski, A., Morris, B., Stein, C., Arendt, R., Huber, A., Höllt, V. and Herz, A. (1988) Inflammation of the hind limb as a model of unilateral, localized pain: influence on multiple opioid systems in the spinal cord of the rat. *Pain,* 35: 299–312.

North, R.A. (1986) Opioid receptor types and embrane ion channels. *Trends Neurosci.,* 114–117.

Parsons, C.G. and Herz, A. (1990) Peripheral opioid receptors mediating antinociception in inflammation. Evidence for activation by enkephalin-like opioid peptides after cold water swim stress. *J. Pharmacol. Exp. Ther,.* 255: 795–802.

Parsons, C.G., Członkowski, A., Stein, C. and Herz, A. (1990) Peripheral opioid receptors mediating antinociception in inflammation. Activation by endogenous opioids and role of the pituitary-adrenal axis. *Pain,* 41: 81–93.

Posner, J., Moody, S.G. Peck, A.W. Rutter, D. and Telekes, A. (1990) Analgesic, central, cardiovascular and endocrine effects of the enkephalin analogue Tyr-D-Arg-Gly-Phe $(4NO_2)$-Pro-NH_2 (443 C 81) in healthy volunteers. *Eur. J. Clin. Pharmcacol.,* 38: 213–218.

Przewłocki, R. (1993) Opioid systems and stress. In: A. Herz (Ed.), *Handbook of Experimental Pharmacology, Vol. Opioids II,* Springer, Heidelberg, pp. 293–324.

Przewłocki, R., Hassan, A.H.S., Lason, W., Epplen, C., Herz, A. and Stein, C. (1992) Gene expression and localization of opioid peptides in immune cells of inflamed tissue. Functional role in antinociception. *Neuroscience,* 48: 491–500.

Raja, S.N., Dirkstein, R.E. and Johnson, C.A. (1992) Comparison of postoperative analgesic effects of intraarticular bupivacaine and morphine following arthroscopic knee surgery. *Anesthesiology,* 77: 1143–1147.

Rogers, H., Birch, P.J., Harrison, S.M., Palmer, E., Manchee,

G.R., Judd, D.B., Naylor, A., Scopes, D.I.C. and Hayes, A.G. (1992) GR94839, a κ-opioid agonist with limited access to the central nervous system, has antinociceptive activity. *Br. J. Pharmacol,.* 106: 783–789.

Russell, N.J.W., Schaible, H.G. and Schmidt, R.F. (1987) Opiates inhibit the discharges of fine afferent units from inflamed knee joint of the cat. *Neurosci. Lett.,* 76: 107–112.

Schäfer, M., Carter, L. and Stein, C. (1994) Interleukin 1β and corticotropin-releasing factor inhibit pain by releasing opioids from immune cells in inflamed tissue. *Proc. Natl. Acad. Sci. USA,* 91: 4219–4223.

Schiller, P.W., Nguyen, T.M.D., Chung, N.N., Dionne, G. and Martel, R. (1990) Peripheral antinociceptive effect of an extremely μ-selective polar dermophin analog (DALDA). In: R. Quirion, K. Jhamandas and C. Gianoulakis (Eds.), *The International Narcotics Research Conference (INRC)*, Alan R. Liss, New York, pp. 53–56.

Schmidt, R.F., Schaible, H.G., Meßlinger, K., Heppelmann, B., Hanesch, U. and Pawlack, M. (1994) In: G.F. Gebhart, D.L. Hammond and T.S. Jensen (Eds.), *Progress in Pain Research and Management, Vol. 2*, IASP Press, Seattle, WA, pp. 213–250.

Selley, E., Breivogel, C.S. and Childers, S. (1993) Modification of G-protein-coupled functions by low-pH pretreatment of membranes from NG 108–15 cells: increase in opioid efficacy by decreased inactivation of G proteins. *Mol. Pharmacol.,* 44: 731–741.

Steen, K.H., Steen, A.E. and Reeh, P.W. (1995) A dominant role of acid pH in inflammatory excitation and sensitization of nociceptors in rat skin, in vitro. *J. Neurosci.,* 15: 3928–3989.

Stein, C., Millan, M.J., Shippenberg T.S. and Herz, A. (1988a) Peripheral effect of fentanyl upon nociception in inflamed tissue of the rat. *Neurosci. Lett.,* 84: 225–228.

Stein, C., Millan, M.J., Yassouridis, A. and Herz, A. (1988b) Antinociceptive effects of μ- and κ-agonists in inflammation are enhanced by a peripheral opioid receptor-specific mechanism. *Eur. J. Pharmacol.,* 155: 255–264.

Stein, C., Millan, M.J., Shippenberg, T.S., Peter, K. and Herz, A. (1989) Peripheral opioid receptors mediating antinociception in inflammation. Evidence for involvement of μ-, δ- and κ-receptors. *J. Pharmacol. Exp. Ther.,* 248: 1269–1275.

Stein, C., Gramsch, C. and Herz, A. (1990a) Intrinsic mechanisms of antinociception in inflammation. Local opioid receptors and β-endorphin. *J. Neurosci.,* 10: 1292–1298.

Stein, C., Hassan, A.H.S., Przewłocki, R., Gramsch, C., Peter, K. and Herz, A. (1990b) Opioids from immunocytes interact with receptors on sensory nerves to inhibit nociception in inflammation. *Proc. Natl. Acad. Sci. USA,* 87: 5935–5939.

Stein, C., Comisel, K., Haimerl, E., Yassouridis, A., Lehrberger, K., Herz, A. and Peter, K. (1991) Analgesic effect of intraarticular morphine after arthroscopic knee surgery. *N. Engl. J. Med.,* 325: 1123–1126.

Stein, C., Hassan, A.H.S., Lehrberger, K., Giefing, J. and Yassouridis, A. (1993) Local analgesic effect of endogenous opioid peptides. *Lancet,* 342: 321–324.

Taiwo, Y.O. and Levine, J.D. (1991) κ- and δ-opioids block sympathetically dependent hyperalgesia. *J. Neurosci.,* 11: 928–932.

Yaksh, T.L. (1988) Substance P release from knee joint afferent terminals: modulation by opioids. *Brain Res.,* 458, 319–324.

CHAPTER 8

Alterations in the functional properties of dorsal root ganglion cells with unmyelinated axons after a chronic nerve constriction in the rat

Robert H. LaMotte[1], Jun-ming Zhang[1] and Marlen Petersen[2]

[1]*Department of Anesthesiology, Yale University School of Medicine, New Haven, CT 06510, USA and*
[2]*Department of Physiology, University of Wurzburg, Wurzburg, Germany*

Introduction

A fascinating sensory disturbance resulting from an injury to a peripheral cutaneous nerve, is a persistent allodynia and hyperalgesia to mechanical stimulation of the skin. An innocuous touch such as might be produced by gently rubbing the skin against clothing is painful (allodynia). A mildly painful prickle from a sharp object such as a von Frey filament or a needle can evoke an abnormally intense and long lasting pain (hyperalgesia). In addition, there may be an ongoing pain referred to the skin even in the absence of overt stimulation. In certain patients some or all of these symptoms are exacerbated by activation of the sympathetic nervous system or alleviated by sympathetic blockade. This condition is called reflex sympathetic dystrophy (Evans, 1946) or sympathetically maintained pain (Roberts, 1986). In this case, it is possible that an abnormal responsiveness of certain peripheral sensory neurons to norepinephrine (NE) may play a role. The same symptoms in other patients are not relieved by sympathetic blockade. These patients are said to have sympathetically independent pain.

Experimental studies in animals have found that peripheral sensory neurons in injured nerves can develop ongoing discharge and/or an abnormal response to certain endogenous chemical substances including NE (for review, Devor, 1994).

There is also evidence for abnormal changes in the functional properties of the somata of sensory neurons with myelinated axons after a nerve injury. For example, spontaneous activity in A fibers can originate in the dorsal root ganglion (DRG) after a loose ligation of the sciatic nerve (Kajander et al., 1992). Also, a chemical coupling can occur within the $L_{4,5}$ DRGs between sympathetic postganglionic neurons and afferent neurons with myelinated fibers after a transection of the sciatic nerve (Burchiel, 1984; McLachlan et al., 1993; Devor et al., 1994; Janig et al., 1994).

Physiological studies of functional changes in primary sensory neurons after nerve injury have focused mainly on distal nerve endings or, in the case of DRG involvement, on A fibers and not C fibers (Devor, 1994). There is evidence that activity in a subpopulation of chemosensitive C fibers is responsible for producing neurogenic (secondary) hyperalgesia and allodynia in a large area of skin surrounding a local injury (Baumann et al., 1991; LaMotte et al., 1991; Simone et al., 1991; Torebjork et al., 1992). If a peripheral nerve injury caused such fibers to become chronically active due, for example, to the development of ectopic electrogenesis and a responsiveness to certain endogenous chemicals such as NE, then chronic pain and cutaneous dysesthesias would be a natural consequence.

If abnormal response properties were found to originate at least in part in the somata of chemosensitive C fibers, then a rationale would exist for using freshly dissociated DRG cells and patch-clamp recording techniques to study neuropathological changes in the expression of receptors and ion channels in putative nociceptive neurons that might contribute to sensory disturbances after nerve injury.

In the following, we describe evidence that the DRG may become a source of abnormal nociceptive activity and the DRG cell is a model for studying neuropathological changes in primary sensory neurons at a cellular and eventually molecular level.

The experimental model of pain and hyperalgesia that we are using is one in which rats develop cutaneous hyperalgesia to heat and mechanical stimulation of the ventral foot after a chronic constriction of the ipsilateral sciatic nerve (Bennett and Xie, 1988). Our first objective was to investigate whether the somata of C-fibers, after a loose ligation, particularly those fibers with spontaneous activity and a responsiveness to capsaicin, had an abnormal response to NE topically applied to the DRG. After finding this to be the case, and that the response was blocked by an alpha-2 blocker (Xie et al., 1995), we asked next whether the ectopic discharges and NE sensitivity resulted from alterations in the properties of the soma membrane or, alternatively, from pathological processes extrinsic to the injured neuron (Petersen and LaMotte, 1994; Zhang and LaMotte, 1995).

Electrophysiological recordings from single unmyelinated dorsal root fibers were typically carried out two to four weeks after ligation of the sciatic nerve. Parallel studies were made in normal rats that had not received a nerve injury.

The perineurium and epineurium of the DRG were removed and the nerve fibers at the injury site exposed so that perfusing solutions with or without NE or other active agents could be applied to both locations (Fig. 1). NE was applied in concentrations of 0.0001–1 mM for a duration of 3 min followed by a 20 min washout with the control solution (HEPES).

Most of the neurons with unmyelinated fibers (C fibers) from injured nerves were first identified on the basis of ongoing ("spontaneous") activity (SA). The SA was not likely to have originated

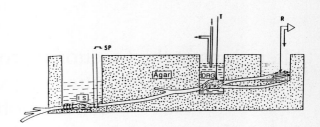

Fig. 1. Schematic of the set-up for microfilament recording in vivo. I.S., injury site; DRG, L_5 dorsal root ganglion; SP, stimulating electrode proximal to the injury site; R, recording electrode; T, thermocouple monitoring the temperature of the perfusing solution (modified reproduction, with permission, from Xie et al., 1995).

from cutaneous receptors since the fibers did not respond to electrical stimulation of the nerve distal to the injury site and/or to noxious mechanical or thermal stimulation of the skin. The hypothesis that SA can originate in the somata, at least for A-beta fibers after a chronic nerve constriction (Kajander et al., 1992) is consistent with our finding that SA persisted after transection of the nerve just proximal to the injury site. Similarly, Zimmerman and his colleagues demonstrated the existence of ectopic generators of SA at various sites in injured primary afferent neurons proximal to an experimentally produced neuroma (Welk et al., 1990).

NE (≤ 0.5 mM) did not evoke activity in any of the C fibers from normal, uninjured nerves when applied either to the peripheral nerve or to the DRG. In contrast, when NE was applied to the injury site or to the DRG of injured nerves, 90 and 81% respectively responded with an increase in discharge rate, defined as an increase by 30% or more (Fig. 2). A pattern of SA consisting of bursting and/or irregular activity changed, after application of NE, into a higher frequency with more regular discharges and shorter interspike intervals that continued for several minutes after the onset of washout. The rate and pattern of discharge typically returned to its original state within 5–10 min after the removal of NE from the perfusion. When a 1% solution of capsaicin, the algesic chemical in red peppers, was applied topically to the DRG at the end of an experiment, nearly all of

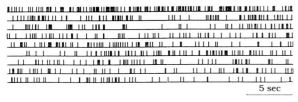
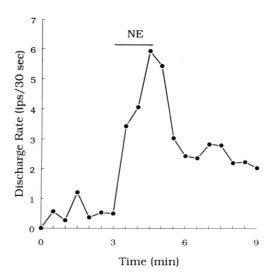

Fig. 2. In vivo recording of the responses of a C fiber to norepinephrine (NE) applied to the DRG of an injured nerve. (A) Action potentials (tick marks) in a C fiber, recorded continually from top left to bottom right, during and after NE (0.1 mM) in HEPES was applied to the DRG for 90 s. HEPES alone was applied before and after NE. (B) Discharge rate plotted for consecutive 30 s intervals.

the uninjured and the NE responsive C fibers tested responded to the drug.

Subsequent experiments were carried out to characterize further the effects of NE application to the DRG on the responses of injured C fibers.

After delivering different concentrations of NE (0.0001–1 mM) to the same fiber, it was determined that 0.01 mM was the lowest dose that evoked a statistically significant response. It was also found that when this concentration was repeatedly delivered, with intervening 20 min washout periods, that the response in each case remained about the same.

The dose of 0.01 mM was then used in an experiment to determine the effects of alpha-adrenergic antagonists on the responses of C fibers to NE applied topically to the DRGs of injured nerves. After obtaining a response to NE and washing out with HEPES alone for 20 min, either yohimbine (10 μm) or prazosin (5 μm), alpha-2 and alpha-1 blockers respectively, were applied in HEPES for 10 min followed by the blocker combined with NE. The response to NE was significantly reduced by yohimbine but not by prazosin (Fig. 3).

Definitive proof for a role of alpha-2 receptors must await further experimentation using more selective alpha adrenergic agonists along with appropriate antagonists. But the results to date are consistent with the hypothesis that alpha-2 adrenergic receptors play an important role in mediating the responses of injured C fibers to NE topically applied to the DRG.

It was also determined that a small but significant increase in the mean SA of injured C fibers occurred during electrical stimulation of the sympathetic trunk (L_2–L_3). The discharge rates of 65% of the fibers tested increased by 30% or more whereas those of 13% decreased by a mean of 45%. We have not yet investigated the effects of alpha-adrenergic antagonists on the sympathetically evoked response.

A conceivable hypothesis, at this point in our research, was that the SA and responses to NE applied to the DRG might not occur as a result of changes in the functional properties of sensory neurons per se after nerve injury but as a conse-

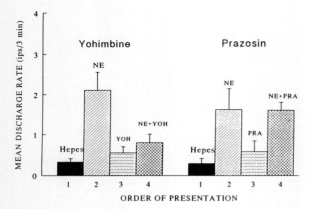

Fig. 3. Effects of alpha-adrenergic antagonists on C fiber activity evoked by NE applied to DRGs of injured nerves. The DRG was perfused with HEPES for 3 min followed by NE (0.01 mM) for 3 min. This was followed by a 20-min washout with HEPES alone (not shown), during which time the discharge rates decreased to approximately normal levels. Next, the alpha-2 blocker, yohimbine (YOH, 10 μM, $n = 9$), or alpha-1 blocker, prazosin (PRA, 5 μM, $n = 5$), was delivered for 10 min, followed by the blocker plus NE (0.01 mM) for 3 min (reproduced, with permission, from Xie et al., 1995).

quence of changes in tissue extrinsic to the neuron (e.g. see Levine et al., 1986; Gonzales et al., 1991). Therefore, the next step in our research was to determine whether these abnormal responses still occurred after the peripheral nerve, including the intact ganglion and dorsal root were removed from the animal but kept viable in an in vitro preparation. In this case, recordings from dorsal root C fibers were carried out while artificial cerebrospinal fluid perfused the ganglion and nerve - the latter including the constriction injury site (Zhang and LaMotte, 1995). Instead of searching primarily for fibers with spontaneous activity, we now searched non-selectively for C fibers by electrically stimulating the peripheral nerve. As a result, a greater proportion of fibers with little or no SA were identified. About 30% of the C fibers isolated for study had no SA, as observed over a 3 min observation period prior to experimentation. Although the mean discharge rates of fibers with SA was lower in vitro (0.4 impulses/s) than in vivo (0.9 impulses/s) the range in ongoing discharge rates was about the same. The SA was typically still present after a nerve transection 1 cm proximal to the ganglion. We interpret these results as suggesting that the SA originated in the DRG and was not greatly changed after the removal of the ganglion from the animal, e.g. removal from the effects of blood borne factors and the effects of activity in the sympathetic nervous system.

Different concentrations of NE and clonidine, a partial alpha-2 agonist, were topically superfused over the DRG. As was the case for the in vivo studies, none of the C fibers from uninjured, normal nerves exhibited either SA or responses to NE. In contrast, sixty percent of the C fibers from injured nerve tested with NE, and 70% given clonidine, responded. In general, the pattern of increased activity and the presence of afterdischarge were findings similar to those obtained in vivo (Fig. 4). The lowest concentrations of NE and clonidine evoking a threshold response (arbitrarily defined as rates of 30% or more above base rates of SA) were 10 and 1 μM respectively. Some of the fibers responsive to either NE or clonidine had no SA. Conversely, some fibers had SA but did not respond to NE or clonidine. Thus, different mechanisms may be responsible for the development of SA and the responsiveness to adrenergic chemicals.

These results support the conclusion that the abnormal SA and responsiveness to NE in C fibers are not dependent on factors extrinsic to the nerve and ganglion, for example, non-adrenergic chemicals delivered from the terminals of sympathetic efferent fibers or chemicals delivered from the blood stream. However, they do not prove that these abnormal characteristics are intrinsic to the soma membrane.

To investigate this possibility, whole cell patch-clamp recordings under current clamp were made from acutely dissociated L_4 and L_5 DRG cells obtained from rats with normal sciatic nerves or nerves that had received a constriction injury two or three weeks earlier (Petersen and LaMotte, 1994). Recordings were made preferentially from small sized cells (mean area of 524 μm^2) within eight hours after plating in culture. A response was defined as a depolarization or hyperpolarization of ≥3 mV.

Fourteen percent of the cells from injured nerves superfused with 10 μM of NE and 52%

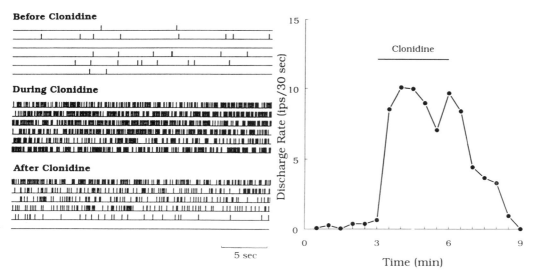

Fig. 4. Responses of a C fiber to clonidine applied to the whole DRG in vitro. The DRG was continuously perfused with oxygenated artificial cerebrospinal fluid (ACSF). Clonidine (0.1 mM) in oxygenated ACSF was then applied to the DRG for 90 s followed by a 20-min washout with oxygenated ACSF again (same format as in Fig. 2).

given 500 μM responded with a membrane depolarization accompanied in some cases by the generation of action potentials (Fig. 5). In contrast, only one of fifteen cells tested from uninjured nerves responded to 100 and 500 μM of NE and did so with a depolarization of only 3 mV and without action potentials. In addition, 13% of the cells from ligated nerves but none from uninjured nerves were spontaneously active. The pattern of activity was irregular. All cells responsive to NE and also tested with a superfusion of 300 nM of capsaicin for ≤30 s responded to capsaicin.

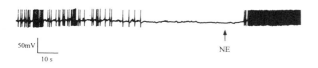

Fig. 5. Polygraph record of a patch-clamp recording, in current clamp mode, of spontaneous activity and a response to NE (0.5 mM) in a DRG cell acutely dissociated 19 days after nerve injury. The resting membrane potential of the cell was −45 mV.

Discussion

Peripheral axonal injury produces biochemical, structural and physiological changes in both injured and neighboring intact neurons including regenerative and collateral sprouting and alterations in excitability (for reviews, e.g. Titmus and Faber, 1990; Walters, 1994). These events are accompanied in DRG cells by the expression of transcription factors such as Jun and Jun D (Herdegan and Zimmermann, 1994), increases in mRNA for NO synthase (Verge et al., 1992) ornithine decarboxylase activity (Soifer et al., 1988) and growth-associated protein GAP-43 (Hoffman, 1989; Woolf et al., 1990).

There is ample evidence that axotomy can increase the excitability of peripheral sensory neurons possibly due to changes in the synthesis of subunit proteins for voltage and ligand gated channels in the soma and their transport into and down the axon (for review, Titmus and Faber, 1990). For example, alterations in the densities and distributions of certain calcium, potassium and sodium channels in the soma membrane might result in ectopic spontaneous activity and conceivably produce a hyperexcitability such that currents generated by a variety of chemical and mechanical stimuli are more effective in depolarizing the cell (Devor, 1994).

A partial peripheral nerve injury can also change the functional properties of intact neurons with axons in the same nerve. An abnormal responsiveness to NE can develop not only in sensory neurons with injured (e.g. transected) axons but also in neurons whose axons are apparently uninjured and which conduct normally across the injury site. After a partial cut of a peripheral nerve, about 40% of the polymodal nociceptors with intact, unmyelinated axons became responsive to close arterial injection of NE or to sympathetic stimulation -effects probably mediated by alpha-2 adrenergic receptors (Sato and Perl, 1991). It was also shown that a partial nerve cut resulted in an up-regulation of alpha-2 receptors in certain small and medium sized cell bodies of neurons with injured axons as well as those with uninjured axons (Nishiyama et al., 1993; Perl, 1994).

Interestingly, it was also found that these results did not occur solely as a result of injury to the sensory nerve fibers since similar effects occurred after a sympathectomy (Perl, 1994). Although weaker responses to NE were obtained after sympathectomy and then in only 20% of the C nociceptors, one effect of a partial nerve cut, as hypothesized by Perl, may be a denervation supersensitivity resulting from the transection of sympathetic fibers that reduces the amount of NE available to C nociceptors. Although it is not known what role NE receptors normally have in uninjured sensory nerves, it is possible that a shortage of NE is at least one cause of an up-regulation of alpha-2 adrenergic receptors in C-polymodal nociceptive neurons after a nerve injury that transects a sufficient number of sympathetic efferent fibers.

The demonstration of an up-regulation of alpha-2 adrenoceptors in injured and non-injured small and medium sized somata of sensory neurons after a partial cut of a peripheral nerve (Perl, 1994) is consistent with the results we obtained after a chronic nerve constriction. It is complemented by our finding that NE can act directly on sensory neurons with injured C fibers, possibly via alpha-2 receptors, to depolarize them and that the abnormal responsiveness to NE and spontaneous activity in at least some nerve injured C fibers are manifested in the soma as well as in peripheral terminal endings. Thus, the DRG can be a source of abnormal nerve impulses in chemosensitive C fibers after a peripheral nerve injury.

In some patients, the pain and/or hyperalgesia resulting from a nerve injury can extend well outside the distribution of the injured nerve suggesting possible changes in sensory processing in the central nervous system (Gracely et al., 1992). In experimental studies of pain, a localized cutaneous injury can produce allodynia and hyperalgesia within a large area of normal skin surrounding the injury as well as within the injured area itself (secondary and primary hyperalgesia respectively) (Lewis, 1936; Hardy et al., 1950). Evidence from studies of secondary hyperalgesia induced by an intradermal injection of capsaicin suggests that chemosensitive C-fibers can initiate and possibly maintain, via sensitization of spinothalamic tract cells, the resulting allodynia and hyperalgesia to mechanical stimulation of the skin (LaMotte et al., 1991; Simone et al., 1991). Presumably, once a state of central sensitization exists, evoked activity in low- as well as high-threshold mechanoreceptive neurons with myelinated axons can evoke pain, hyperalgesia and allodynia. Thus, it is possible that after certain peripheral nerve injuries, the same kinds of chemosensitive C-fibers can become chronically spontaneously active or respond abnormally to normally ineffective endogenous stimuli such that a state of secondary hyperalgesia and allodynia persists.

Acknowledgements

Supported by USPHS Grants NS14624 and NS 10174.

References

Baumann, T.K., Simone, D.A., Shain, C.N. and LaMotte, R.H. (1991) Neurogenic hyperalgesia: the search for the primary cutaneous afferent fibers that contribute to capsaicin-induced pain and hyperalgesia. *J. Neurophysiol.*, 66: 212–227.

Bennett, G.J. and Xie, Y.-K. (1988) A peripheral mononeuropathy in rat that produces disorders of pain sensation like those seen in man. *Pain*, 33: 87–107.

Burchiel, K.J. (1984) Spontaneous impulse generation in normal and denervated dorsal root ganglia: sensitivity to alpha-adrenergic stimulation and hypoxia. *Exp. Neurol.*, 85: 257–272.

Devor, M. (1994) The pathophysiology of damaged peripheral nerve. In: P.D. Wall and R. Melzack (Eds.), *Textbook of Pain*, 3rd edn. Churchill-Livingston, London, pp. 79–100.

Devor, M., Janig, W. and Michaelis, M. (1994) Modulation of activity in dorsal root ganglion neurons by sympathetic activation in nerve-injured rats. *J. Neurophysiol.*, 71: 38–47.

Evans, J.A. (1946) Reflex sympathetic dystrophy. *Surg. Clin. North Am.*, 26: 780–90.

Gonzales R., Sherbourne C.D., Goldyne M.E. and Levine J.D. (1991) Noradrenaline-induced prostaglandin production by sympathetic postganglionic neurons is mediated by 2-adrenergic receptors. *J. Neurochem.*, 57: 1145–1150.

Gracely, R.H., Lynch, S.A. and Bennett, G.J. (1992) Painful neuropathy: altered central processing maintained dynamically by peripheral input. *Pain*, 51: 175–194.

Hardy, J.D., Wolff, H.G. and Goodell, H. (1950) Experimental evidence on the nature of cutaneous hyperalgesia. *J. Clin. Invest.*, 29: 115–140.

Herdegen, T. and Zimmermann, M. (1994) Expression of c-Jun and Jun D transcription factors represent specific changes in gene expression following axotomy. *Prog. Brain Res.*, 103: 153–171.

Hoffman, P.N. (1989) Expression of GAP-43, a rapidly transported growth-associated protein, and class II beta tublin, a slowly transported cytoskeletal protein, are coordinated in regeerating neurons. *J. Neurosci.*, 9: 893–897.

Janig, W., Devor, M., McLachlan, E.M. and Michaelis, M. (1994) A new site of coupling between sympathetic neurones and afferent neurones following peripheral nerve lesions. In: J.M. Besson, G. Guilbaud and H. Ollat (Eds.), *Peripheral Neurons in Nociception: Physiopharmacological Aspects*. John Libbey Eurotext, Paris.

Kajander, K.C., Wakisaka, S. and Bennett, G.J. (1992) Spontaneous discharge originates in the dorsal root ganglion at the onset of a painful peripheral neuropathy in the rat. *Neurosci. Lett.*, 138: 225–228.

LaMotte, R.H., Shain,C.N., Simone, D.A. and Tsai, E. (1991) Neurogenic hyperalgesia: psychophysical studies of underlying mechanisms. *J. Neurophysiol.*, 66: 190–211.

Levine, J.D., Taiwo, Y.O., Collins, S.D. and Tam, J.K. (1986) Noradrenaline hyperalgesia is mediated through interaction with sympathetic postganglionic neuron terminals rather than activation of primary afferent nociceptors, *Nature*, 323: 158–160.

Lewis, T. (1936) Experiments relating to cutaneous hyperalgesia and its spread through somatic nerves. *Clin Sci*., 2: 373–423.

McLachlan, E.M., Janig, W., Devor, M. and Michaelis, M. (1993) Peripheral nerve injury triggers noradrenergic sprouting within dorsal root ganglia. *Nature*, 363: 543–545.

Nishiyama, K., Brighton, B.W., Bossut, D.F. and Perl, E.R. (1993) Peripheral nerve injury enhances $_2$-adrenergic receptor expression by some DRG neurons. *Soc. Neurosci. Abstr.*, 19: 499.

Perl, E.R. (1994) Causalgia and reflex sympathetic dystrophy revisited. In: J. Boivie, P. Hanson, and U. Lindblom (Eds.), *Touch, Temperature, and Pain in Health and Disease: Mechanisms and Assessments*. Progress in Pain Research and Management, Vol. 3, IASP Press, Seattle, WA.

Petersen, M. and LaMotte, R.H. (1994) Morphological and physiological changes in isolated dorsal root ganglion cells after nerve injury. *Soc. Neurosci. Abstr.*, 20: 1569.

Roberts, W.J. (1986) A hypothesis on the physiological basis for causalgia and related pain. *Pain*, 24: 297–311.

Sato, J. and Perl, E.R. (1991) Adrenergic excitation of cutaneous pain receptors induced by peripheral nerve injury. *Science*, 251: 1608–1610.

Simone, D.A., Sorkin, L.S., Oh, U., Chung, J.M., Owens, C., LaMotte, R.H. and Willis, W.D. (1991) Neurogenic hyperalgesia: central neural correlates in responses of spinothalamic tract neurons. *J. Neurophysiol.*, 66: 228–246.

Soiefer, A.I., Moretto, A., Spencer, P.S. and Sabri, M.I. (1988) Axotomy-induced ornithine decarboxylase activity in the mouse dorsal root ganglion is inhibited by the vinca alkaloids. *Neurochem. Res.*, 13: 1169–1173.

Titmus, M.J. and Faber, D.S. (1990) Axotomy-induced alterations in the electrophysiological characteristics of neurons. *Prog. Neurobiol.*, 35: 1–51, 1990.

Torebjork, H.E., Lundberg, L.E.R. and LaMotte, R.H. (1992) Central changes in processing of mechanoreceptive input in capsaicin-induced secondary hyperalgesia in humans. *J. Physiol.*, 448: 763–780.

Verge, V.M., Xu, Z., Xu, X.J., Wiesenfeld-Hallin, Z. and Hokfelt, T. (1992) Marked increase in nitric oxide synthase in RNA in rat dorsal root ganglion after peripheral axotomy: in situ hybridization and functional studies. *Proc. Natl. Acad. Sci. USA*, 89: 11617–11621.

Walters, E.T. (1994) Injury-related behavior and neuronal plasticity: an evolutionary perspective on sensitization, hyperalgesia, and analgesia. *Int. Rev. Neurobiol.*, 36: 325–426.

Welk, E., Leah, J.D. and Zimmermann, M. (1990) Characteristics of A- and C-fibers ending in a sensory nerve neuroma in the rat. *J. Neurophysiol.*, 63: 759–766.

Woolf, C.J., Reynolds, M.L., Molander, C., O'Brien, C., Lindsay, R.M. and Benowitz, LI. (1990) The growth-associated protein GAP-43 appears in dorsal root ganglion cells and in the dorsal horn of the rat spinal cord following peripheral nerve injury. *Neuroscience*, 34: 465–478.

Xie, Y.-K., Zhang, J.-M. and LaMotte, R.H. (1995) Functional changes in dorsal root ganglion cells after chronic nerve constriction in the rat. *J. Neurophysiol.*, 73: 1811–1820.

Zhang, J.-M. and LaMotte, R.H. (1995) Abnormal spontaneous activity and adrenergic sensitivity in intact nerve injured dorsal root ganglion cells in vitro. *Soc. Neurosci. Abstr.*, 21: 383, 1995.

CHAPTER 9

Plasticity of messenger function in primary afferents following nerve injury - implications for neuropathic pain

Zsuzsanna Wiesenfeld-Hallin and Xiao-Jun Xu

Karolinska Institute, Department of Medical Laboratory Sciences and Technology, Section of Clinical Neurophysiology, Huddinge University Hospital, Sweden

Introduction

Two major categories of substances are possible candidates for neurotransmitters/neuromodulators in primary sensory neurons - excitatory amino acids (EAA) and peptides. The EAAs glutamate and aspartate have been identified biochemically in primary afferents (see Salt and Hill, 1983 for review) and more recently glutamate has been located immunohistochemically in dorsal root ganglion cells and in the dorsal horn (Battaglia and Rustioni, 1988; De Biasi and Rustioni, 1988, 1990). The physiological function of glutamate in the spinal cord has been examined with specific antagonists acting at a variety of receptors for glutamate. There is considerable evidence that activation of the N-methyl-D-aspartate (NMDA) receptor is of great importance for the mediation of nociceptive stimuli (Davies and Lodge, 1987; Dickenson and Sullivan, 1990; Woolf and Thompson, 1991). Of the neuropeptides, substance P (SP) was the first to be identified immunohistochemically in a subpopulation of small primary afferents (Hökfelt at al., 1975). The role of SP in nociceptive transmission was suggested in early studies (Otsuka and Konishi, 1976; Jessell and Iversen, 1977). Since then a large number of other peptides have been identified in primary afferents and their role in normal and pathological circumstances have been examined (Hökfelt et al., 1980, 1994). In this review the function of EAA and neuropeptides in animals with intact peripheral nerves and after deafferentation will be summmarized and the possible role of functional plasticity of neurotransmitters in neuropathic pain will be discussed.

The use of spinal reflexes in studies of spinal nociceptive mechanisms

Spinal reflexes have been extensively used in studies of pain mechanisms in animals and man. The flexor reflex has been used in our laboratory to study the role of EAA and neuropeptides in modulating spinal cord excitability. The magnitude of the nocifensive hindpaw flexor reflex is positively correlated with the activity of dorsal horn neurons that are activated by both innocuous and noxious stimuli, the so-called wide dynamic range cells (Schouenborg and Sjölund, 1983; Carstens et al., 1990). Furthermore, the magnitude of the flexor reflex evoked by noxious stimulation on the foot and recorded from the biceps femoris muscle is well correlated to painful sensations in adult humans (Willer, 1977; Chang and Dallaier, 1989). The flexor reflex is a graded response and is very suitable for both physiological and pharmacological studies. The preparation used is the decerebrate, spinalized rat as described by Wall and Woolf (1984). Since the animals are decerebrated under the influence of a short lasting barbiturate, the responses recorded during the course of the experiment, which starts at least 1 h after the preparation is finished, are not influenced by the presence of a general anaesthetic. The spinalization procedure removes the tonic descending control exerted by inhibitory pathways originating in the brainstem (Liebeskind et al., 1973; Willis et al., 1977; Basbaum et. al, 1978). Thus, a very brisk flexor reflex can be evoked by electrical or natural (mechanical or thermal) stimulation of the skin and electrical stimulation of muscle nerves in the hindlimb.

The effect of conditioning stimulation of unmyelinated afferents on the flexor reflex

Activation of dorsal horn interneurons and the flexor reflex are under modulatory control, involving both excitation and inhibition (Melzack and Wall, 1965; Besson and Chaouch, 1987). Repetitive activation of C-afferents leads to a gradual increase in the response of dorsal horn interneurons, which has been termed "wind-up" (Mendell, 1966). In their pioneering study Wall and Woolf (1984) demonstrated that a brief conditioning stimulus (CS) train that activates unmyelinated (C) cutaneous afferents leads to wind-up and facilitation of the flexor reflex that considerably exceeds the duration of stimulation. Thus, 20 electric shocks at 1 Hz applied to the cutaneous sural nerve facilitated the flexor reflex for 5–10 min, whereas the same CS applied to muscle afferents in the gastrocnemious nerve facilitated the flexor reflex for about 1 h. Such changes in reflex excitability occurred independently of changes in the excitability of afferent terminals or motorneurons (Cook et al., 1986) Thus, prolonged changes in spinal cord excitability may be due to sensitization of dorsal horn interneurons. Corresponding changes in spinal cord excitability may be responsible for some painful conditions in humans, such as hyperalgesia, allodynia and tenderness (Woolf, 1983; Woolf and Chong, 1993). The flexor reflex does not become facilitated if the intensity of the CS is weaker and only activates rapidly conducting (Aβ) or slowly conducting (Aδ) myelinated afferents (Wall and Woolf, 1984). This difference between the consequences of activation of myelinated and unmyelinated afferents has lead us to consider whether sensory neuropeptides may be involved in the sensitization of the flexor reflex following C-fiber activation.

Assessment of the role of neuropeptides and glutamate on spinal nociceptive mechanisms with the flexor reflex when peripheral nerves are intact

The mechanisms of action of a large number of peptides, peptide agonists and antagonists and an NMDA antagonist have been examined with the rat flexor reflex model.

The role of the tachykinins substance P and neurokinin A in flexor reflex hyperexcitability

SP-like immunoreactivity (SP-LI) and neurokinin A-LI (NKA-LI) have been found to be colocalized in small dorsal root ganglion cells (Dalsgaard et al., 1985). SP and NKA applied on the spinal cord excite dorsal horn interneurons and may function in nociceptive neurotransmission (Henry, 1976; Fleetwood-Walker et al., 1990). Picomolar quantities of SP or NKA injected intrathecally (i.t.) onto the lumbar spinal cord facilitate the flexor reflex (Wiesenfeld-Hallin, 1986; Woolf and Wiesenfeld-Hallin, 1986; Xu and Wiesenfeld-Hallin, 1992). Interestingly, the duration of the facilitatory effect of i.t. SP is similar to that evoked by a CS applied to the sural nerve (Woolf and Wiesenfeld-Hallin, 1986). If NKA and SP are coadministered the facilitation of the flexor reflex is synergistically increased (Fig. 1), exceeding the additive effect of the two peptides (Xu and Wiesenfeld-Hallin., 1992; Wiesenfeld-Hallin and Xu, 1993). The synergistic interaction of SP and NKA may indicate that neuropeptides that are colocalized may interact functionally (also see below).

The role of endogenous SP and NKA in the CNS were difficult to assess before suitable antagonists became available. Spantide II, a nonselective antagonist of tachykinin receptors (Folkers et al., 1990), proved to be the first nontoxic tachykinin antagonist with a clear effect in the CNS (Wiesenfeld-Hallin et al., 1990a,b). Spantide II applied i.t. antagonized the scratching/biting behaviour evoked by i.t. SP (Wiesenfeld-Hallin et al., 1990b) and blocked the sensitization of the flexor reflex by SP, as well as following CS of both cutaneous (Wiesenfeld-Hallin

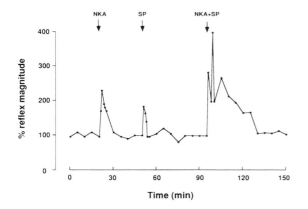

Fig.1. The excitatory effect of intrathecal neurokinin A (7 pmol), substance P (7 pmol) and a mixture of 3.5 pmol of both tachykinins on the flexor reflex in a decerebrate spinalized rat. The reflex was evoked by subcutaneous electrical shocks (1/min) that activated A and C fibers in the sural innervation area. Note the synergistic interaction of the two peptides. Control reflex magnitude is defined as 100%. (from Xu and Wiesenfeld-Hallin, 1992)

et al., 1990a) and muscle (Wiesenfeld-Hallin et al., 1991a) afferents. These results indicated that tachykinins may mediate spinal cord hyperexcitability when released into the spinal cord following activation of both cutaneous and muscle afferents.

More recently, specific antagonists of both NK_1 and NK_2 receptors for SP and NKA, respectively, have become available. With the use of specific antagonists it has been possible to demonstrate the differential functions of SP and NKA in the spinal cord (Wiesenfeld-Hallin and Xu, 1993). CP 96 345, a non-peptide antagonist of the NK_1 receptor that can be applied systemically (Snider et al., 1991), effectively blocked facilitation of the flexor reflex following activation of both cutaneous (Xu et al., 1992a) and muscle (Wiesenfeld-Hallin and Xu, 1993) afferents. In contrast, Menarini 10207, a specific antagonist of the NK_2 receptor (Maggi et al., 1990) blocked spinal hyperexcitability following CS of muscle, but not cutaneous, afferents (Xu et al., 1991a; Wiesenfeld-Hallin and Xu, 1993). These results indicate that SP released after activation of cutaneous afferents may be responsible for the brief facilitation of the flexor reflex and co-release of SP and NKA may underly the prolonged facilitation of the flexor reflex following activation of muscle afferents. This differential effect of the activation of NK_1 and NK_2 receptors on spinal reflex hyperexcitability has been recently confirmed with the use of other specific antagonists (Ma and Woolf, 1995).

The interaction of SP with other neuropeptides present in primary afferents

Peptidergic primary afferents exhibit a complex pattern of coexistence where a number of peptides derived from different precursors can be localized in the same afferent (see Hökfelt et al., 1980, 1994 for review). The tachykinins SP and NKA are colocalized in primary afferents and synergistically potentiate each other (see above). Tachykinins have been found to be colocalized with other peptides derived from very different precursors. In rats with intact peripheral nerves SP is colocalized with calcitonin gene-related peptide (CGRP) (Wiesenfeld-Hallin et al., 1984). CGRP-LI, which is very abundant in primary afferents and is found in both small and large dorsal root ganglion cells, may be localized in all cells exhibiting SP-LI. The possible functional consequence of the coexistence of these two peptides was examined in behavioural and physiological studies (Wiesenfeld-Hallin et al., 1984). SP injected onto the lumbar spinal cord in rats with implanted i.t. catheters evoked a brief caudally directed biting/scratching response lasting less than 5 min. CGRP by itself evoked no response at even high doses. CGRP co-administered with SP potentiated the effect of SP, evoking a much more prolonged and intense biting/scratching behavior. Parallel results were obtained in flexor reflex studies (Woolf and Wiesenfeld-Hallin, 1986). By itself i.t. CGRP only weakly facilitated the flexor reflex, but synergistically increased the excitatory effect of i.t. SP. The mechanism of the interaction between CGRP and SP has been investigated and CGRP has been found to inhibit the degradation of SP (Le Grevés et al., 1985). The synergistic interaction of SP and CGRP is a further example of functional interaction between coexisting neuropeptides.

SP is also colocalized with the neuropeptide galanin in rat dorsal root ganglion cells and the three peptides, SP, CGRP and galanin, have been identified in the same cell (Ch'ng et al., 1985; Ju et al., 1987). Normally few cells exhibit galanin-LI in rat, but the expression of this peptide increases remarkably after peripheral nerve injury (Hökfelt t al., 1987; Villar et al., 1989). We have examined the effect of galanin on the flexor reflex both in rats with intact and sectioned sciatic nerves (see below). Galanin has a complex effect on the reflex and has primarily an inhibitory function (Wiesenfeld-Hallin et al., 1989a). Galanin functions as a physiological SP antagonist when the two peptides are coadministered i.t. (Xu et al., 1989, 1990).

The interaction of SP and glutamate

SP is also colocalized with the EAA glutamate in dorsal root ganglion cells and their terminals in the spinal cord (Battaglia and Rustioni, 1988; De Biasi and Rustioni, 1988, 1990). Glutamate appears to have an important role in wind-up (Davies and Lodge, 1987) and sensitization of the flexor reflex. The prolonged facilitation of the flexor reflex following C-afferent CS was reduced by blockade of the NMDA receptor (Woolf and Thompson, 1991). We examined the interaction of the NK_1 receptor antagonist CP-96,345 and the NMDA antagonist MK-801 on wind-up and on the facilitation of the reflex by CS of cutaneous afferents (Xu et al., 1992b). MK-801 depressed the baseline reflex and reduced wind-up and post-stimulus facilitation. CP 96 345 was less effective in reducing wind-up, but blocked the post-CS facilitation as effectively as MK 801. When subthreshold doses of the NK_1 and NMDA antagonists were coadministered both wind-up and reflex facilitation were synergistically reduced. Thus, glutamate and SP may be co-released upon C-fiber stimulation and interact synergistically to induce central sensitization.

The function of galanin and vasoactive intestinal polypeptide on flexor reflex hyperexcitability in rats with intact peripheral nerves

SP, NKA, somatostatin (SOM) and CGRP are the major peptides normally identified in rat primary afferents, whereas galanin-LI and VIP -LI are much less prominent in normal dorsal root ganglia (Hökfelt et al., 1994). Although i.t. VIP facilitates spinal cord reflex excitability in rats with intact peripheral nerves (Wiesenfeld-Hallin, 1987, 1989; Xu and Wiesenfeld-Hallin, 1991), endogenous VIP does not seem to have a role in C-fiber activity-induced spinal sensitization under normal conditions since a VIP antagonist had no effect on CS-induced reflex facilitation in rats with intact sciatic nerves (Wiesenfeld-Hallin et al., 1990c; Xu and Wiesenfeld-Hallin, 1991).

In rats with intact nerves galanin has a complex effect on the flexor reflex with a brief facilitatory effect at low doses, facilitation followed by inhibition at higher doses and a purely inhibitory effect at the highest dose (Wiesenfeld-Hallin et al., 1988, 1989a). Galanin's interaction with C-fiber CS of cutaneous and muscle afferents was also examined. I.t. galanin inhibited the facilitation of the flexor reflex by CS of both cutaneous (Wiesenfeld-Hallin et al., 1989a) and muscle (Xu et al., 1991b) afferents. Since reflex facilitation by CS of cutaneous and muscle afferents appears to involve the release of tachykinins and CGRP by primary afferents, we tested the interaction of galanin with SP and CGRP. Galanin preadministration antagonized the excitatory effect of SP and CGRP (Xu et al., 1989, 1990). Thus, galanin appears to function as an antagonist of excitatory neuropeptides with which it coexists. No interaction between galanin and SOM or VIP was observed, which corresponds to a lack of coexistence between galanin and VIP or SOM in normal dorsal root ganglion cells. An inhibitory function of exogenous galanin was also indicated in behavioral and electrophysiological studies where it potentiated the analgesic effect of morphine (Wiesenfeld-Hallin et al., 1990d).

We have been able to demonstrate an inhibitory role for endogenous galanin with some recently developed galanin receptor antagonists (Bartfai et al., 1991, 1993). The galanin antagonist M-35 significantly potentiated CS-induced reflex facilitation in rats with intact nerves (Wiesenfeld-Hallin et al., 1992), although the effect was much stronger after sciatic nerve section (Fig 2).

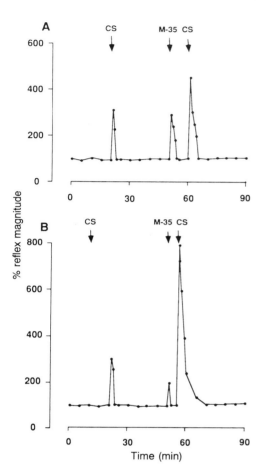

Fig. 2. Potentiation of the facilitation of the flexor reflex induced by a conditioning stimulus (CS) train (20 shocks at 0.9 Hz) that activated C-fibers in the sural nerve in a rat with with intact sciatic nerves (A) and 12 days after nerve section (B). The CS facilitated the flexor reflex similarly when the nerve was intact or sectioned. The cut nerve was stimulated proximal to the lesion. The galanin antagonist M-35 [galanin-(1–13)-bradykinin-(2–9)-amide] injected intrathecally (300 pmol) prior to a second CS potentiated the facilitatory effect of the stimulus train. Note, however, that the potentiation was much stronger in the axotomized rat, indicating that galanin exerted a more potent tonic inhibition of spinal hyperexcitability after nerve injury than under normal conditions. (from Wiesenfeld-Hallin et al., 1992)

These results indicate that nociceptive transmission is under tonic galaninergic inhibitory influence, which is significantly increased after nerve injury.

The effects of cholecystokinin on the flexor reflex

Cholecystokinin (CCK) is not present in normal dorsal root ganglion cells in the rat (Marley et al., 1982; Ju et al., 1986), but is localized in dorsal horn interneurons or tracts descending from higher centers (Skirboll et al., 1983). CCK may function as an endogenous opioid antagonist since CCK reduces the analgesic effect of morphine and β-endorphin (Itoh et al., 1982; Faris et al., 1983). Antagonists of the CCK-B receptor, which predominates in the rat spinal cord, potentiate opioid analgesia and prevent the development of morphine tolerance (Dourish et al., 1990; Wiesenfeld-Hallin et al., 1990d). I.t. CCK dose-dependently facilitates the flexor reflex (Wiesenfeld-Hallin and Duranti, 1987b). CCK injected i.t. after morphine did not reverse morphine-induced reflex depression. However, if CCK was injected prior to i.t. morphine it enhanced the initial excitatory effect of morphine and reduced the subsequent inhibitory effect of the opioid. Thus, CCK antagonizes the analgesic effect of morphine on the flexor reflex by increasing morphine-induced excitation, which may be due to the release of tachykinins from primary afferent terminals by the opioid (Wiesenfeld-Hallin et al., 1991b). Furthermore, CCK also reduced morphine induced depression of the flexor reflex. We have tested the role of endogenous CCK with an antagonist of the CCK-B receptor, CI-988 (previously PD134308) (Wiesenfeld-Hallin et al., 1990d). CI-988 by itself caused a moderate naloxone-reversible depression of the flexor reflex. Since CI-988 has negligible affinity to opioid receptors, it was presumably acting indirectly, indicating that the endogenous CCK system tonically antagonized the endogenous opioid system. CI-988 potentiated the analgesic effect of morphine both on the flexor reflex and in behavioural tests in intact animals, further indicating that endogenous CCK is an opioid antagonist. In view of the fact that galanin potentiated the analgesic effect of morphine (Wiesenfeld-Hallin et al., 1990e), the effect of CI-988 combined with galanin was also evaluated. CI-988 intensely potentiated the inhibitory effect of galanin. Finally, a profound, long lasting depression of the flexor reflex occurred

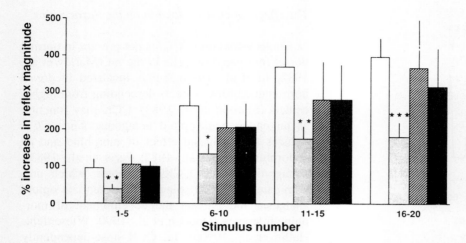

Fig. 3. The effect of the NMDA antagonist MK-801 (0.5 mg/kg i.v.) on wind-up of the flexor reflex during a CS train in rats with intact ($n = 8$, open columns) and sectioned ($n = 7$, hatched columns) sciatic nerves. Note that wind-up developed similarly in both groups. MK-801 partially reduced the wind-up in normal (dotted columns), but not in axotomized rats (filled columns). *$P < 0.05$, **$P < 0.01$ and ***$P < 0.005$ compared to control response. Wind-up is expressed as average increase in reflex magnitude over baseline response before or after MK-801 administration for groups of 5 stimuli in the 20 shock CS train. Variability is expressed as standard error. (from Xu et al., 1995)

when CI-988 was coadministered with galanin and morphine (Wiesenfeld-Hallin et al., 1990d) Such results indicate that a convergence of the effect of opioids, galanin and CCK may occur in the dorsal horn.

The functional implications of messenger plasticity in primary sensory neurons following axotomy.

Plasticity of glutamate function

A loss of synaptic vesicles close to the synapse has been found following axotomy in studies of the fine structure of peptide-containing glomeruli of primary afferents in lamina II of rat dorsal horn (Zhang et al., 1995a,b). These results indicate that the release of glutamate may be reduced following axotomy, leading to attenuation of glutamatergic synaptic transmission. We tested this hypothesis by examining the effect of the NMDA antagonist MK-801 in blocking reflex hyperexcitablity following CS of the axotomized nerve (Xu et al., 1995). After unilateral section of the sciatic nerve MK-801 did not reduce the wind-up of the flexor reflex during CS of C-fibers in the axotomized nerve (Fig. 3). Furthermore, MK-801 was ineffective in blocking the peak facilitation of the flexor reflex following the CS (Fig. 4). In contrast, in rats with intact nerves MK-801 effectively blocked both effects of a C-fiber stimulus train (Xu et al., 1992b). These findings indicate that the involvement of NMDA receptors in mediating activity-dependent spinal hyperexcitability is substantially reduced after peripheral nerve section, possibly reflecting a reduced release of glutamate by primary sensory neurons.

Plasticity of neuropeptide function

Following peripheral axotomy, long term changes occur in the expression of neuropeptides and their receptors in primary sensory neurons (Table 1). For example, SP-LI, SOM-LI and CGRP-LI are downregulated and galanin-LI and VIP-LI, which are normally found in few rat primary afferents, are upregulated. Furthermore, CCK-LI and NPY-LI and receptor protein for CCK and NPY, two neuropeptides normally not found in rat primary afferents, are expressed following peripheral nerve section. Clear physiological correlates to some of these changes in neu-

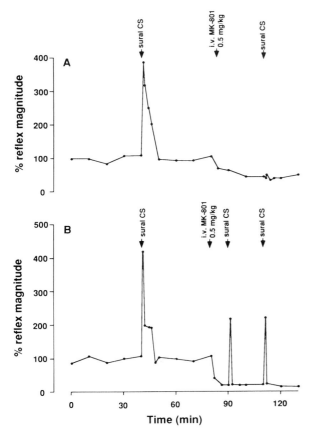

Fig. 4. The effect of 0.5 mg/kg MK-801 on the facilitation of the flexor reflex induced by a CS train to the sural nerve in a rat with intact sciatic nerves (A) and in another rat 14 days after ipsilateral sciatic nerve section (B). Note that MK-801 depressed the baseline reflex similarly in both experiments. However, while MK-801 totally blocked reflex hyperexcitability in the rat with intact nerves, it failed to do so in the axotomized rat. (from Xu et al., 1995)

ropeptidergic phenotype have been demonstrated with the flexor reflex model.

After peripheral nerve section myelinated (Wall and Devor, 1981) and unmyelinated (Wall et al., 1981) afferents from the cut nerve are still able to excite central cells. However, the sensitization of the flexor reflex following CS of C-afferents is altered after axotomy (Wall and Woolf, 1986). The brief reflex hyperexcitability following CS of cutaneous afferents is maintained, but the prolonged effect of CS of muscle afferents is reduced. In view of the importance of SP and CGRP in the mediation of prolonged spinal cord hyperexcitability (Woolf and Wiesenfeld-Hallin, 1986), it is interesting to consider the role of neuropeptides following nerve injury. Within 2 weeks following sciatic nerve section tachykinins totally lose their role in mediating spinal reflex hyperexcitability (Wiesenfeld-Hallin et al., 1990c). The time course of this effect parallels the decline of SP-LI in dorsal root ganglion cells (Jessell et al., 1979). VIP, which normally has no role in spinal reflex sensitization (Xu and Wiesenfeld-Hallin, 1991), is upregulated after peripheral nerve injury (Shehab and Atkinson, 1986) and becomes a major excitatory mediator following section of the sciatic nerve (Wiesenfeld-Hallin, 1989; Wiesenfeld-Hallin et al., 1990c). Thus, there is a switch in the role of excitatory neuropeptides following peripheral nerve injury.

Galanin, which has a demonstrable inhibitory role, is intensely upregulated in dorsal root ganglion cells following peripheral nerve section (Hökfelt et al., 1987). This inhibitory role became enhanced after peripheral nerve section (Wiesenfeld-Hallin et al., 1989b). The reflex depressive effect of i.t. galanin was significantly increased, occurring at lower drug concentrations than in animals with intact peripheral nerves. Furthermore, the magnitude of reflex depression was significantly stronger with a more rapid onset after nerve section. The enhanced inhibitory role of

TABLE 1

Changes in neuropeptide and peptide receptor (R) expression following sciatic nerve section in small and large dorsal root ganglion cells in rat. (adapted from Hökfelt et al., 1994)

	Small cells	Large cells
Downregulated	CGRP	CGRP
	SP	
	NPY-R	
	SOM	
Upregulated	VIP	VIP
	GAL	GAL
	CCK	
	CCK-R	CCK-R
	NPY	NPY
		NPY-R

galanin following nerve section was also demonstrated with the selective anagonist M-35 (Wiesenfeld-Hallin et al., 1992). In rats with intact nerves M-35 moderately potentiated spinal cord sensitization following CS of a cutaneous nerve. This potentiation was significantly more pronounced after peripheral nerve section (Fig. 2). These results indicate that the moderate tonic galaninergic control of nociceptive input to the spinal cord is enhanced after nerve injury. Thus, galanin agonists may be useful analgesics for the treatment of neuropathic pain following nerve injury.

Galanin is a functional antagonist of the excitatory effect of SP and CGRP, but not VIP and SOM when peripheral nerves are intact (Xu et al., 1989, 1990). We examined the interaction of galanin with these neuropeptides after peripheral nerve section (Xu et al., 1990). Galanin's antagonism of the excitatory effect of SP was totally abolished after axotomy and its antagonism of the excitatory effect of CGRP was significantly reduced. Just as in rats with intact nerves, galanin did not interact with SOM. In contrast, after nerve section galanin antagonized the excitatory effect of VIP. Interestingly, a strong coexistence between galanin-LI and VIP-LI was observed in dorsal root ganglion cells following axotomy. These results reinforce the conclusion that peptides that are colocalized have a functional interaction.

We have also examined the functional significance of the upregulation of CCK-LI and CCK-B receptor mRNA in rat DRG (Xu et al., 1993; Zhang et al., 1993). In the clinic, pain arising after nerve injury is difficult to treat and is usually insensitive to opioid analgesics (Arnér and Meyerson, 1988). In agreement with the clinical observation, chronic i.t. morphine administration was ineffective in blocking autotomy, a behavioural model of experimental neuropathic pain following peripheral nerve injury (Xu et al., 1993). Since CCK is an endogenous opioid antagonist, we examined whether morphine insensitivity could involve up-regulation of CCK. Chronic coadministration of the CCK-B receptor antagonist CI-988 with morphine significantly reduced autotomy behavior (Xu et al., 1993). The effect of CI-988 and

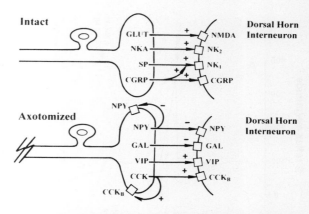

Fig. 5. Schematic illustration of the major neutotransmitters/modulators released from intact and axotomized primary afferents. +, excitation, −, inhibition.

morphine on the flexor reflex in rats with sectioned sciatic nerves was evaluated (Xu et al., 1994). Morphine was significantly less effective in depressing the flexor reflex after axotomy than when peripheral nerves were intact. However, the depressive effect of CI-988 was significantly enhanced in axotomized rats, which probably signifies increased tonic endogenous CCK activity. Combination of morphine and CI-988 resulted in significant potentiation of the analgesic effect of morphine, just as in rats with intact peripheral nerves (Wiesenfeld-Hallin et al., 1990d). Thus, coadministration of CCK antagonists in combination with opioids may offer a new approach for treating neuropathic pain.

Conclusions

Nerve injury leads to complex changes in the phenotype of primary afferents in the rat. This plasticity, which can be demonstrated with histochemical techniques, also has physiological consequences in the processing of sensory information (Fig 5). Thus, neuropathic pain following nerve injury may arise as a result of plasticity of the nervous system, in the conduction of afferent information from the periphery into the spinal cord, intraspinally, as well as at other levels as discussed in other chapters in this volume. Only by systematic analysis of these alterations in neural function following injury can we hope to develop treatment strategies for treating pain that arises as a consequence nerve pathology. It must be also kept in mind that there are considerable species differences in the reaction of nervous tissue to injury (Hökfelt et al., 1994). Drugs developed for the treatment of neuropathic pain need to be tested on carefully selected, homogeneous groups of patients with well defined pathology.

Acknowledgements

This work was supported by the Bank of Sweden Tercentenary Foundation, the Swedish Medical Research Council (project no. 07913), and Astra Pain Control AB.

References

Arnér, S. and Meyerson, B.A. (1988) Lack of analgesic effect of opioids on neuropathic and idiopathic forms of pain. *Pain*, 33: 11–23.

Bartfai, T., Bedecs, K., Land, T., Langel, U., Bertorelli, R., Girotti, P., Consolo, S., Xu, X.-J., Wiesenfeld-Hallin, Z., Nilsson, S., Pieribone, V.A. and Hökfelt, T. (1991) M-15: High affinity chimeric peptide that blocks the neuronal actions of galanin in the hippocampus, locus coeruleus, and spinal cord. *Proc. Natl. Acad. Sci. USA*, 88: 10961–10965.

Bartfai, T., Langel, U., Bedecs, K., Andell, S., Land, T., Gregersen, S., Ahrén, B., Girotti, P., Consolo, S., Corwin, R., Crawley, J., Xu, X.-J., Wiesenfeld-Hallin, Z. and Hökfelt, T. (1993) Galanin-receptor ligand M40 peptide distinguishes between putative galanin-receptor subtypes. *Proc. Natl. Acad. Sci. USA*, 90: 11287–11291.

Basbaum, A.I., Clanton, C.H. and Fields, H.L. (1978) Three bulbospinal pathways from the rostral medulla of the cat: an autoradiographic study of pain modulating systems. *J. Comp. Neurol.*, 178: 209–224.

Battaglia, G. and Rustioni, A. (1988) Coexistence of glutamate and substance P in dorsal root ganglion neurons of the rat and monkey. *J. Comp. Neurol.*, 277: 297–312.

Besson, J.-M. and Chaouch, A. (1987) Peripheral and spinal mechanisms of nociception. *Physiol. Rev.*, 67: 67–186.

Carstens, E., Hartung, M., Stelzer, B. and Zimmermann, M. (1990) Suppression of a hind limb flexion withdrawal reflex by microinjection of glutamate or morphine into the periaqueductal gray in the rat. *Pain*, 1990: 105–112.

Chang, M.M. and Dallaier, M. (1989) Subjective pain sensation is linearly correlated with the flexion reflex in man. *Brain Res.*, 479: 145–150.

Ch'ng, J.L.C., Christofides, N.D, Anand, P., Gibson, S.J., Allen, Y.S., Su, H.C., Tatemoto, K., Morrison, J.F.B., Polak, J.M. and Bloom, S.R. (1985) Distribution of galanin immunoreactivity in the central nervous system and the response of galanin-containing neuronal pathways to injury. *Neuroscience*, 16: 343–354.

Cook, A.J., Woolf, C.J. and Wall, P.D. (1986) Prolonged C-fibre mediated facilitation of the flexion reflex in the rat is not due to changes in afferent terminal or motoneuron excitability. *Neurosci. Lett.*, 70: 91–96.

Dalsgaard, C.-J., Haegerstrand, A., Theodorsson-Norheim, E., Brodin, E. and Hökfelt, T. (1985) Neurokinin-A like immunoractivity in rat primary sensory neurons: coexistence with substance P. *Histochemistry*, 83: 37–40.

Davies, S.N. and Lodge, D. (1987) Evidence for involvement of N-methylaspartate receptors in "wind-up" of class 2 neurones in the dorsal horn of the rat. *Brain Res.*, 424: 402–406.

De Biasi, S. and Rustioni, A. (1988) Glutamate and substance P coexist in primary afferent terminals in the superficial laminae of spinal cord. *Proc. Natl. Acad. Sci. USA*, 85: 7820–7824.

De Biasi, S. and Rustioni, A. (1990) Ultrastructural immunohistochemical localization of excitatory amino acids in the somatosensory system. *J. Histochem. Cytochem.*, 38: 1745–1754.

Dickenson, A.H. and Sullivan. A.F. (1990) Differential effects of excitatory amino acid antagonists on dorsal horn nociceptive neurones in the rat. *Brain Res.*, 506: 31–39.

Dourish, C.T., O'Neill, M.F., Coughlaan, J., Kitchenek, S.J., Hawley, D. and Iversen, S.D. (1990) The selective CCK-B receptor antagonist L-365,260 enhances morphine analgesia and prevents morphine tolerance in the rat. *Eur. J. Pharmacol.*, 176: 35–44.

Faris, P.L., Komisaruk, B.R., Watkins, L.R. and Mayer, D.L. (1983) Evidence for the neuropeptide cholecystokinin as an antagonist of opiate analgesia. *Science*, 219: 211–222.

Fleetwood-Walker, S.M., Mitchell, R., Hope, P.J., El-Yassir, N., Molony, V. and Bladon, C.M. (1990) The involvement of neurokinin receptor subtypes in somatosensory processing in the superficial dorsal horn of the cat. *Brain Res.*, 519: 169–182.

Folkers, K., Feng, D.-M., Asano, N., Håkanson, R., Wiesenfeld-Hallin, Z. and Leander, S. (1990) Spantide II, an effective tachykinin antagonist having high potency and negligible neurotoxicity. *Proc. Natl. Acad. Sci. USA*, 87: 4833–4835.

Henry, J.L. (1976) Effects of substance P on functionally identified units in cat spinal cord. *Brain Res.*, 114: 439–451.

Hökfelt, T., Kellerth, J.-O., Nilsson, G. and Pernow, B. (1975) Experimental immunohistochemical studies on the localization and distribution of substance P in cat primary sensory neurons. *Brain Res.*, 100: 235–252.

Hökfelt, T., Johansson, O., Ljungdahl, Å., Lundberg, J.M. and Schultzberg, M. (1980) Peptidergic neurons. *Nature*, 287: 515–521.

Hökfelt, T., Wiesenfeld-Hallin, Z., Villar, M.J. and Melander, T. (1987) Increase of galanin-like immunoreactivity in rat dorsal root ganglion cells after peripheral axotomy. *Neurosci. Lett.*, 83: 217–220.

Hökfelt, T., Zhang, X. and Wiesenfeld-Hallin, Z. (1994) Messenger plasticity in primary sensory neurons following axotomy and its functional implications. *Trends Neurosci.*, 17: 22–30.

Itoh, S., Katsuura, G. and Maeda, Y. (1982) Caerulein and cholecystokinin suppress endorphin-induced analgesia in the rat. *Eur. J. Pharmacol.*, 80: 421–425.

Jessell, T.M. and Iversen, L.L. (1977) Opiate analgesics inhibit substance P release from rat trigeminal nucleus *Nature*, 268: 549–551.

Jessell, T., Tsunoo, A., Kanazawa, I. and Otsuka, M. (1979) Substance P: depletion in the dorsal horn of rat spinal cord after section of the peripheral processes of primary sensory neurons. *Brain Res.*, 168: 247–259.

Ju, G., Hökfelt, T., Fischer, J.A., Frey, P., Rehfeld, J.F. and Dockray, G.J. (1986) Does cholecystokinin-like immunoractivity in rat primary sensory neurons represent calcitonin-gene related peptide? *Neurosci. Lett.*, 68: 305–310.

Ju, G., Hökfelt, T., Brodin, E., Fahrenkrug, J., Fischer, J.A., Frey, P., Elde, R.P. and Brown, J.C. (1987) Primary sensory neurons of the rat showing calcitonin gene-related peptide immunoreactivity and their relation to substance P-, somatostatin-, galanin- vasoactive intestinal polypeptide- and cholecystokinin-immunoreactive ganglion cells. *Cell Tissue Res.*, 247: 417–431.

Le Grevés, P., Nyberg, F., Terenius, L. and Hökfelt, T. (1985) Calcitonin gene-related peptide is a potent inhibitor of substance P degradation. *Eur. J. Pharmacol.*, 115: 309–311.

Liebeskind, J.C., Guilbaud, G, Besson, J.-M. and Oliveras, J.-L. (1973) Analgesia from electrical stimulation of the periaqueductal gray matter in the cat: behavioral observations and inhibitory effects on spinal cord interneurons. *Brain Res.*, 50: 441–446.

Ma, O.P. and Woolf, C.J. (1995) Involvement of neurokinin receptors in the induction but not the maintenance of mechanical allodynia in rat flexor motoneurons. *J. Physiol.*, 486: 769–777.

Maggi, C.A., Patacchini, R., Giuliani, S., Rovero, P., Dion, S., Regoli, D., Giachetti, A. and Meli, A. (1990) Competitive antagonists discriminate between NK_2 tachykinin receptor subtypes. *Br. J. Pharmacol.*, 100: 588–592.

Marley, P.D., Nagy, J.E., Emson, P.C. and Rehfeld, J.F. (1982) Cholecystokinin in the rat spinal cord: distribution and lack of effect on neonatal capsaicin treatment and rhizotomy. *Brain Res.*, 238: 494–498.

Melzack, R. and Wall, P.D. (1965) Pain mechanisms: a new theory. *Science*, 150: 971–979.

Mendell, L.M. (1966) Physiological properties of unmyelinated fibre projections to the spinal cord. *Exp. Neurol.*, 16: 316–332.

Otsuka, M. and Konishi, S. (1976) Release of substance P-like immunoreactivity from isolated spinal cord of newborn rat. *Nature*, 264: 83–84.

Salt, T.E. and Hill, R.G. (1983) Neurotransmitter candidates of somatosensory primary afferent fibres. *Neuroscience*, 10: 1083–1103.

Schouenborg, J. and Sjölund, B.H. (1983) Activity evoked by A- and C-afferent fibers in rat dorsal horn neurons and its relation to a flexion reflex. *J. Neurophysiol.*, 50: 1108–1121.

Shehab, S.A.S. and Atkinson, M.E. (1986) Vasoactive intestinal polypeptide (VIP) increases in the spinal cord after peripheral axotomy of the sciatic nerve originate from primary afferent neurones. *Brain Res.*, 372: 37–44.

Skirboll, L., Hökfelt, T., Dockray, G. and Rehfeld, J. (1983) Evidence for periaqueductal cholecystokinin-substance P neurons projecting to the spinal cord. *J. Neurosci.*, 3: 1151–1157.

Snider, M., Constantine, J.W., Lowe, III, J.A., Longo, K.P., Lebel, W.S., Woody, H.A., Drozda, S.E., Desai, M.C., Vinick, F.J., Spencer, R.W. and Hess, H.-J. (1991) A potent nonpeptide antagonist of the substance P (NK1) receptor. *Science*, 251: 435–437.

Villar, M.J., Cortés, R., Theodorsson, E., Wiesenfeld-Hallin, Z., Schalling, M., Fahrenkrug, J., Emson, P.C. and Hökfelt, T. (1989) Neuropeptide expression in rat dorsal root ganglion cells and spinal cord after peripheral nerve injury with special reference to galanin. *Neuroscience*, 33: 587–604.

Wall, P.D. and Devor, M. (1981) The effect of peripheral nerve injury on dorsal root potentials and in transmission of afferent signals into the spinal cord. *Brain Res.*, 209: 95–111.

Wall, P.D. and Woolf, C.J. (1984) Muscle but not cutaneous C-afferent input produces prolonged increases in the excitability of the flexion reflex in the rat. *J. Physiol.*, 356: 443–458.

Wall, P.D. and Woolf, C.J. (1986) The brief and the prolonged facilitatory effects of unmyelinated afferent input on the rat spinal cord are independently influenced by peripheral nerve section. *Neuroscience*, 17: 1199–1205.

Wall, P.D., Fitzgerald, M. and Gibson, S.J. (1981) The response of rat spinal cord cells to unmyelinated afferents after peripheral nerve section and after changes in substance P levels. *Neuroscience*, 6: 2205–2215.

Wiesenfeld-Hallin, Z. (1986) Substance P and somatostatin modulate spinal cord excitability via physiologically different sensory pathways. *Brain Res.*, 372: 172–175.

Wiesenfeld-Hallin, Z. (1987) Intrathecal vasoactive intestinal polypeptide modulates spinal reflex excitability primarily to cutaneous thermal stimuli in rats. *Neurosci. Lett.*, 80: 293–297.

Wiesenfeld-Hallin, Z. (1989) Nerve section alters the interaction between C-fibre activity and intrathecal neuropeptides on the flexor reflex in rat. *Brain Res.*, 489: 129–136.

Wiesenfeld-Hallin, Z. and Duranti, R. (1987) Intrathecal cholecystokinin interacts with morphine but not substance P in modulating the nociceptive flexion reflex in the rat. *Peptides*, 8: 153–158.

Wiesenfeld-Hallin, Z. and Xu, X.-J. (1993) The differential roles of substance P and neurokinin A in spinal cord hyperexcitability and neurogenic inflammation. *Regul. Peptides*, 46: 165–173.

Wiesenfeld-Hallin, Z., Hökfelt, T., Lundberg, J.M., Forsmann, W.G., Reinecke, M., Tschapp, F.A. and Fischer, J.A. (1984) Immunoractive calcitonin gene-related peptide and substance P co-exist in sensory neurons to the spinal cord and interact in spinal behavioural responses of the rat. *Neurosci. Lett.*, 52: 199–204.

Wiesenfeld-Hallin, Z., Villar, M.J. and Hökfelt, T. (1988) Intrathecal galanin at low doses increases spinal reflex excitability in rats more to thermal than mechanical stimuli. *Exp. Brain Res.*, 71: 663–666.

Wiesenfeld-Hallin, Z., Villar, M.J. and Hökfelt, T. (1989a) The effects of intrathecal galanin and C-fiber stimulation on the flexor reflex in the rat. *Brain Res.*, 486: 205–213.

Wiesenfeld-Hallin, Z., Xu, X.-J., Villar, M.J. and Hökfelt, T. (1989b) The effects of intrathecal galanin on the flexor reflex in rat: increased depression after sciatic nerve section. *Neurosci. Lett.*, 105: 149–154.

Wiesenfeld-Hallin, Z., Xu, X.-J., Håkanson, R., Feng, D.-M. and Folkers, K. (1990a) The specific antagonistic effect of intrathecal spantide II on substance P- and C-fiber conditioning stimulation-induced facilitation of the nociceptive flexor reflex in rat. *Brain Res.*, 526: 284–290.

Wiesenfeld-Hallin, Z., Xu, X.-J., Kristensson, K., Håkanson, R., Feng, D.-M. and Folkers, K. (1990b) Antinociceptive and substance P antagonistic effects of intrathecally injected spantide II in rat: no signs of motor impairment or neurotoxicity. *Regul. Peptides*, 29: 1–11.

Wiesenfeld-Hallin, Z., Xu, X-J., Håkanson, R., Feng, D.-M. and Folkers, K. (1990c) Plasticity of the peptidergic mediation of spinal reflex facilitation after peripheral nerve section in the rat. *Neurosci. Lett.*, 116: 293–298.

Wiesenfeld-Hallin, Z., Xu, X.-J., Hughes, J., Horwell, D.C. and Hökfelt, T. (1990d) PD134308, a selective antagonist of cholecystokinin type B receptor, enhances the analgesic effect of morphine and synergistically interacts with intrathecal galanin to depress spinal nociceptive reflexes. *Proc. Natl. Acad Sci. USA*, 87: 7105–7109.

Wiesenfeld-Hallin, Z., Xu, X.-J., Villar, M.J. and Hökfelt, T. (1990e) Intrathecal galanin potentiates the spinal analgesic effect of morphine: electrophysiological and behavioural studies. *Neurosci. Lett.*, 109: 217–221.

Wiesenfeld-Hallin, Z., Xu, X.-J., Håkanson, R., Feng, D.-M. and Folkers, K. (1991a) Tachykinins mediate changes in spinal reflexes after activation of unmyelinated muscle afferents in the rat. *Acta Physiol. Scand.*, 141: 57–61.

Wiesenfeld-Hallin, Z., Xu, X.-J., Håkanson, R., Feng, D.-M. and Folkers, K. (1991b) Low-dose intrathecal morphine facilitates the spinal flexor reflex by releasing different neuropeptides in rats with intact and sectioned peripheral nerves. *Brain Res.*, 551: 157–162.

Wiesenfeld-Hallin, Z., Xu, X.-J., Langel, U., Bedecs, K., Hökfelt, T. and Bartfai, T. (1992) Galanin-mediated control of pain: enhanced role after nerve injury. *Proc. Natl. Acad. Sci. USA*, 89: 3334–3337.

Willer, J.C. (1977) Comparative study of perceived pain and nociceptive flexion reflex in man. *Pain*, 3: 69–80.

Willis, W.D., Haber, L.H. and Martin, R.F. (1977) Inhibition of spinothalamic tract cells and interneurons by brain stem stimulation in the monkey. *J. Neurophysiol.*, 40; 968–981.

Woolf, C.J. (1983) Evidence for a central component of post-injury pain hypersensitivity. *Nature*, 306: 686–688.

Woolf, J.C. and Chong, M.-S. (1993) Preemptive analgesia - Treating postoperative pain by preventing the establishment of central sensitization. *Anesth. Analg.*, 77: 362–379.

Woolf, C.J. and Thompson, S.W.N. (1991) The induction and maintenance of central sensitization is dependent on N-methyl-D-aspartic acid receptor activation: implications for the treatment of post-injury pain hypersensitivity states. *Pain*, 44: 293–300.

Woolf, C. and Wiesenfeld-Hallin, Z. (1986) Substance P and calcitonin gene-related peptide synergistically modulate the gain of the nociceptive flexor withdrawal reflex in the rat. *Neurosci. Lett.*, 66: 226–230.

Xu, X.-J. and Wiesenfeld-Hallin, Z. (1991) An analogue of growth hormone releasing factor (GRF), (Ac-Try^1, D-Phe^2) GRF-(1–29), specifically antagonizes the facilitation of the flexor reflex induced by intrathecal vasoactive intestinal peptide in rat spinal cord. *Neuropeptides*, 18: 129–135.

Xu, X.-J. and Wiesenfeld-Hallin, Z. (1992) Intrathecal neurokinin A facilitates the spinal nociceptive flexor reflex evoked by thermal and mechanical stimuli and synergistically interacts with substance P. *Acta Physiol. Scand.*, 144: 163–168.

Xu, X.-J., Wiesenfeld-Hallin, Z., Villar, M.J. and Hökfelt, T. (1989) Intrathecal galanin antagonizes the facilitatory effect of substance P on the nociceptive flexor reflex in the rat. *Acta Physiol. Scand.*, 137: 463–464.

Xu, X.-J., Wiesenfeld-Hallin, Z., Villar, M.-J., Fahrenkrug, J. and Hökfelt, T. (1990) On the role of galanin, substance P and other neuropeptides in primary sensory neurons of the rat: studies on spinal reflex excitability and peripheral axotomy. *Eur. J. Neurosci.*, 2: 733–743.

Xu, X.-J., Maggi, C.A. and Wiesenfeld-Hallin, Z. (1991a) On the role of the NK-2 tachykinin receptors in the mediation of spinal reflex excitability in the rat. *Neuroscience*, 44: 483–490.

Xu, X.-J., Wiesenfeld-Hallin, Z. and Hökfelt, T. (1991b) Intrathecal galanin blocks the prolonged increase in spinal cord flexor reflex excitability induced by conditioning stimulation of unmyelinated muscle afferents in the rat. *Brain Res.*, 541: 350–353.

Xu, X.-J., Dalsgaard, C.-J. and Wiesenfeld-Hallin, Z. (1992a) Intrathecal CP-96,345 blocks reflex facilitation induced in rats by substance P and C-fiber-conditioning stimulation. *Eur. J. Pharmacol.*, 216: 337–344.

Xu, X.-J., Dalsgaard, C.-J. and Wiesenfeld-Hallin, Z. (1992b) Spinal substance P and N-methyl-D-aspartate receptors are coactivated in the induction of central sensitization of the nociceptive flexor reflex. *Neuroscience*, 51: 641–648.

Xu, X.-J., Puke, M.J.C., Verge, V.M.K., Wiesenfeld-Hallin, Z., Hughes, J. and Hökfelt, T. (1993) Up-regulation of cholecystokinin in primary sensory neurons is associated with morphine insensitivity in experimental neuropathic pain in the rat. *Neurosci. Lett.,* 152: 129–132.

Xu, X.-J., Hökfelt, T., Hughes, J. and Wiesenfeld-Hallin, Z. (1994) The CCK-B antagonist CI988 enhances the reflex-depressive effect of morphine in axotomized rats. *NeuroReport,* 5: 718–720.

Xu, X.-J., Zhang, X., Hökfelt, T. and Wiesenfeld-Hallin, Z. (1995) Plasticity in spinal nociception after peripheral nerve section: reduced effectiveness of the NMDA receptor antagonist MK-801 in blocking wind-up and central sensitization of the flexor reflex. *Brain Res.,* 670: 342–346.

Zhang, X., Dagerlind, Å., Elde, R.P., Castel, M.-N., Broberger, C., Wiesenfeld-Hallin, Z. and Hökfelt, T. (1993) Marked increase in cholecystokinin B receptor messenger RNA levels in rat dorsal root ganglia after peripheral axotomy. *Neuroscience,* 57: 227–233.

Zhang, X., Bean, A.J., Wiesenfeld-Hallin, Z., Xu, X.-J. and Hökfelt, T. (1995a) Ultrastructural studies on peptides in the dorsal horn of the spinal cord. III. Effects of peripheral axotomy with special reference to galanin. *Neuroscience,* 64: 893–915.

Zhang, X., Bean, A.J., Wiesenfeld-Hallin, Z. and Hökfelt, T. (1995b) Ultrastructural studies on peptides in the dorsal horn of the rat spinal cord. IV. Effects of peripheral axotomy with special reference to neuropeptide Y and vasoactive intestinal polyppetide/peptide histidine isoleucine. *Neuroscience,* 64: 917–941.

CHAPTER 10

The possible role of substance P in eliciting and modulating deep somatic pain

S. Mense, U. Hoheisel and A. Reinert

Institut für Anatomie und Zellbiologie, Im Neuenheimer Feld 307, D-69120 Heidelberg, Germany

Introduction

Neuropeptides have been and still are controversely discussed as transmitters/modulators of neuronal activity in the peripheral and central nervous systems. Substance P (SP) is of particular interest for the understanding of pain mechanisms, since it is assumed to be involved in the processing of nociceptive information at various levels of the nervous system (Nicoll et al., 1980; Dubner and Bennett, 1983; Duggan et al., 1988). The exact role of the neuropeptide is uncertain, however, and some authors question an association between SP release and pain (Frenk et al., 1988).

In this article, recent results are presented and discussed concerning the involvement of SP in the mediation of deep somatic pain (from muscle, fascia, tendon, joint). SP-related events in skeletal muscle and spinal cord are addressed.

SP and primary afferent fibres in muscle

After its synthesis in the soma of primary afferent neurones the bulk of SP is transported to the receptive endings in the periphery and not to the presynaptic endings in the spinal cord (Brimijoin et al., 1980). The activation of nociceptive endings is assumed to be associated with the release of SP from these endings, the neuropeptide triggering a cascade of events that eventually lead to neurogenic inflammation (Lembeck and Holzer, 1979).

At the primary afferent level, the presence of SP in nociceptive endings is still not firmly established. In a study in which dorsal root ganglion cells of the cat were functionally identified and intracellularly stained, Leah et al. (1985) found that the majority of nociceptive neurones did not show SP immunoreactivity (SP-IR).

There is evidence indicating that the SP content of skeletal muscle is less than that of the skin, and a hypothesis has been put forward that – teleologically – the low concentration of SP in muscle is necessary for preventing tissue necrosis during inflammation (McMahon et al., 1984). If the SP content of the muscle was high, a myositis would cause the release of large amounts of this neuropeptide. The ensuing tissue oedema could lead to excessively high intramuscular pressure in muscles that have a tight fascia. A comparison of the SP content of muscle and skin is difficult to make, however, since muscle is a three-dimensional structure and skin approximately two-dimensional.

There is no doubt, however, that SP is present in a large proportion of nerve endings in skeletal muscle (see below). Many of the dorsal root ganglion cells projecting in a muscle nerve exhibit SP-IR (Molander et al., 1987; O'Brien et al., 1989).

In the literature, information concerning the influence of SP on the discharges of identified single nerve endings is scarce. Kumazawa and Mizumura (1979) examined SP effects on receptive endings of the testicular surface in dogs and found the neuropeptide to have a weak excitatory action in micromolar concentrations. In another in vitro study SP had a conditioning effect on the excitation of

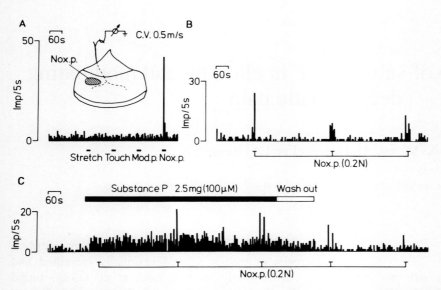

Fig. 1. Time histogram of the impulse activity of a single HTM receptor of the rat diaphragm in vitro. (A) The unit had a conduction velocity (C.V.) of 0.5 m/s; it could be activated only by noxious local pressure (Nox. p.) exerted on the hatched region which marks the receptive field in the insert. Mod. p., moderate (innocuous) pressure. (B) Repeated mechanical stimulation performed with a stimulating apparatus at a noxious intensity (0.2 N) and at intervals of 6 min elicited reproducible responses. (C) During the period marked by the filled bar above the histogram, SP 100 μM was added to the organ bath; this led to an increase in background discharge but not in response magnitude to mechanical stimulation. During wash-out (open bar) the background activity returned to the original level.

cutaneous afferent nerve endings by a combination of inflammatory mediators (Kessler et al., 1992).

In our laboratory, the effects of SP on primary afferent neurones were studied using a rat hemidiaphragm-phrenic nerve preparation in vitro. The hemidiaphragm was mounted in an organ bath, and fine filaments were split by hand from the phrenic nerve for the recording of electrical activity in single non-myelinated (group IV- or C-) afferent units (conduction velocity less than 1.5 m/s). The receptors were classified as low-threshold mechanosensitive (LTM, presumably non-nociceptive) or high-threshold mechanosensitive (HTM, presumably nociceptive) units according to their responsiveness to mechanical stimulation of the diaphragm. LTM units could be activated by weak stimuli such as touching the tissue with an artist's brush and increased their discharge rate with the intensity of mechanical stimulation. HTM units did not respond to innocuous deformation of the receptive field (RF) but required strong mechanical stimuli such as pinching or noxious local pressure for activation. To the organ bath containing the hemidiaphragm defined concentrations of SP were added to study the influence of the neuropeptide on the discharge behaviour of the endings.

LTM and HTM receptors were differentially affected by SP. In LTM units, the background activity (defined as discharges in the absence of intentional stimulation) was not influenced at all, but the responses to mechanical stimulation showed a decrease in magnitude. The latter effect was particularly strong after the stimulation had been repeated several times.

The HTM receptors showed a marked increase in background discharge under the influence of SP (Fig. 1). The effect had a threshold concentration between 1 and 10 μM and was significant at 100 μM. Simultaneous administration of SP at 100 μM and the SP antagonist (D-Arg[1], D-Trp[7,9], Leu[11])-SP (spantide) at 50 μM resulted in a significant decrease of the activity compared to the effect of SP alone (Fig. 2). In contrast, the mechanical responsiveness of the HTM receptors showed a (statistically insignificant, U test, two-

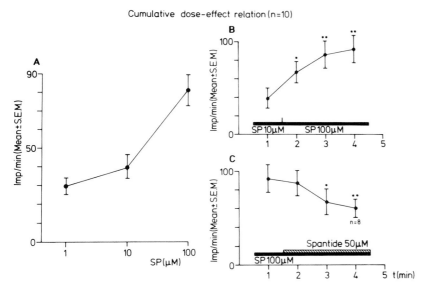

Fig. 2. Dose-effect relations for a population of 10 HTM receptors in vitro. (A) Cumulative dose-response curve for the background discharge (impulses per minute). (B) Background activity plotted versus time to show the sharp increase in activity during the change from SP 10 to 100 μM. (C) Plot as in B to show the effect of a combination of spantide 50 μM and SP 100 μM. Asterisks in (B) and (C) indicate the level of significance: *$P < 0.05$; **$P < 0.01$ in comparison to the first data point of the curve (U test, two-sided).

sided) tendency towards a decrease. Thus, the action of SP on non-myelinated afferent units from the diaphragm was characterized by a combination of increased background activity in nociceptive endings and a decreased mechanical sensitivity in non-nociceptive units.

Recent data from our laboratory suggest that during longer-lasting pathological alterations of skeletal muscle not only the discharge activity in slowly conducting fibers changes but also the IR to neuropeptides in free nerve endings. In these experiments, the number and distribution of nerve endings immunoreactive (ir) to SP and other neuropeptides was examined in intact and inflamed skeletal muscle. In striated muscle free nerve endings are mainly located around blood vessels (Fleming et al., 1989; Popper and Micevych, 1989). In the literature, there are several reports on the neuropeptide content in sensory nerves innervating chronically inflamed tissue (e.g. Lembeck et al., 1981; Kuraishi et al., 1989), but data on changes in SP-IR in inflamed skeletal muscle are lacking.

Two animal models of myositis were used: (1) An acute myositis was induced by infiltrating the muscle with 2% carrageenan. Histological signs of inflammation develop within 2 h; at about the same time the electrical activity of afferent C fibres from the muscle begins to change (Diehl et al., 1993). After a survival time of 10 h these animals were killed with an overdose of anaesthetic. (2) A persistent (subacute) myositis was induced by a single injection of complete Freund's adjuvant into the muscle. The animals were killed after a survival period of 12 days.

After immunohistochemical processing of the muscle, all ir terminal nerve fibres (endings and preterminal axons showing clear varicosities) were counted light microscopically in transverse sections of the muscle. The number of fibres per cm^2 section at a constant section thickness of 12 μm was determined as a measure of innervation density.

In both normal and inflamed muscle, SP-ir fibres were less numerous than units that were ir for calcitonin gene-related peptide (CGRP). The predominant location of SP-ir fibres was the wall of larger arteries, whereas CGRP-ir units were most numerous around smaller vessels (arterioles or venules). After 12 days of persistent inflammation (induced by Freund's adjuvans) the total number of SP-ir fibres had increased by a factor of about 2 (Fig. 3). This increase was particularly pronounced in endings around arteries. In contrast, CGRP-ir fibres exhibited only a non-significant trend towards a higher innervation density. The density of fibres showing IR for nerve growth factor (NGF) and growth-associated protein 43 (GAP-43/B-50) was likewise significantly increased. These effects of the subacute myositis appeared to start early in the development of the inflammation, since a strong tendency to higher figures was present already at the end of the 10 h period of acute (carrageenan-induced) myositis.

Effects of SP on dorsal horn neurones

Dorsal horn neurones were electrophysiologically studied in anaesthetized Sprague-Dawley rats. The spinal cord was surgically exposed and the lumbar segments 4 and 5 superfused with artificial cerebrospinal fluid that formed a thin layer around the spinal cord. The perfusion system could be switched from this fluid to another one that contained SP in defined concentrations. The superfusion rate was kept at a minimum; therefore, the exchange in the fluid film was slow (judging from tests with a dye solution, a complete exchange took approximately 3 min). The rationale of this technique was to simulate the slow rise and long-lasting elevation of the intrathecal SP concentration that probably occurs in the course of a peripheral tissue lesion. With the method used the neuropeptide had to diffuse via the extracellular space into the spinal cord to reach its presumable site of action (predominantly cells in laminae I and III, Bleazard et al., 1994). Therefore, the superfusion is assumed to simulate the extrasynaptic actions of SP which are an important aspect in the function of SP in volume transmission (Vizi and Labos, 1991; Beck et al., 1995; Sandkühler, this volume). In fact – and in contrast to the reported SP effects after intrathecal or iontophoretical injection – SP ap-

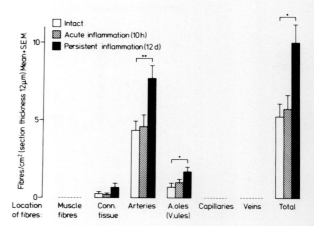

Fig. 3. Inflammation-induced changes in innervation density of SP-ir fibres in the rat gastrocnemius-soleus muscle. Three experimental conditions are compared: intact muscle (open bars), acute myositis of 10 h duration (hatched bars) and persistent (subacute) myositis of 12 days duration (filled bars). On the ordinate the numerical area density (number of fibres per cm² of tissue section) is plotted. The abscissa shows the distribution of the fibres over the various components of muscle tissue. Arterioles (A.oles) could not be clearly distinguished from venules (V.ules); therefore, fibres around these vessels were evaluated together. $*P < 0.05$; $**P < 0.01$ (U test, two-sided).

plied this way did not activate the dorsal horn neurones studied (only 1 cell out of 24 tested increased its discharge rate during superfusion with SP 100 µM; Chapter 9).

The impulse activity of single dorsal horn neurones was recorded with microelectrodes introduced into the spinal cord. To determine the responsiveness of the dorsal horn neurones, the major spinal nerves of the hindlimb (common peroneal n., tibial n., sural n.) were stimulated electrically and the response characteristics measured (threshold, latency, following frequency, input convergence from the stimulated nerves). The study concentrated on neurones that responded to mechanical stimulation of receptors in deep somatic tissues (muscle, fascia, tendon, joint). Cells that had a mechanosensitive RF in the skin were excluded from the study.

Superfusion of the spinal cord with SP had heterogeneous effects on the electrical responsiveness of dorsal horn neurones. The main SP action was a

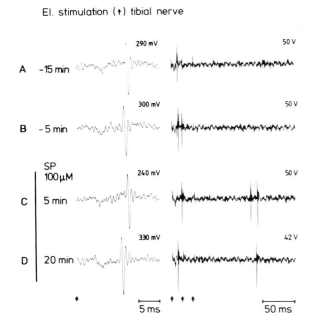

Fig. 4. Appearance of a new response to C-fibre input during superfusion with SP 100 μM. (A,B) 15 and 5 min before administration of SP, the recorded neurone had exclusive input from A-fibres in the tibial nerve. Stimulation of the nerve at an intensity supramaximal for C fibres (50 V) did not elicit a response at C-fibre latency. (C,D) during SP superfusion a response at C-fibre latency appeared which showed a decrease in threshold after 15 min of superfusion (50 V in (B) and 42 V in (C) correspond to 1.2 times threshold; the C-fibre stimulus consisted of a train of three pulses at 50 Hz). The cell had no background activity in the absence of intentional stimulation and was not excited by SP (from Hoheisel et al., 1995).

change in the input a given neurone responded to. An example is shown in Fig. 4. Before administration of SP, the cell could be driven only by A fibres in the tibial nerve; there was no response to the electrical activation of C fibres in that nerve (right half of the upper two panels). During superfusion with SP 100 μM, a response at C-fibre latency appeared in addition to the A-fibre response (two lower panels). The electrical threshold of the nerve to evoke a C-fibre response dropped considerably within the first 15 min of the superfusion, whereas the threshold of the A-fibre response showed little variation.

Other cells exhibited an additional A-fibre input from other peripheral nerves under the influence of SP, i.e. the neuropeptide increased the degree of input convergence onto the neurones. A general feature of the SP-induced increase in responsiveness was that it outlasted the period of superfusion. Even after relatively long wash-out periods (30–60 min) the additional A-fibre input was not reversible. Another effect of the spinal superfusion with SP was that an existing A- or C-fibre input disappeared. In contrast to the additional input described above, this effect was reversible after a few minutes of wash-out time.

Again other neurones showed a reduction in amplitude of the action potential which often was associated with a loss of electrical excitability. In some cases the amplitude reduction was so marked that the neurone could no longer be recorded from. Since only neurones were selected for study which gave a stable recording for at least 10 min prior to SP administration, and since in the control population such a behaviour did not occur, this effect must be attributed to an action of SP. Some of the cells that showed this effect reappeared during wash-out.

The actions of SP superfused on the spinal cord were concentration dependent. At a concentration of 100 μM, SP elicited one of the above effects in 75 % of the neurones tested. This proportion dropped to 50 % at a concentration of 10 μM, and 1 μM was completely ineffective.

In some neurones with high-threshold mechanosensitive RFs in deep tissues, the influence of SP superfusion on the mechanical threshold and configuration of the RFs was tested. The aim of these experiments was to find out whether the changes in electrical excitability described above were associated with detectable changes in RF properties. In about half of the neurones there was a marked increase in RF size during SP superfusion (Fig. 5).

This effect may be due to an increase in synaptic efficacy of afferent fibres that connect the neurones with body regions outside the original RF. Intracellular recordings in dorsal horn neurones showed that under the influence of SP subthreshold excitatory post-synaptic potentials (EPSPs) may become suprathreshold. The EPSP effects all

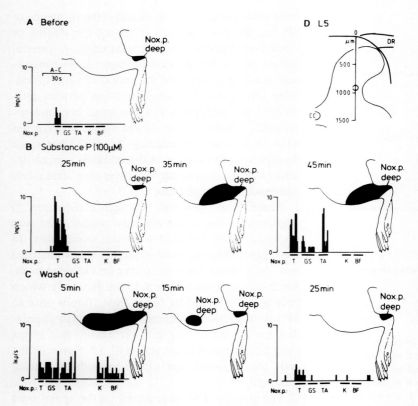

Fig. 5. Changes in receptive field (RF) size of a dorsal horn neurone during superfusion with 100 μM SP. The recording site is indicated in panel D (CC, central canal; DR, dorsal root). Size and location of the RF (black area) on the hindlimb are shown before SP superfusion (A), during superfusion (B) and during wash-out with CSF (C). The peristimulus time histograms underneath the outlines of the hindlimb show the neurone's responses to noxious deep pressure stimulation (Nox.p.deep) of the calcaneal tendon (T), gastrocnemius-soleus muscle (GS), tibialis anterior muscle (TA), knee joint (K) and biceps femoris muscle (BF). Before SP superfusion, the neurone had exclusive input from nociceptors in the calcaneal tendon (A). During SP superfusion (B, 35, 45 min after switching to SP) and in the early phase of the wash-out period (C, 5 min) the RF expanded. At the time of the greatest RF expansion noxious pressure stimulation of GS, TA, K and BF was effective in driving the neurone. The effect was reversible after a wash-out period of 25 min (C, from Hoheisel et al., 1995).

occurred at A-fibre latency, but it is conceivable that SP has a similar action also on C-fibre synapses.

The SP concentration used for intrathecal administration can not be considered to be extremely high if the following facts are taken into account: (1) Determinations of the SP concentrations in the rat dorsal horn have yielded values of approximately 0.5 μM SP (Ogawa et al., 1985). (2) Intrathecal SP concentrations of 1–10 μM are required to elicit effects on the discharges of dorsal horn neurones (Kellstein et al., 1990; Liu and Sandkühler, 1995) and to influence the release of amino acids from the spinal cord (Maehara et al., 1995). (3) An unspecific damage of the neurones caused by SP appears unlikely because in our experiments the first effects occurred at concentrations between 1 and 10 μM, therefore, 100 μM was not far supramaximal. (4) Some of the nociceptive neurones that were tested for SP-induced changes in RF size did not show alterations in behaviour indicative of an unspecific damage (e.g. change in background activity, irregular responses to mechanical stimuli). (5) There was no change in systemic blood pressure during SP superfusion at 100 μM.

Perspectives, interpretations and speculations

SP-induced discharges of primary afferent units in vitro

In the hemidiaphragm-phrenic nerve preparation used, SP was a powerful excitant of HTM (presumably nociceptive) receptors. In fact, the effective SP concentration was in the same order of magnitude as that of bradykinin which had proven to be one of the most effective stimulants for muscle nociceptors in previous studies (Fock and Mense, 1976). In contrast to bradykinin which is known to excite not only nociceptive but also non-nociceptive muscle afferent units (Mense and Meyer, 1985), SP appears to have a selective excitatory action on nociceptive endings. Again in contrast to bradykinin (Mense and Meyer, 1988), SP activates muscle nociceptors without sensitizing them to mechanical stimulation.

The data suggest that SP released from free nerve endings in skeletal muscle influences the discharge behaviour of the various types of ending in different ways. In nociceptors, it increases the background activity which may be one reason for the spontaneous pain that often accompanies such a lesion. The simultaneous decrease in mechanical responsiveness in LTM receptors might contribute to the pain by reducing the pain-inhibiting segmental input to the spinal cord (Handwerker et al., 1975). The tenderness of a damaged muscle is probably not due to SP release, since there were no signs of a SP-induced sensitization to mechanical stimuli. Taken together the neurophysiological findings at the primary afferent level suggest that activation of a nociceptive ending is not necessarily associated with its sensitization. Both effects are controlled by different mechanisms and can occur independently. If in fact the reason for spontaneous pain at the primary afferent level is background activity in nociceptors, and the reason for tenderness is sensitization of these endings, then these two forms of pain sensations have to be regarded as independent phenomena and may require different forms of treatment.

Inflammation-induced changes in innervation density in SP-ir fibres

The increase of SP-ir nerve fibres during a subacute inflammation found in our study is in accordance with data from Weihe et al. (1988) who induced an inflammation in the subcutaneous tissue of the rat paw and observed an increase in SP-ir fibres (and an even larger increase in CGRP-ir ones) after 1 week of inflammation. Apparently, this effect is not present in all types of inflammation and species, because in patients with rheumatoid arthritis a depletion of SP- and CGRP-ir nerve fibres in the synovia has been described (Mapp et al., 1990).

The increase in the density of SP-ir fibres can be interpreted in various ways. It may be due to (1) a decrease in SP-release which leads to an elevated intraaxonal concentration of the neuropeptide so that SP can be visualized in a larger proportion of fibres, (2) increased synthesis of SP in dorsal root ganglion cells (which would have the same effect on the histological visualization), or (3) sprouting of ir nerve fibres. The first mechanism may not hold true, since noxious stimuli are known to release SP from peripheral nerve endings (Helme et al., 1986). Theoretically, this release could be associated with an increase in intraaxonal peptide concentration, since in peripheral nerves innervating chronically inflamed tissue the transport rate of neuropeptides was found to be higher (Donnerer et al., 1992). As to the second mechanism, SP synthesis is probably upregulated under the conditions of our study (Donaldson et al., 1992). Generally, noxious stimuli are known to induce preprotachykinin-A gene expression in the rat dorsal root ganglion (Noguchi et al., 1988).

The increase in NGF expression in the inflamed muscle may be a supporting factor, since NGF regulates peptide gene expression in dorsal root ganglion cells (Lindsay and Harmar, 1989).

Another effect of an increased NGF level in inflamed tissue could be to induce sprouting of nerve fibres. Sprouting is generally associated with an increased IR to GAP-43/B-50, because this protein is found in particularly large quantities in the axonal growth cones during embryonic and postnatal development (Skene, 1989) and also in regenerating axon fibres following nerve injury (Tetzlaff et al., 1989). The increase in GAP-43/B-50-ir fibres in the persistently inflamed muscle suggests that fibre sprouting contributes to the observed increase in the density of SP-ir fibres.

The functional consequences of an increase in the innervation density of SP-ir nerve endings – probably combined with an increase in SP release – may be manyfold. SP is known to be a potent vasodilator of peripheral blood vessels, including those in skeletal muscle (Öhlen et al., 1987). It also causes plasma protein extravasation; this effect is enhanced under inflammatory conditions (Scott et al., 1992). SP-ir fibres show a close relationship to mast cells (Skofitsch et al., 1985), and recently NK-receptors have been found on the membrane of these cells (Krumins and Broomfield, 1992).

Besides these pro-inflammatory effects, SP and other tachykinins have potent immunomodulatory actions (McGillis et al., 1990; Mantyh, 1991). Receptors for SP have been reported to exist on T lymphocytes (Payan et al., 1987), and in human T lymphocytes the neuropeptide induces a rise in the intracellular calcium concentration (Kavelaars et al., 1993). During longer-lasting lesions the mitogenic effect of SP could also be of importance, because it leads to neovascularization and proliferation of endothelial cells (Ziche et al., 1990).

The physiological function of the SP- and CGRP-ir fibres that exhibited an increase in numerical density is unknown. If the effect is present also in nociceptors, pain sensations from an inflamed muscle could be due not only to an increased discharge activity in these endings but also to a recruitment of active nociceptive endings. This latter effect could lead to an enhancement of pain because of an increased spatial summation at the first synapses in the spinal cord.

Dorsal horn neurones and substance P

The experiments employing superfusion of the spinal cord demonstrate that elevated levels of intrathecal SP are associated with a marked change in the responsiveness of dorsal horn neurones. One possible explanation for the observed unmasking of additional input is that SP strengthens synaptic connections with afferent fibres that originally were ineffective in driving the cell. This means that under normal conditions a considerable proportion of dorsal horn neurones must have afferent connections that are ineffective, in addition to the effective synapses that connect the neurone with the original RF. This interpretation is supported by the presence of somatotopically inappropriate projections from primary afferent units to the spinal cord (Meyers and Snow, 1984).

The release of SP from the spinal terminals of nociceptive afferent fibres may increase the efficacy of the ineffective synapses and thus change the connectivity in the dorsal horn. The loss of input or of electrical excitability observed in other neurones are further aspects of this functional reorganization. The data are not sufficient to tell whether those neurones that showed a loss of excitability were inhibitory interneurones. If so, their disappearance could be causally related to the increase in excitability observed in other cells.

In the present study only neurones were included that had no mechanosensitive RFs in the skin. A more detailed functional identification was not attempted. Therefore, no information is available concerning a possible relationship between functional properties and type of SP effect. It is likely, however, that such a relationship exists, because in nociceptive neurones the only SP effect observed was an increase in RF size or number of RFs. Whether all nociceptive neurones react to SP in this way is unknown.

The SP-induced increase in RF size or responsiveness of dorsal horn neurones to electrical nerve stimulation was not associated with a change in background activity. This finding is in agreement with results obtained from muscle primary afferent units showing that SP had a differential action on background activity and excitability. However, at the primary afferent level the SP effect was inverse in that the background activity was increased with little change in mechanical excitability.

A possible explanation for the reduction in spike size which often was accompanied by a loss of excitability is a SP-induced depolarization of the neurones (cf. Randic and Miletic, 1977; Zieglgänsberger and Tulloch, 1979). Actually, intracellular recordings from a limited number of cells in the present series demonstrated that such a reduction in spike amplitude was associated with a marked depolarization of the cell membrane.

Because of the heterogeneity of the SP effects, the subjective sensations possibly elicited by the neuropeptide following its intrathecal release under (patho)physiological conditions is hard to assess. The expansion of the RFs and the increase in responsiveness in neurones that probably mediate deep pain suggests that a given noxious stimulus will elicit more pain if SP is present. An increase in RF size of nociceptive neurones as observed in our experiments has been discussed as a possible mechanism for the hyperalgesia in patients (Dubner, 1992). This mechanism may be of importance for fibromyalgia patients who have elevated levels of SP in the cerebrospinal fluid (Vaeroy et al., 1988). It has to be emphasized, however, that the concentration of SP measured in fibromyalgia patients is many orders of magnitudes lower than that applied in the present study.

In our experiments, intrathecal SP concentrations that markedly changed the responsiveness of dorsal horn neurones had no influence on their background activity. If the view is accepted that an increase in excitability reflects allodynia or hyperalgesia, whereas an increase in background activity is related to spontaneous pain, then at the spinal level SP could be involved in the development of allodynia/hyperalgesia but not in spontaneous pain. At the primary afferent level, however, it could contribute to spontaneous pain without producing tenderness.

Apart from modulating the processing of nociceptive information in the spinal cord, SP has been shown to trigger long-term changes by inducing the expression of immediate/early genes such as c-fos (Williams et al., 1989; Zimmerman and Herdegen, this volume). Thus, SP may not only be involved in spinal allodynia/hyperalgesia but also in neuroplastic changes that eventually may lead to the transition from acute to chronic pain.

Acknowledgements

The authors wish to thank Ms. B. Quenzer and Ms. M. Pauli for expert technical assistance. This work was supported by the Deutsche Forschungsgemeinschaft.

References

Beck, H., Schröck, H. and Sandkühler, J. (1995) Controlled superfusion of the rat spinal cord for studying non-synaptic transmission: an autoradiographic analysis. *J. Neurosci. Methods*, 58: 193–202.

Bleazard, L., Hill, R.G. and Morris, R. (1994) A comparison of the distribution of NK_1 receptors in the dorsal horn of the spinal cord with that of neurones responsive to agonists acting at this receptor, in the rat. *J. Physiol.*, 476: 434P.

Brimijoin, S. Lundberg, J.M., Brodin, E., Hökfelt, T. and Nilsson, G. (1980) Axonal transport of substance P in the vagus and sciatic nerves of the guinea pig. *Brain Res.*, 191: 443–457.

Diehl, B., Hoheisel, U. and Mense, S. (1993) The influence of mechanical stimuli and of acetylsalicylic acid on the discharges of slowly conducting afferent units from normal and inflamed muscle in the rat. *Exp. Brain Res.*, 92: 431–440.

Donaldson, L.F., Harmar, A.J., Mc Queen, D.S. and Seckl, J.R. (1992) Increased expression of preprotachykinin, calcitonin gene-related peptide, but not vasoactive intestinal peptide messenger RNA in dorsal root ganglia during the development of adjuvant monoarthritis in the rat. *Mol. Brain Res.*, 16: 143–149.

Donnerer, J., Schuligoi, R. and Stein, C. (1992) Increased content and transport of substance P and calcitonin gene-related peptide in sensory nerves innervating inflamed tissue: evidence for a regulatory function of nerve growth factor in vivo. *Neuroscience*, 49: 693–698.

Dubner, R. (1992) Hyperalgesia and expanded receptive fields. *Pain*, 48: 3–4.

Dubner, R. and Bennett, G.J. (1983) Spinal and trigeminal mechanisms of nociception. *Annu. Rev. Neurosci.*, 6: 381–418.

Duggan, A.W., Hendry, I.A., Morton, C.R., Hutchison, W.D. and Zhao, Z.Q. (1988) Cutaneous stimuli releasing immunoreactive substance P in the dorsal horn of the cat. *Brain Res.*, 451: 261–273.

Fleming, B.P., Gibbins, I.L., Morris, J.L. and Gannon, B.J. (1989) Noradrenergic and peptidergic innervation of the extrinsic vessels and microcirculation of the rat cremaster muscle. *Microvasc. Res.*, 38: 255–268.

Fock, S. and Mense, S. (1976) Excitatory effects of 5-hydroxytryptamine, histamine and potassium ions on muscular group IV afferent units: a comparison with bradykinin. *Brain Res.*, 105: 459–469.

Frenk, H., Bossut, D., Urca, G. and Mayer, D.J. (1988) Is substance P a primary afferent neurotransmitter for nociceptive input? I. Analysis of pain-related behaviors resulting from intrathecal administration of substance P and 6 excitatory compounds. *Brain Res.*, 455: 223–231.

Handwerker, H.O., Iggo, A. and Zimmermann, M. (1975) Segmental and supraspinal actions on dorsal horn neurons responding to noxious and non-noxious skin stimuli. *Pain*, 1: 147–165.

Helme, R.D., Koschorke, G.M. and Zimmermann, M. (1986) Immunoreactive SP release from skin nerves in the rat by noxious thermal stimulation. *Neurosci. Lett.*, 63: 295–299.

Hoheisel, U., Mense, S. and Ratkai, M. (1995) Effects of spinal cord superfusion with substance P on the excitability of rat dorsal horn neurons processing input from deep tissues. *J. Musculoskel. Pain*, in press.

Kavelaars, A., Jeurissen, F., von Frijtag-Drabbe-Kunzel, J., Herman van Roijen, J., Rijkers, G.T. and Heijnen, C.J. (1993) Substance P induces a rise in intracellular calcium concentration in human T lymphocytes in vitro: evidence of a receptor-independent mechanism. *J. Neuroimmunol.*, 42: 61–70.

Kellstein, D.E., Price, D.D., Hayes, R.L. and Mayer, D.J. (1990) Evidence that substance P selectively modulates C-fiber-evoked discharges of dorsal horn nociceptive neurones. *Brain Res.*, 526: 291–298.

Kessler, W., Kirchhoff, C., Reeh, P.W. and Handwerker, H.O. (1992) Excitation of cutaneous afferent nerve endings in vitro by a combination of inflammatory mediators and conditioning effect of substance P. *Exp. Brain Res.*, 91: 467–476.

Krumins, S.A. and Broomfield, C.A. (1992) Evidence of NK1 and NK2 tachykinin receptors and their involvement in histamine release in a murine mast cell line. *Neuropeptides*, 21: 65–72.

Kumazawa, T. and Mizumura, K. (1979) Effects of synthetic substance P on unit-discharges of testicular nociceptors of dogs. *Brain Res.*, 170: 553–557.

Kuraishi, Y., Nanayama, T., Ohno, H., Fujii, N., Otaka, A., Yajima, H. and Satoh, M. (1989) Calcitonin gene-related peptide increases in the dorsal root ganglia of adjuvant arthritic rat. *Peptides*, 10: 447–452.

Leah, J.D., Cameron, A.A. and Snow, P.J. (1985) Neuropeptides in physiologically identified mammalian sensory neurones. *Neurosci. Lett.*, 56: 257–263.

Lembeck, F. and Holzer, P. (1979) Substance P as neurogenic mediator of antidromic vasodilation and neurogenic plasma extravasation. *Naunyn-Schmiedeberg's Arch. Pharmacol.*, 310: 175–183.

Lembeck, F., Donnerer, J. and Colpaert, F.C. (1981) Increase of substance P in primary afferent nerves during chronic pain. *Neuropeptides*, 1: 175–180.

Lindsay, R.M. and Harmar, A.J. (1989) Nerve growth factor regulates expression of neuropeptide genes in adult sensory neurons. *Nature*, 337: 362–364.

Liu, X.-G. and Sandkühler, J. (1995) The effects of extrasynaptic substance P on nociceptive neurons in laminae I and II in rat lumbar spinal dorsal horn. *Neuroscience*, 68: 1207–1218.

Maehara, T., Suzuki, H., Yoshioka, K. and Otsuka, M. (1995) Characteristics of substance P-evoked release of amino acids from neonatal rat spinal cord. *Neuroscience*, 68: 577–584.

Mantyh, P.W. (1991) Substance P and the inflammatory immune response. *Ann. N. Y. Acad. Sci.*, 632: 263–271.

Mapp, P.I., Kidd, B.L., Gibson, S.J., Terry, J.M., Revell, P.A., Ibrahim, N.B.N., Blake, D.R. and Polak, J.M. (1990) Substance P-, calcitonin gene-related peptide- and C-flanking peptide of neuropeptide Y-immunoreactive fibres are present in normal synovium but depleted in patients with rheumatoid arthritis. *Neuroscience*, 37: 143–153.

McGillis, J.P., Mitsuhashi, M. and Payan, D.G. (1990) Immunomodulation by tachykinin neuropeptides. *Ann. N. Y. Acad. Sci.*, 594: 85–94.

McMahon, S.B., Sykova, E., Wall, P.D., Woolf, C.J. and Gibson, S.J. (1984) Neurogenic extravasation and substance P levels are low in muscle as compared to skin in the rat hindlimb. *Neurosci. Lett.*, 52: 235–240.

Mense, S. and Meyer, H. (1985) Different types of slowly conducting afferent units in cat skeletal muscle and tendon. *J. Physiol.*, 363: 403–417.

Mense, S. and Meyer, H. (1988) Bradykinin-induced modulation of the response behaviour of different types of feline group III and IV muscle receptors. *J. Physiol.*, 398: 49–63.

Meyers, D.E.R. and Snow, P.J. (1984) Somatotopically inappropriate projections of single hair follicle afferent fibres to the cat spinal cord. *J. Physiol.*, 347: 59–73.

Molander, C., Ygge, I. and Dalsgaard, C.-J. (1987) Substance P-, somatostatin- and calcitonin gene-related peptide-like immunoreactivity and fluoride resistant acid phosphatase-activity in relation to retrogradely labeled cutaneous, muscular and visceral primary sensory neurons in the rat. *Neurosci. Lett.*, 74: 37–42.

Nicoll, R.A., Schenker, C. and Leeman, S.E. (1980) Substance P as a transmitter candidate. *Annu. Rev. Neurosci.*, 3: 227–268.

Noguchi, K., Morita, Y., Kiyama, H., Ono, K. and Tohyama, M. (1988) A noxious stimulus induces the preprotachykinin-A gene expression in the rat dorsal root ganglion: a quantitative study using in situ hybridisation histochemistry. *Mol. Brain Res.*, 4: 31–35.

O'Brien, C., Woolf, C.J., Fitzgerald, M., Lindsay, R.M. and Molander, C. (1989) Differences in the chemical expression of rat primary afferent neurons which innervate skin, muscle or joint. *Neuroscience*, 32: 493–502.

Öhlen, A., Lindbom, L., Staines, W., Hökfelt, T., Cuello, A.C., Fischer, J.A. and Hedquist, P. (1987) Substance P and calcitonin gene-related peptide: immunohistochemical localization and microvascular effects in rabbit skeletal muscle. *Naunyn-Schmiedeberg's Arch Pharmacol.*, 336: 87–93.

Ogawa, T., Kanazawa, I. and Kimura, S. (1985) Regional

distribution of substance P, neurokinin alpha and neurokinin beta in rat spinal cord, nerve roots and dorsal root ganglia and the effects of dorsal root section or spinal transection. *Brain Res.*, 359: 152–157.

Payan, D.G., McGillis, J.P., Renold, F.K., Mitsuhashi, M. and Goetzl, E.J. (1987) Neuropeptide modulation of leukocyte function. *Ann. N. Y. Acad. Sci.*, 496: 183–191.

Popper, P. and Micevych, P.E. (1989) Localization of calcitonin gene-related peptide and its receptors in a striated muscle. *Brain Res.*, 496: 180–186.

Randic, M. and Miletic, V. (1977) Effect of substance P in cat dorsal horn neurones activated by noxious stimuli. *Brain Res.*, 128: 164–169.

Scott, D.T., Lam, F.Y. and Ferrell, W.R. (1992) Acute inflammation enhances substance P-induced plasma protein extravasation in the rat knee joint. *Regul. Peptides*, 39: 227–235.

Skene, J.H.P. (1989) Axonal growth associated proteins. *Annu. Rev. Neurosci.*, 12: 127–156.

Skofitsch, G., Savitt, J.M. and Jakobowitz, D.M. (1985) Suggestive evidence for a functional unit between mast cells and substance P fibres in the rat diaphragm and mesentery. *Histochemistry*, 82: 5–8.

Tetzlaff, W., Zwiers, H., Lederis, K., Cassar, L. and Bisby, M.A. (1989) Axonal transport and localization of B-50/GAP-43-like immunoreactivity in regenerating sciatic and facial nerves of the rat. *J. Neurosci.*, 9: 1303–1313.

Vaeroy, H., Helle, R., Forre, O., Kass, E. and Terenius, L. (1988) Elevated LSF levels of substance P and high incidence of Raynaud's phenomenon in patients with fibromyalgia: new features of diagnosis. *Pain*, 33: 21–26.

Vizi, E.S. and Labos, E. (1991) Non-synaptic interactions at presynaptic level. *Prog. Neurobiol.*, 37: 145–163.

Weihe, E., Nohr, D., Millan, M.J., Stein, C., Müller, S., Gramsch, C. and Herz, A. (1988) Peptide neuroanatomy of adjuvant-induced arthritic inflammation in rat. *Agents Actions*, 25: 255–259.

Williams, S., Pini, A., Evan, G. and Hunt, S.P. (1989) Molecular events in the spinal cord following sensory stimulation. In: F. Cervero, G.J. Bennett and P.M. Headley (Eds.), *Processing of Sensory Information in the Superficial Dorsal Horn of the Spinal Cord*, Plenum Press, New York, pp. 273–283.

Ziche, M., Morbidelli, L., Pacini, M., Geppetti, P., Alessandri, G. and Maggi, C.A. (1990) Substance P stimulates neovascularization in vivo and proliferation of cultured endothelial cells. *Microvasc. Res.*, 40: 264–278.

Zieglgänsberger, W. and Tulloch, I.F. (1979) Effects of substance P on neurones in the dorsal horn of the spinal cord of the cat. *Brain Res.*, 166: 273–282.

CHAPTER 11

Studies of the release of immunoreactive galanin and dynorphin $A_{(1-8)}$ in the spinal cord of the rat

A.W. Duggan and R.C. Riley

Department of Preclinical Veterinary Sciences, Royal (Dick) School of Veterinary Studies, University of Edinburgh, Summerhall, Edinburgh EH9 1QH, UK

Introduction

In current therapy, centrally acting analgesics are exemplified by μ-agonist opioids and α_2-adrenomimetics. In both cases the administered drugs are agonists at receptors believed to be important in inhibiting the transmission of nociceptive information and there is considerable evidence that part of that action occurs in the spinal cord (Duggan et al., 1976; Le Bars et al., 1980; Advokat, 1988; Clarke et al., 1988; Takano and Yaksh, 1992). Thus mimicking inhibition as a strategic approach to pain control has been successful but the resultant drugs do have significant side effects when administered systemically. These effects, however, can be minimized by epidural or intrathecal administration.

Recent research has identified a number of other compounds which may also function as inhibitory to the transmission of nociceptive information. Dynorphin and galanin are two such compounds although their roles, even in the spinal cord, have been the subject of dissent amongst investigators and this will be subsequently discussed. Despite this, galanin and dynorphin may be lead compounds for newer analgesics and hence work in our laboratory has been directed at defining the conditions which produce a spinal release of these neuropeptides.

The experiments have employed antibody microprobes to detect release of neuropeptides. Antibody microprobes are fine glass micropipettes having a siloxane polymer on their outer surfaces which permits immobilization, on these surfaces, of antibodies to a neuropeptide under study. When inserted into the central nervous system a focal release of the relevant neuropeptide, if adjacent to a microprobe, may result in the binding of a proportion of the released molecules to a localized area of the microprobe (Duggan and Hendry, 1986). After withdrawal from the brain or spinal cord, microprobes are incubated in a radiolabelled form of the peptide under study and then autoradiographs obtained. Focal binding of released molecules are revealed as deficits in the binding of the tracer and this is analysed quantitatively by image analysis. The amounts detected are very small. For example, complete inhibition of the binding of ^{125}I-SP to 100 μm of the length of microprobes has been estimated to correspond to 10^{-17} mole of unbound label. The concentrations in the vicinity to produce comparable inhibition of binding within 30 min at 37°C vary from 10^{-7} to 10^{-9}M with in vitro tests (Duggan, 1991)

The amount of time microprobes need to remain in the nervous system to detect a neuropeptide varies according to local neuropeptide concentrations and the sensitivity of the particular microprobes, but it has ranged from 5 to 30 min (Duggan et al., 1988; Schaible et al., 1990; Duggan et al., 1990). It is usual to average results from defined groups of microprobes (Hendry et al., 1988). The mean image analyses have as the ordinate an integral (grey scale) obtained by analysing microprobe autoradiographs in 10 μm^2 areas on a density scale of 0–255 and then performing transverse integrations across the images. This analysis with an interval of 10 μm is beyond the biological resolution of the method (see Duggan 1991) and hence the mean of three successive intervals is usually performed giving a biological resolution of 30 μm. Since a working definition of release is an increase in extracellular levels produced by a stimulus, the differences (in 30 μm intervals) in the mean optical densities of pre-stimulus and post-stimulus microprobes can be assigned statistical significance and hence can define areas where significant release has followed a defined stimulus. The method has been used successfully to define the stimuli producing central release of substance P (Duggan et al., 1988; Schaible et al., 1990), somatostatin (Morton et al., 1989) and calcitonin gene-related peptide (Schaible et al., 1994).

Galanin

Galanin was originally isolated from porcine intestine (Tatemoto et al., 1983) but subsequently shown to be present in neurones of the peripheral and central nervous systems. In the rat spinal cord galanin occurs in dorsal root ganglion neurones, intrinsic neurones of the spinal cord and in fibres of supraspinal origin (Ch'ng et al., 1985; Melander et al., 1986; Tuchscherer and Seybold, 1989; Klein et al., 1990). There is dispute among investigators on the abundance of galanin-containing neurones in dorsal root ganglia of the rat with estimates varying from rare (<5%) (Villar et al., 1990) to more common than neurones containing calcitonin gene-related peptide (Klein et al., 1990). This difference has not been resolved but it indicates the need to determine whether galanin is readily released in the spinal cord when impulses invade the central terminal of primary afferent fibres.

Following nerve crush in the rat, galanin levels in dorsal root ganglion cells (along with those of vasoactive intestinal polypeptide and neuropeptide Y) increase markedly (Villar et al., 1990, 1991; Xu et al., 1990), although the functional significance of these alterations is uncertain. By immunocytochemical techniques, galanin-contain-ing neural structures in the spinal cord have been shown to be relatively heavily concentrated in laminae I and II of the dorsal horn but not in deeper laminae IV and V nor in the dorsal columns (Tuchscherer and Seybold, 1989).

In most studies of the possible functions of galanin in the spinal cord this neuropeptide has been applied topically to the spinal cord surface. When galanin has been administered in this way the effects have not been uniform. Following intrathecal administration to rats in relatively low doses, galanin has been variously found to produce hyperalgesia to mechanical but not thermal noxious stimuli (Cridland and Henry, 1988), to result in vocalisation to innocuous mechanical peripheral stimuli (Cridland and Henry, 1988) and to enhance reflexes to peripheral noxious thermal but not mechanical stimuli (Wiesenfeld-Hallin et al., 1988). In contrast to these enhanced responses, a number of depressant actions of intrathecal galanin have also been observed although following the administration of larger amounts than those producing increased responses. These depressant actions in the rat include decreased responses in hot plate and tail flick tests (Post et al., 1988) and decreased flexor reflexes produced by peripheral nerve stimulation (Wiesenfeld-Hallin et al., 1989). In addition galanin has been found to reduce the potentiation of flexor reflexes produced by conditioning stimulation of a peripheral nerve (Wiesenfeld-Hallin et al., 1989) and to markedly potentiate the depression of these reflexes by intrathecal morphine (Wiesenfeld-Hallin et al., 1990).

Compounds released from the central endings of primary afferent fibres are normally thought of as important in excitatory transmission. These depressant actions of galanin do not appear compatible with such a role although galanin could be co-released with an excitant amino acid and/or tachykinin and have independent inhibitory actions. Alternatively these depressant effects could represent the action of galanin released from intrinsic neurones of the spinal cord or from extrinsic fibres such as those descending from the brain stem.

Galanin inhibits the release of other compounds. In the central nervous system this has been shown for acetylcholine release in the rat ventral hippocampus (Fisone et al., 1987), dopamine release in the rat median eminence (Nordstrom et al., 1987) and noradrenaline release from rat spinal cord slices (Reimann and Schneider, 1993). Release of galanin itself has been relatively little studied but Morton and Hutchison (1989) used antibody microprobes in experiments on galanin release in the spinal cord of the anaesthetized cat. In this work a basal presence of immunoreactive (ir)-galanin was detected in the region of lamina II and at the spinal cord surface. Electrical stimulation of both A and C fibres of the tibial nerve (10 Hz) failed to produce a spinal release of ir-galanin. Similarly ineffective were peripheral noxious thermal and mechanical stimuli. This is not the pattern one would expect of a neuroactive compound present and released from an abundance of primary afferent fibres as suggested by the work of Klein et al. (1990). Indeed a significant basal presence not increased by peripheral nerve stimulation in spinal cats suggests release predominantly from intrinsic spinal elements.

Recent experiments in this laboratory have examined spinal release of galanin in normal rats and those with peripheral inflammation. Apart from a direct interest in galanin per se, these experiments were part of continued interest in changes in the central release of neuropeptides with developing inflammation in peripheral tissues (Hope et al., 1990; Schaible et al., 1990). Little is known of the role of galanin in inflammation but Kuraishi et al. (1991) found that intrathecal administration of an antiserum to galanin induced hypoalgesia to mechanical stimulation of an inflamed but not a normal hind paw of the rat.

The antibody microprobes used in these experiments employed an antiserum to rat galanin (Peninsula Laboratories). Data from the manufacturer indicated negligible cross-reactivity with porcine galanin, neuropeptide Y and substance P. With microprobes bearing anti-rat galanin antibodies, however, pre-incubation in porcine galanin (10^{-7}M) for 30 min at 37°C resulted in greater than 60% suppression of ^{125}I-rat galanin. Pre-incubation with rat galanin (10^{-7}M) under the same conditions resulted in approximately 90% suppression of ^{125}I-rat galanin binding.

The experiments on galanin release were performed on normal rats and those with a peripheral inflammation. Both groups were urethane anaesthetized. Peripheral inflammation was induced by subdermal injections of Freund's adjuvant around the ankle joint to induce a unilateral inflammation. This injection results in a unilateral swelling of the ankle region and previous studies (Hanesch et al., 1993) have shown that this involves both the joint and periarticular tissues, e.g. connective tissue, tendons and overlying skin.

Because the most likely stimulus producing pain from inflamed skin, joints or tendons is mechanical, resulting either from direct physical contact with an environmental object or movement of the damaged tissues, these experiments employed a series of repeated mechanical stimuli of increasing intensity. These consisted of flexing the inflamed joint area and then mechanically compressing the inflamed tissues. Comparable stimuli were also applied to the uninflamed hind limbs of normal animals. Flexion of the ankle was applied by holding the hind paw gently and pulling the limb out to its full extent, then pushing the paw gently until it was in close apposition to the knee. This flexion of the ankle joint was repeated 10 times per minute throughout the period that microprobes remained in the spinal cord. The second (more severe) stimulus consisted of laterally compressing the ankle with a modified spring clip for 10 s of each minute for the period that microprobes were present in the spinal cord.

Comparisons between the mean image analyses of in vitro microprobes not exposed to galanin and those inserted into the spinal cord in the absence of any active peripheral stimulus showed a basal presence of ir-galanin in both sides of the spinal cords of normal rats and those with unilateral ankle inflammation. Further analysis showed that there were no significant differences in basal extracellular levels of ir-galanin in the dorsal horns and dorsal columns of normal and arthritic animals. The ipsilateral and contralateral sides of the spinal cords of animals with peripheral ankle inflammation also showed no significant differences in basal levels of ir-galanin. Thus comparable basal levels of ir-galanin were found

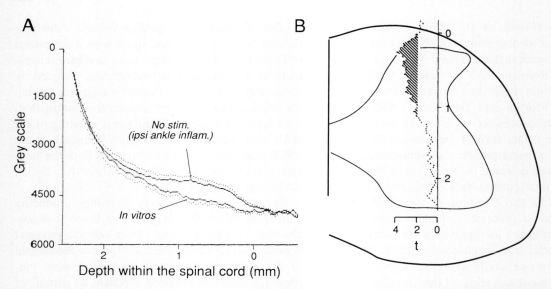

Fig. 1. Basal presence of immunoreactive (ir) galanin in the side of the spinal cord ipsilateral to ankle inflammation. (A) The mean image analyses of two groups of microprobes are plotted with respect to length: those which were not inserted into the spinal cord but simply incubated in ^{125}I-galanin (in vitros $n = 25$) and those remaining in the spinal cord in the absence of added peripheral stimulation (no stim. ipsilateral ankle inflammation $n = 29$). With each mean image analysis the mean grey scale was determined in 30 μm intervals and a line joins these points. At each point ± S.E.M. is also plotted. (B) A plot of the t statistics derived from the differences between the two mean image analyses of (A) is related to an outline of the spinal cord at the area sampled. The hatched areas indicate where these differences are significant at the $P < 0.05$ level. Reproduced with permission from Hope et al. (1994)

in normals and animals with unilateral peripheral inflammation.

Fig. 1A compares the mean image analysis of microprobes inserted into the side of the spinal cord ipsilateral to peripheral inflammation but in the absence of any added stimulation with that of in vitro microprobes. Fig. 1B plots, in 30 μm intervals, the differences between the mean image analyses of the two groups and the hatched area indicates where these differences are significant at the $P < 0.05$ level. This shows that there was a basal presence of ir-galanin at significant levels throughout the whole of the dorsal horn and dorsal columns but not within the ventral horn.

A diffuse basal presence of a compound gives little clue to the releasing structures. A localized presence in the superficial dorsal horn would favour release in part from primary afferent fibres but even release from this source could result in a diffuse presence if the compound were slowly degraded after release. Relevant to this is the finding that, when incubated with membrane preparations of rat hypothalamus, the half-lives of galanin 1–29 and galanin 1–16 were 100 min and 28 min respectively (Land et al., 1991). If this reflects the rate of degradation of galanin under in vivo conditions then diffusion away from sites of release is to be expected. It is equally possible, however, that the basal presence of ir-galanin in the superficial dorsal horn represents release from intrinsic neurones of the spinal cord or release from terminals of supraspinal origin.

Flexing the ankle of normal animals failed to produce a central release of ir-galanin and flexing the inflamed ankle only did so after prolonged repetitive stimulation. Indeed it was only after a third period of 20 min of intermittent flexion of the inflamed ankle that increased levels of ir-galanin were detected in the dorsal horn by microprobes. Increasing the intensity of mechanical stimulation of inflamed tissues by laterally compressing the inflamed ankle gave the unexpected result of decreasing extracellular levels of ir-galanin. This decrease was not significant during the first period of noxious ankle compression but was very evident during the second application of this stimulus.

Thus Fig. 2A compares the mean image analysis of microprobes present in the spinal cord during the second period of ankle flexion and that of microprobes present in the absence of active peripheral stimulation. Fig. 2B shows that significant differences occurred between these groups at several sites in the nucleus proprius of the dorsal horn. Severe compression of the foot of the normal rat also failed to alter basal levels of ir-galanin in the ipsilateral spinal cord.

The relative difficulty with which galanin was released by peripheral stimulation argues against this neuropeptide being a simple transmitter released by a particular population of afferent fibres. Thus these experiments do not support the proposal of Klein et al. (1990) that galanin is present in more primary afferent fibres than calcitonin gene related peptide. The results are supportive of those of Kuraishi et al. (1991) in suggesting that galanin has a functional role in the spinal cord with the development of peripheral inflammation. Release studies alone cannot determine function and, in the present instance, whether released galanin facilitates or inhibits transfer of information from nociceptive primary afferents. It remains to be determined if galanin reduces transmitter release from the central terminals of primary afferent fibres but, as cited previously, an inhibition of transmitter release has been shown in areas of the brain and spinal cord. Following peripheral nerve damage the synthesis of both NPY and of galanin commences in a population of both large and small dorsal root ganglion neurones (Wakisaka et al., 1992; Noguchi et al., 1993; Kashiba et al., 1994). It has been shown that NPY reduces substance P release form the central terminals of small diameter primary afferents (Duggan et al., 1991) and from cultured dorsal root ganglion neurones (Walker et al., 1988; Herdegen et al., 1993). It is possible that both NPY and galanin function to silence primary afferents during a regeneration phase and that the decreased levels of ir-galanin produced by severe ankle compression represents an inhibition of both basal and evoked release of galanin by the activity of afferents brought into play by such stimuli. An alternative explanation is that repeated stimuli deplete the stores of releas-

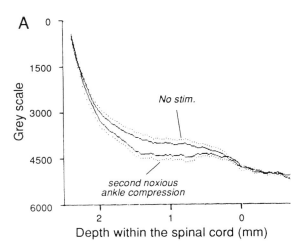

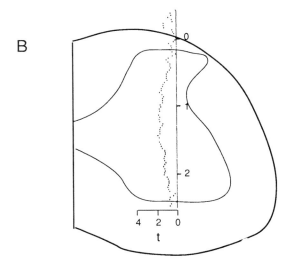

Fig. 2. Noxious lateral compression of the inflamed ankle region and spinal release of ir-galanin. In (A) the mean image analysis of 29 microprobes present in the spinal cord in the absence of any added stimulus (no stim.) is compared with those present during the second ($n = 12$) period of noxious lateral ankle compression. (B) The differences between the two groups of microprobes in A are plotted with respect to depth in the spinal cord. Because only isolated points attain significance at the $P < 0.05$ level there is not hatching as in Fig. 1. Reproduced with permission from Hope et al. (1994)

able galanin to the extent that basal levels decline below those present prior to any active stimulus. The report of Klein et al. (1992) that 30 min of peripheral nerve stimulation in the rat resulted in reduced levels of ir-galanin measured by immunocytochemistry of the dorsal horn supports such a proposal.

Dynorphin

The dynorphins are a family of neuropeptides derived from the precursor molecule prodynorphin. The peptide commonly called dynorphin $A_{(1-17)}$ is prodynorphin 209–225 and another cleavage product of this region is dynorphin $A_{(1-8)}$ (prodynorphin 209–216). Other dynorphins are: α-neoendorphin (prodynorphin 175–184), β-neoendorphin (prodynorphin 175–183), leumorphin (prodynorphin 228–256) and dynorphin $B_{(1-13)}$ (prodynorphin 228–240) (Day et al., 1993). Although it is a common statement that dynorphins are endogenous ligands at κ opioid receptors, this is an oversimplification, since at best there is only a six-fold difference in the affinity of the dynorphins for the κ receptor than for the μ receptor (Day et al., 1993).

From a functional viewpoint dynorphins are enigmatic neuropeptides. At the single cell level, activation of kappa receptors appears to inhibit a voltage sensitive calcium channel and this may function to reduce transmitter release at nerve endings (North, 1993). When administered systemically, κ-agonist drugs depress some spinal reflexes and indeed κ receptors were first proposed from the effects of ketocyclazocine on spinal reflexes in the chronic spinal dog (Martin et al., 1976).

Administered iontophoretically both dynorphin $A_{(1-13)}$ and the kappa receptor agonist U50 488H depressed the excitation of spinocervical tract neurones of the cat by noxious peripheral stimuli (Fleetwood-Walker et al., 1988). Applied topically to the spinal cord, however, dynorphin and kappa agonists have produced variable effects on neuronal firing with mixed excitatory, inhibitory actions (Hylden et al., 1991). Necrosis of the spinal cord has followed topical application although this is not an opioid effect in the sense that it is not prevented by opioid antagonists (Faden and Jacobs, 1983; Bakshi et al., 1992; Long et al., 1994).

Renewed interest in dynorphin and the spinal cord came with the finding that, when inflammation develops in the periphery of the rat, the spinal levels of dynorphins are greatly increased. This has been shown for dynorphin $A_{(1-8)}$ (Iadarola et al., 1988; Nahin et al., 1992), dynorphin $A_{(1-17)}$ (Millan et al., 1986) and α-neoendorphin (Przewlocka et al., 1992). Preceding these rises in neuropeptide levels, mRNA for prodynorphin has also been found to be increased (Iadarola et al., 1988; Weihe et al., 1989; Draisci and Iadarola, 1989) Under these conditions measurements of ir-dynorphin-containing neurones by immunocytochemical means admits of several interpretations. A depletion may represent enhanced release not compensated by increased synthesis but equally could result from decreased synthesis and decreased release. Conversely, increased numbers of ir-dynorphin-containing neurones may represent either decreased or increased release. The simplest explanation for the increased RNA message for dynorphin A is that this is a response to prior release since neuropeptides are synthesized in cell bodies and transported thence to terminals.

There is thus a need to examine release of dynorphin A and related neuropeptides in the spinal cord both in normal animals and those with developing peripheral inflammation. Hutchison et al. (1990) studied release of dynorphin $A_{(1-17)}$ in the spinal cord of the normal cat. They found a basal presence of ir-dynorphin in lamina I of the dorsal horn which was increased by electrical stimulation of small diameter afferents of the ipsilateral tibial nerve with frequencies of 50–100 Hz. Evoked release was abolished by spinal transection suggesting that such release came from intrinsic spinal neurones activated by supraspinally derived fibres. Possibly relevant to this proposal is the finding that adding corticotrophin-releasing factor to a perfusate of the isolated mouse spinal cord produced a release of ir-dynorphin (Song and Takemori, 1992).

The experiments described in the present account have used microprobes bearing antibodies to porcine dynorphin $A_{(1-8)}$ (Peninsula Laboratories). This was chosen since much of the original work from the laboratory of Iadarola et al. (1988) studied this peptide. Data from the manufacturer indicated negligible cross-reactivity with porcine pro-

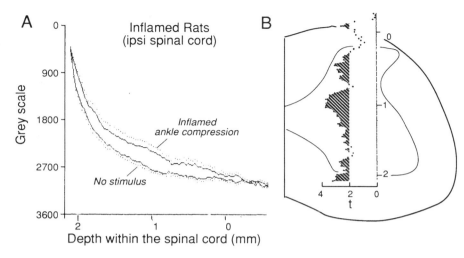

Fig. 3. Spinal release of ir-dynorphin $A_{(1-8)}$ by lateral compression of the inflamed ankle region. (A) The mean image analysis of microprobes present in the spinal cord in the absence of any peripheral stimulus (no stimulation) is compared with that of microprobes present in the side of the spinal cord ipsilateral to an inflamed ankle subject to intermittent lateral compression. (B) The differences between the two groups of microprobes in (A) are plotted with respect to depth in the spinal cord. The hatched areas indicate where these differences are significant at the $P < 0.05$ level. Reproduced with permission from Riley et al. (1996)

dynorphin 209–240, dynorphin $A_{(1-17)}$, dynorphin $A_{(1-13)}$, dynorphin B, α-neoendorphin, β-endorphin and both leu/met enkephalins. With microprobes bearing anti-porcine dynorphin $A_{(1-8)}$ antibodies preincubation in porcine dynorphin $A_{(1-8)}$ (10^{-7}M) for 30 min at 37°C resulted in greater than 55% suppression of ^{125}I-porcine dynorphin $A_{(1-8)}$. As with galanin, experiments were performed on normal rats and those with inflammation produced by injecting Freund's adjuvant around one ankle.

In normal animals not subjected to any peripheral stimulus, microprobes detected ir-dynorphin $A_{(1-8)}$ in the dorsal horn of the spinal cord. With the development of inflammation of the ankle of 3 or 4 days duration, however, these basal levels of ir-dynorphin$_{(1-8)}$ were detected in both the dorsal and the ventral horn of the both sides of the spinal cord. Lateral compression of the ankles of the normal rats did not release ir-dynorphin $A_{(1-8)}$ during the period of stimulation, but unexpectedly this peptide was detected in increased amounts in the ventral horn following the stimulus. By contrast, compression of inflamed ankles markedly elevated levels of ir-dynorphin $A_{(1-8)}$ during the period of stimulus application at three major sites in the ipsilateral spinal grey matter (the mid dorsal horn, the deep dorsal/upper ventral horn and the lower ventral horn) as shown in Fig. 3. These levels persisted for at least one hour after the period of stimulation.

The finding of both basal and evoked release of ir-dynorphin $A_{(1-8)}$ in several areas of the spinal cord suggests an involvement of dynorphins in several aspects of spinal cord processing, with release in the superficial horn potentially reflecting a role in antinociception, whilst that in the deep dorsal horn/ventral horn may indicate an influence of motor behaviour. Although the similar basal presence of ir-dynorphin $A_{(1-8)}$ in the dorsal horn of normal and inflamed rats could be the result of surgery, an enhanced afferent input was almost certainly responsible for the increased levels in the ventral horn of inflamed rats, particularly as active manipulations of the inflamed region increased these levels still further. The delayed increase in the ventral horn levels of ir-dynorphin $A_{(1-8)}$ following ankle compression may be indicative of a rapid synthesis of prodynorphin, since there is evidence that expression of the prodynorphin gene requires prior expression of the immediate early gene product c-fos (Hunter et al., 1995) and the latter is known to be induced rapidly following peripheral noxious stimuli (Hunt et al., 1987). Similarly, whilst the prolonged elevation of ir-dynorphin $A_{(1-8)}$ for at least one hour after compression of the inflamed ankle could represent persistence following release during the stimulus, there could also be a contribution from de novo synthesis. Since these experiments did not employ rapidly repeated noxious stimuli, it is unknown if dynorphin apparently depleted in the manner of spinal galanin. Nevertheless, it is clear that when inflamed peripheral tissues are subject to mechanical manipulation, dynorphins are released centrally and it is likely that this will influence the central processing of nociceptive information.

Although opiates are normally thought of as inhibitory to such processing (Duggan and North, 1984), and by inference to the perception of pain, functional studies have produced disparate proposals as to the roles of dynorphins in the spinal cord when inflammation develops peripherally. Several investigators have found that the receptive fields of many neurons of the dorsal horn of the rat and cat expand as inflammation develops in the periphery (Simone et al., 1991; Grubb et al., 1993; Hoheisel et al., 1993; Urban et al., 1993). Hylden et al. (1991) observed that topically applied dynorphin $A_{(1-8)}$ and dynorphin $A_{(1-17)}$ produced a similar effect with 5 of 15 dorsal horn neurones of normal rats. This receptive field expansion, however, was not reversed by systemic naloxone. As cited previously, topical dynorphin has been found to produce spinal cord necrosis and this is also not preventable by opioid antagonists. Stiller et al. (1993) administered the kappa opioid antagonist nor-binaltorphamine microiontophoretically near single spinal neurones of rats with peripheral inflammation and the predominant effect was an expansion of receptive fields. This implies the tonic presence of a ligand active at kappa receptors acting to reduce neuronal excitability and thereby to reduce responses to peripheral stimuli. These discrepancies among investigators have not been resolved.

There is now considerable evidence to indicate the development of a highly complex central nervous system response when inflammation develops peripherally. Fifteen years ago noxious stimuli and tissue damage were commonly proposed as invoking analgesia probably through opioid peptides. A decade ago evidence began to accumulate that the antithesis, hyperalgesia and allodynia could follow peripheral insults and this generated attempts to prevent this state by stopping nociceptive impulses ever reaching the central nervous system - preemptive analgesia. Whether these two apparently opposing processes develop together or at differing rates and times as inflammation follows injury is unknown. Release studies have shown that mechanical stimuli applied to inflamed joints indeed release tachykinins (probably hyperalgesia producing) and galanin and dynorphin (probably analgesic) in the spinal cord. This is a very complex picture but it does suggest that if the endogenous response involves a host of compounds then therapy could equally logically mimic or block the action of more than one substance.

References

Advokat, C. (1988) The role of descending inhibition in morphine-induced analgesia. *Trends Pharmacol. Sci.*, 9: 330–334.

Bakshi, R., Ni, R.-X. and Faden, A.I. (1992) *N*-Methyl-D-aspartate (NMDA) and opioid receptors mediate dynorphin-induced spinal cord injury: behavioral and histological studies. *Brain Res.*, 580: 255–264.

Ch'ng, J.L.C., Christofides, P., Anand, S.J., Gibson, Allen, H.C., Su, K., Tatemoto, J.F.B., Morrison, J.M., Polak, J.M. and Blomm, S.R. (1985) Distribution of galanin immunoreactivity in the central nervous system and responses of galanin-containing neuronal pathways to injury. *Neuroscience*, 16: 343–354.

Clarke, R.W., Ford, T.W. and Taylor, J.S. (1988) Adrenergic and opioidergic modulation of a spinal reflex in the decerebrated rabbit. *J. Physiol.*, 404: 407–417.

Cridland, R.A. and Henry, J.L. (1988) Effects of intrathecal administration of neuropeptides on a spinal nociceptive reflex in the rat: VIP, Galanin, CGRP, TRH, Somatostatin and Angiotensin II. *Neuropeptides*, 11: 23–32.

Day, R., Trujillo, K.A. and Akil, H. (1993) Prodynorphin biosynthesis and posttranslational processing. In: A. Herz (Ed.), *Opioids*, Vol. 1, Springer Verlag, Berlin, pp. 449–470.

Draisci, G. and Iadarola, M.J. (1989) Temporal analysis of increases in c-fos, preprodynorphin and preproenkephalin mRNAs in rat spinal cord. *Mol. Brain Res.*, 6: 31–37.

Duggan, A.W. (1991) Antibody Microprobes. In: J. Stamford (Ed.), *Monitoring Neuronal Activity: A Practical Approach*, Oxford University Press, Oxford, UK, pp. 181–202.

Duggan, A.W. and North, R.A. (1984) Electrophysiology of opioids. *Pharmacol. Rev.*, 35: 219–281.

Duggan, A.W. and Hendry, I.A. (1986) Laminar localization of the sites of release of immunoreactive substance P in the dorsal horn with antibody coated microelectrodes. *Neurosci. Lett.*, 68: 134–140.

Duggan, A.W., Hall, J.G. and Headley, P.M. (1976) Morphine, enkephalin and the substantia gelatinosa. *Nature*, 264: 456–458.

Duggan, A.W., Hendry, I.A., Green, J.L., Morton, C.R. and Zhao, Z.Q. (1988) Cutaneous stimuli releasing immunoreactive substance P in the dorsal horn of the cat. *Brain Res.*, 451: 261–273.

Duggan, A.W., Hope, P.J., Jarrott, B., Schaible, H. and Fleetwood-Walker, S.M. (1990) Release, spread and persistence of immunoreactive neurokinin A in the dorsal horn of the cat following noxious cutaneous stimulation. Studies with antibody microprobes. *Neuroscience*, 35: 195–202.

Duggan, A.W., Hope, P.J. and Lang, C.W. (1991) Microinjection of neuropeptide Y into the superficial dorsal horn reduces stimulus evoked release of immunoreactive substance P in the anaesthetized cat. *Neuroscience*, 44: 733–740.

Faden, A.I. and Jacobs, T.P. (1983) Dynorphin induces partially reversible paraplegia in the rat. *Eur. J Pharmacol.*, 91: 321–324.

Fisone, G., Wu, C.F., Consolo, S., Nordstrom, O., Brynne, N., Bartfai, T., Melander, T. and Hokfelt, T. (1987) Galanin inhibits acetylcholine release in the ventral hippocampus of the rat: Histochemical, autoradiographic, in vivo, and in vitro studies. *Proc. Natl. Acad. Sci. USA*, 84: 7339–7343.

Fleetwood-Walker, S.M., Hope, P.J., Mitchell, R., El-Yassir, N. and Molony, V. (1988) The influence of opioid receptor subtypes on the processing of nociceptive inputs in the spinal dorsal horn of the cat. *Brain Res.*, 451: 213–226.

Grubb, B.D., Stiller, R.U. and Schaible, H.-G. (1993) Dynamic changes in the receptive field properties of spinal cord neurons with ankle input in rats with chronic unilateral inflammation in the ankle region. *Exp. Brain Res.*, 92: 441–452.

Hanesch, U., Pfrommer, U., Grubb, B.D. and Schaible, H.-G. (1993) Acute and chronic phases of unilateral inflammation in rat's ankle are associated with an increase in the proportion of calcitonin gene related peptide-immunoreactive dorsal root ganglion cells. *Eur. J. Neurosci.*, 5: 154–161.

Hendry, I.A., Morton, C.R. and Duggan, A.W. (1988) Analysis of antibody microprobe autoradiographs by computerized image processing. *J. Neurosci. Methods*, 23: 249–256.

Herdegen, T., Fiallos-Estrada, C.E., Bravo, R. and Zimmermann, M. (1993) Colocalisation and covariation of c-jun transcription factor with galanin in primary afferent neurons and with CGRP in spinal motoneurons following transection of rat sciatic nerve. *Mol. Brain Res.*, 17: 147–154.

Hoheisel, U., Mense, S., Simons, D.G. and Yu, X.-M. (1993) Appearance of new receptive fields in rat dorsal horn neurons following noxious stimulation of skeletal muscle: a model for referral of muscle pain. *Neurosci. Lett.*, 153: 9–12.

Hope, P.J., Jarrott, B., Schaible, H., Clarke, R.W. and Duggan, A.W. (1990) Release and spread of immunoreactive neurokinin A in the cat spinal cord in a model of acute arthritis. *Brain Res.*, 533: 292–299.

Hope, P.J., Lang, C.W., Grubb, B.D. and Duggan, A.W. (1994) Release of immunoreactive galanin in the spinal cord of rats with ankle inflammation: studies with antibody microprobes. *Neuroscience*, 60: 801–807.

Hunt, S.P., Pini, A. and Evan, G. (1987) Induction of c-fos like protein in spinal cord neurons following sensory stimulation. *Nature*, 328: 632–634.

Hunter, J.C., Woodburn, V.L., Durieux, C., Pettersson, E.K.E., Poat, J.A. and Hughes, J. (1995) C-fos antisense oligodeoxynucleotide increases formalin-induced nociception and regulates preprodynorphin expression. *Neuroscience*, 65: 485–492.

Hutchison, W.D., Morton, C.R. and Terenius, L. (1990) Dynorphin A: in vivo release in the spinal cord of the cat. *Brain Res.*, 532: 299–306.

Hylden, J.L.K., Nahin, R.L., Traub, R.J. and Dubner, R. (1991) Effects of spinal kappa-opioid receptor agonists on the responsiveness of nociceptive superficial dorsal horn neurons. *Pain*, 44: 187–193.

Iadarola, M.J., Brady, L.S., Draisci, G. and Dubner, R. (1988) Enhancement of dynorphin gene expression in spinal cord following experimental inflammation: stimulus specificity, behavioural parameters and opioid receptor binding. *Pain*, 35: 313–326.

Kashiba, H., Noguchi, K., Ueda, Y. and Senba, E. (1994) Neuropeptide Y and galanin are coexpressed in rat large type A sensory neurons after peripheral transection. *Peptides*, 15: 411–416.

Klein, C.M., Westlund, K.N. and Coggeshall, R.E. (1990) Percentages of dorsal root axons immunoreactive for galanin are higher than those immunoreactive for calcitonin gene-related peptide. *Brain Res.*, 519: 97–101.

Klein, C.M., Coggeshall, R.E., Carlton, S.M. and Sorkin, L.S. (1992) The effects of A- and C-fiber stimulation on patterns of neuropeptide immunostaining in the rat superficial dorsal horn. *Brain Res.*, 580: 121–128.

Kolhekar, R., Meller, S.T. and Gebhart, G.F. (1993) Characterization of the role of spinal N-methyl-D-aspartate receptors in thermal nociception in the rat. *Neuroscience*, 57: 385–395.

Kuraishi, Y., Kawamura, M., Yamaguchi, T., Houtani, T., Kawabata, S., Futaki, S., Fujii, N. and Satoh, M. (1991) Intrathecal injections of galanin and its antiserum affect nociceptive response of rat to mechanical, but not thermal, stimuli. *Pain*, 44: 321–324.

Land, T., Langel, Ü. and Bartfai, T. (1991) Hypothalamic degradation of galanin(1–29) and galanin(1–16): Identification and characterization of the peptidolytic products. *Brain Res.*, 558: 245–250.

Le Bars, D., Guilbaud, G., Chitour, D. and Besson, J.M.

(1980) Does systemic morphine increase descending inhibitory controls of dorsal horn neurons involved in nociception? *Brain Res.*, 202: 223–228.

Long, J.B., Rigamonti, D.D., Oleshansky, M.A., Wingfield, C.P. and Martinez-Arizala, A. (1994) Dynorphin A-induced rat spinal cord injury: evidence for excitatory amino acid involvement in a pharmacological model of ischemic spinal cord injury. *J. Pharmacol. Exp. Ther.*, 269: 358–366.

Martin, W.R., Eades, C.G., Thompson, J.A., Huppler, R.E. and Gilbert, P.E. (1976) The effects of morphine- and nalorphine-like drugs in the nondependent and morphine-dependent chronic spinal dog. *J. Pharmacol. Exp. Ther.*, 197: 517–532.

Melander, T.T., Hokfelt, T. and Rokaeus, A. (1986) Distribution of galanin-like immunoreactivity in the rat central nevous system. *J. Comp. Neurol.*, 248: 475–517.

Millan, M.J., Millan, M.H., Czlonkowski, A., Pilcher, C.W., Hollt, V., Colpaert, F.C. and Herz, A. (1986) Functional response of multiple opioid systems to chronic arthritic pain in the rat. *Ann. N. Y. Acad. Sci.*, 467: 182–193.

Morton, C.R. and Hutchison, W.D. (1989) Release of sensory neuropeptides in the spinal cord: studies with calcitonin gene-related peptide and galanin. *Neuroscience*, 31: 807–815.

Morton, C.R., Hutchison, W.D., Hendry, I.A. and Duggan, A.W. (1989) Somatostatin: evidence for a role in thermal nociception. *Brain Res.*, 488: 89–96.

Nahin, R.L., Hylden, J.L.K. and Humphrey, E. (1992) Demonstration of dynorphin A 1–8 immunoreactive axons contacting spinal cord projection neurons in a rat model of peripheral inflammation and hyperalgesia. *Pain*, 51: 135–143.

Noguchi, K., De León, M., Nahin, R.L., Senba, E. and Ruda, M.A. (1993) Quantification of axotomy-induced alteration of neuropeptide mRNAs in dorsal root ganglion neurons with special reference to neuropeptide Y mRNA and the effects of neonatal capsaicin treatment. *J. Neurosci. Res.*, 35: 54–66.

Nordstrom, O., Melander, T., Hokfelt, T., Bartfai, T. and Goldstein, M. (1987) Evidence for an inhibitory effect of the peptide galanin and dopamine release from the rat median eminence. *Neurosci. Lett.*, 73: 21–26.

North, R.A. (1993) Opioid actions on membrane ion channels. In: A. Herz (Ed.), *Opioids*, Vol. 1, Springer Verlag, Berlin pp. 773–798.

Post, C., Alari, L. and Hokfelt, T. (1988) Intrathecal galanin increases the latency in the tail flick and hot plate tests in the mouse. *Acta Physiol. Scand.*, 132: 583–584.

Przewlocka, B., Lason, W. and Przewlocki, R. (1992) Time-dependent changes in the activity of opioid systems in the spinal cord of monoarthritic rats - a release and in situ hybridization study. *Neuroscience*, 46: 209–216.

Reimann, W. and Schneider, F. (1993) Galanin receptor activation attenuates norepinephrine release from rat spinal cord slices. *Life Sci.*, 52: PL251–PL254.

Riley, R.C., Zhao, Z.Q. and Duggan, A.W. (1996) Spinal release of immunoreactive dynorphin $A_{(1-8)}$ with the development of peripheral inflammation. *Brain Res.*, in press.

Schaible, H., Jarrott, B., Hope, P.J. and Duggan, A.W. (1990) Release of immunoreactive substance P in the spinal cord during development of acute arthritis in the knee joint of the cat: a study with antibody microprobes. *Brain Res.*, 529: 214–223.

Schaible, H.-G., Freudenberger, U., Neugebauer, V. and Stiller, R.U. (1994) Intraspinal release of immunoreactive calcitonin gene-related peptide during development of inflammation in the joint in vivo - a study with antibody microprobes in cat and rat. *Neuroscience*, 62: 1293–1305.

Simone, D.A., Sorkin, L.S., Oh, U., Chung, J.M., Owens, C., LaMotte, R.H. and Willis, W.D. (1991) Neurogenic hyperalgesia: central neural correlates in responses of spinothalamic tract neurons. *J. Neurophysiol.*, 66: 228–246.

Song, Z.H. and Takemori, A.E. (1992) Stimulation by corticotropin-releasing factor of the release of immunoreactive dynorphin A from mouse spinal cords in vitro. *Eur. J. Pharmacol.*, 222: 27–32.

Stiller, R.U., Grubb, B.D. and Schaible, H.-G. (1993) Neurophysiological evidence for increased kappa opioidergic control of spinal neurons in rats with unilateral inflammation at the ankle. *Eur. J. Neurosci.*, 5: 1520–1527.

Takano, Y. and Yaksh, T.L. (1992) Characterization of the pharmacology of intrathecally administered *Alpha*-2 agonists and antagonists in rats. *J. Pharmacol. Exp. Ther.*, 261: 764–772.

Tatemoto, K., Rokaeus, A., Jornvall, H., McDonald, T.J. and Mutt, V. (1983) Galanin-A novel biologically active peptide from porcine intestine. *FEBS Lett.*, 164: 124–128.

Tuchscherer, M.M. and Seybold, V.S. (1989) A qauntitative study of the coexistence of peptides in the varicosities within the superficial laminae of the dorsal horn of the rat spinal cord. *J. Neurosci.*, 9: 195–205.

Urban, L., Perkins, M.N., Campbell, E. and Dray, A. (1993) Activity of deep dorsal horn neurons in the anaesthetized rat during hyperalgesia of the hindpaw induced by ultraviolet irradiation. *Neuroscience*, 57: 167–172.

Van Gilst, W.H., Tio, R.A., Van Wijngaarden, J., De Graeff, P.A. and Wesseling, H. (1992) Effects of converting enzyme inhibitors on coronary flow and myocardial ischemia. *J. Cardiovasc. Pharmacol.*, 19(Suppl. 5): S134–S139.

Villar, M.J., Cortes, R., Theodorsson, E., Wiesenfeld-Hallin, Z., Schalling, M., Fahrenkrug, J., Emson, P.C. and Hokfelt, T. (1990) Neuropeptide expression in rat dorsal root ganglion cells and spinal cord after peripheral nerve injury with special reference to galanin. *Neuroscience*, 33: 587–604.

Villar, M.J., Wiesenfeld-Hallin, Z., Xu, X.-J., Theodorsson, E., Emson, P.C. and Hokfelt, T. (1991) Further studies on galanin-, substance P-, and CGRP-like immunoreactivities in primary sensory neurons and spinal cord: effects of dorsal rhizotomies and sciatic nerve lesions. *Exp. Neurol.*, 112: 29–39.

Wakisaka, S., Kajander, K.C. and Bennett, G.J. (1992) Effects of peripheral nerve injuries and tissue inflammation on the levels of neuropeptide Y-like immunoreactivity in rat primary afferent neurons. *Brain Res.*, 598: 349–352.

Walker, M.W., Ewald, D.A., Perney, T.M. and Miller, R.J. (1988) Neuropeptide Y modulates neurotransmitter release and Ca++ currents in rat sensory neurons. *J. Neurosci.*, 8: 2438–2446.

Weihe, E., Millan, M.J., Hollt, V., Nohr, D. and Herz, A. (1989) Induction of the gene encoding pro-dynorphin by experimentally induced arthritis enhances staining for dynorphin in the spinal cord of rats. *Neuroscience*, 31: 77–95.

Wiesenfeld-Hallin, Z., Villar, M.J. and Hokfelt, T. (1988) Intrathecal galanin at low doses increases spinal reflex excitability in rat more to thermal than mechanical stimuli. *Exp. Brain Res.*, 71: 663–666.

Wiesenfeld-Hallin, Z., Villar, M.J. and Hokfelt, T. (1989) The effects of intrathecal galanin and C-fibre stimulation as the flexor reflex in the rat. *Brain Res.*, 486: 205–213.

Wiesenfeld-Hallin, Z., Xu, X.J., Hughes, J., Horwell, D.C. and Hokfelt, T. (1990) A selective antagonist of cholecystokinin type B receptor, enhances the analgesic effect of morphine and synergistically interacts with intrathecal galanin to depress spinal nociceptive reflexes. *Proc. Natl. Acad. Sci. USA*, 87: 7105–7109.

Xu, X., Wiesenfeld-Hallin, Z., Villar, M.J., Fahrenkrug, J. and Hokfelt, T. (1990) On the role of galanin, substance P and other neuropeptides in primary sensory neurons of the rat: studies on spinal reflex excitability and peripheral axotomy. *J. Neurosci.*, 2: 733–743.

Section III

Neurotransmitters and Plasticity

CHAPTER 12

Cooperative mechanisms of neurotransmitter action in central nervous sensitization

W.D. Willis[1], K.A. Sluka[2], H. Rees[3] and K.N. Westlund[1]

[1]Marine Biomedical Institute, Medical Research Building, Galveston, TX 77555-1069, USA, [2]2600 Steindler Building, Iowa City, IA 52242, USA and [3]67 Nant Talwg Way, Barry, South Glamorgan, CF62 6LZ, UK

Introduction

During the past 5 years, our group has investigated the mechanisms of central sensitization of nociceptive pathways using 2 animal models: intradermal injection of capsaicin (Dougherty and Willis, 1992; Dougherty et al., 1992a; Simone et al., 1991; reviewed in Willis, 1994) and induction of acute arthritis (Dougherty et al., 1992b; Sluka et al., 1992; Sluka and Westlund, 1992, 1993a–d; Sorkin et al., 1992; Westlund et al., 1992; see review by Sluka et al., 1995b). We have also explored the possibility of provoking sensitization more directly by administering drugs into the dorsal horn, either by iontophoresis or by microdialysis (Dougherty and Willis, 1991; Dougherty et al., 1993, 1995; Palecek et al., 1993a,b). Agents administered have included neurotransmitter agonists and antagonists, as well as drugs affecting second messenger systems. Central sensitization is characterized in our experiments by the following: (1) changes in the responses of primate spinothalamic tract (STT) neurons to stimulation of the skin, including increased responses to innocuous and marginally noxious stimuli and an enlarged receptive field; (2) behavioral responses to stimulation of the skin in rats indicative of allodynia and/or hyperalgesia; (3) morphological alterations in the dorsal horn in rats and monkeys; (4) and release of neurotransmitters into the dorsal horn, as determined by microdialysis and high performance liquid chromatography (HPLC) in rats and monkeys.

Intradermal injection of capsaicin

When human subjects are given an intradermal injection of capsaicin (100 mg in 10 ml), severe pain occurs immediately, and primary hyperalgesia develops near the injection site. Secondary mechanical hyperalgesia and allodynia also develop over an extended area surrounding the site of primary hyperalgesia (Simone et al., 1989; LaMotte et al., 1991, 1992). A zone of analgesia just at the injection site is presumed to result from inactivation of nociceptive terminals where the capsaicin concentration is high (LaMotte et al., 1992). There is an area of primary heat hyperalgesia around the injection site in which the threshold for heat pain is reduced from 45 to 32°C. However, there is no area of secondary heat hyperalgesia. The pain caused by capsaicin injection lasts about 15–20 min. The time course of the secondary mechanical hyperalgesia and allodynia is much longer: the peak effect occurs within about 15–20 min, and these sensory changes last at least 1–2 h (Simone et al., 1989; LaMotte et al., 1991). A flare response also occurs, extending nearly to the limits reached by the secondary mechanical hyperalgesia.

In human subjects, microneurographic recordings from primary afferent C fibers supplying the area of secondary mechanical hyperalgesia and allodynia do not show enhanced responses following a capsaicin injection. A reduced excitability has been found for C nociceptors whose receptive fields were in the analgesic zone; an increased excitability was seen in several units after topical application of capsaicin (LaMotte et al., 1992). An electrical stimulus applied through the microneurography electrode that initially caused a purely tactile sensation projected to an area away from the injection site became painful when the area of secondary mechanical hyperalgesia and allodynia expanded to include the projected receptive field (Torebjörk et al., 1992). The pain disappeared and only touch was felt when the area of secondary mechanical hyperalgesia and allodynia regressed.

In more recent human studies by Kilo et al. (1994), different types of stimuli were used to assess hyperalgesia and allodynia following topical capsaicin, and nerve blocks were used to decipher which fiber types mediated a particular response. Hypersensitivity to von Frey filaments (>400 mN bending force) applied to the receptive field, presumably reflecting mechanical hyperalgesia, was shown to be mediated by C-fiber afferents. Painful responses to innocuous brushing of the skin were mediated by low threshold mechano-sensitive Aβ fibers. Koltzenburg et al. (1994) further suggest that continuous C-fiber input is necessary to maintain Aβ fiber-mediated mechanical hyperalgesia, since stimulation of C-fibers results in both central sensitization and brush-evoked pain.

These studies suggest that activation of several peripheral fiber types is necessary for the perception of hyperalgesia and allodynia. C-fiber activation has a pivotal role both in establishing central sensitization and in allowing previously innocuous stimuli to produce pain. Activation of Aβ fibers results in allodynia when sensitization is present due to the effects of C nociceptor input to the spinal cord.

Parallel results have been obtained in animal experiments. Baumann et al. (1991) found in primates that only a few types of nociceptive fibers were activated sufficiently strongly by capsaicin injections to account for the pain that is produced by this procedure. Nociceptors supplying the skin surrounding a capsaicin injection site did not become sensitized to mechanical or heat stimulation after injection of capsaicin. Nociceptors innervating the injection site had reduced responses.

Our group has used the responses of primate spinothalamic tract (STT) cells as a model system to predict human sensory changes that would be produced by experimental manipulations of spinal cord function that can only be done in animal subjects. Intradermal injection of capsaicin (100 μg in 10 μl) into the skin in the receptive field of STT cells causes robust discharges that peak within seconds and then taper off over 15–20 min (Simone et al., 1991). The responsiveness of the neurons to mechanical and thermal stimulation of the skin changes in a characteristic manner. There may be an increase or a decrease in responses to mechanical stimuli applied at the injection site, and there is a lowered threshold for noxious heat pulses applied to the skin immediately around the injection site. However, the threshold for noxious heat does not change in the surrounding skin. Responses to weak tactile stimuli and to poking the skin with a von Frey filament that is painful on normal skin become considerably increased (Simone et al., 1991). Similar results are seen when a large dose of capsaicin (100 μl of a 3% solution of capsaicin) is injected intradermally and the innocuous mechanical stimuli are brushing the skin (BRUSH) or application of a large arterial clip to a fold of skin (PRESS) (Fig. 1; Dougherty and Willis, 1992). In such experiments, the responses of STT cells to noxious stimuli, such as application of a small arterial clip (PINCH) to the skin, squeezing the skin or application of noxious heat, are often unchanged. The time course of the enhanced responses is similar to that of allodynia and secondary hyperalgesia in human skin, 15 min for a peak effect and 1.5 h duration; furthermore, the receptive fields of individual STT cells increase in size (Figs. 1 and 2A; Dougherty and Willis, 1992).

In experiments in which GLU receptor agonists were released in the vicinity of STT cells by iontophoresis (Dougherty and Willis, 1992), we showed that following intradermal injection of capsaicin, STT cells became more responsive to one or more of the following excitatory amino acids (EAAs): glutamate (GLU), aspartate (ASP), N-methyl-D-aspartate (NMDA), quisqualic acid (QUIS), α-amino-3-hydroxy-5-methyl-isoxazoleproprionic

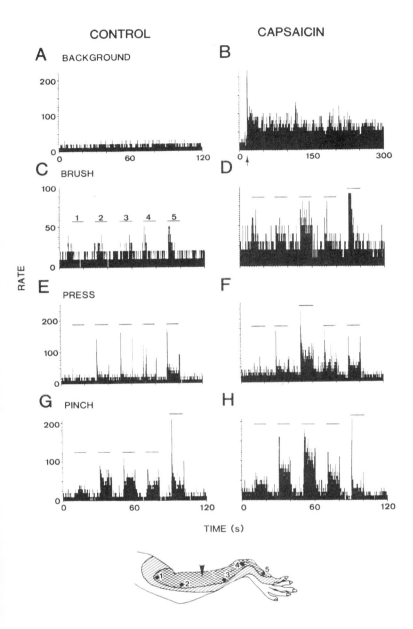

Fig. 1. Responses of a primate STT neuron to mechanical stimulation of the skin before and after in intradermal injection of capsaicin. (A) shows the background activity before capsaicin and (B) the response to the injection (time indicated by the arrow). (C), (E) and (G) are the responses to mechanical stimulation of the skin at the 5 sites indicated in the drawing at the bottom, and (D), (F) and (H) are the responses to the same stimuli after the injection. The enlargement of the receptive field is indicated by the singly hatched area on the drawing. (From Dougherty and Willis, 1992)

acid (AMPA), and kainic acid (KAIN). The time course of the increased sensitivity to EAAs (Fig. 2B) parallels that of the alterations in sensitivity to mechanical stimulation of the skin (Fig. 2A). These observations indicate that a capsaicin injection causes a long-lasting increase in the excitability of STT neurons to locally applied EAAs, but it does not rule out the possibility that the enhanced

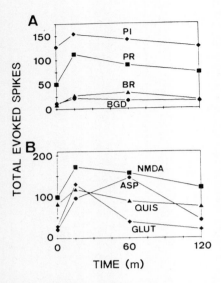

Fig. 2. (A) shows the time course of the increased responsiveness of a primate STT neuron to mechanical stimulation of the skin (BGD, background; BR, brush; PR, pressure; PI, pinch) and (B) to iontophoretically released EAAs (GLUT, glutamate; QUIS, quisqualate; ASP, aspartate; NMDA, N-methyl-D-aspartate) following an intradermal injection of capsaicin. (From Dougherty and Willis, 1992)

excitability of these neurons depends upon a continued input from the periphery. However, recordings from peripheral nerve fibers have so far not revealed changes in the activity or excitability of sensory receptors supplying the skin in the area of secondary hyperalgesia (Baumann et al., 1991; cf., LaMotte et al., 1992).

Evidence that sensitization of central neurons can occur for periods of up to a few h without maintained peripheral input comes from both in vivo and in vitro experiments. Our group has found that local coadministration of an EAA, such as NMDA or QUIS, with substance P (SP) can result in the enhancement of the responses of STT cells to later applications of the EAA or to mechanical stimulation of the skin (Dougherty and Willis, 1991; Dougherty et al., 1993). Repeated applications of an EAA, such as NMDA, do not cause sensitization; instead, the responses tend to habituate (Dougherty and Willis, 1991). Nor does SP alone cause sensitization. The sensitization of STT cells produced by coadministration of NMDA or QUIS and SP may last hours (Fig. 3). Randic's group has found that neurons isolated from the dorsal horn and placed in an in vitro bath and then exposed to NMDA and SP show enhanced responses that persist for up to 50 min (Randic et al., 1990). Thus, dorsal horn neurons can be sensitized by exposure to the combination of an EAA and SP without any input from a peripheral nerve and without processing by dorsal horn circuits. However, this does not imply that sensitization is unaffected by an additional peripheral input.

These observations suggested to us that the sensitizing effect of an intradermal injection of capsaicin may depend on the co-release of EAAs and SP in the dorsal horn. A similar co-release of EAAs and SP may be responsible for the hyperalgesia seen in a number of animal models (see review by Urban et al., 1994). Therefore, we have investigated the effects of administering substances that antagonize EAA receptors and neurokinin receptors on the sensitization of STT cells by capsaicin injection (Dougherty et al., 1992a, 1994). As a control experiment, it was important to demonstrate that a second injection of capsaicin, given after the STT cell had recovered from the effects of the first injection, would produce a comparable sensitization. Since this was found to be the case (Dougherty et al., 1994), the design of the experiments was as follows. After testing the responses of an STT cell to mechanical stimulation of the skin, a first injection of capsaicin was given and the sensitized responses to the mechanical stimuli observed. A period of 1.5–2 h was allowed to elapse and then the responses were tested again to show that they had returned to the control level. Following this, an antagonist drug was administered by microdialysis, and the effects of the antagonist on the responses to mechanical stimuli were determined. Finally, a second dose of capsaicin was given (the injection was in a separate region of skin than the first injection) and the responses to mechanical stimulation again tested.

The antagonists that we have used in these experiments and the receptors that they block have included: CNQX, non-NMDA EAA receptors; AP7, NMDA EAA receptors; CP96345 and GR82334, NK1 receptors; and MEN10376 and GR94800, NK2 receptors. The results of these

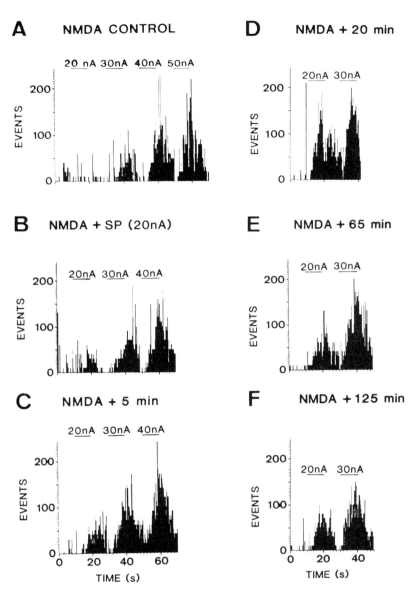

Fig. 3. Prolonged increase in the responsiveness of a primate STT neuron to iontophoretically released NMDA following co-application of NMDA and SP by iontophoresis. (A) shows the control responses to graded iontophoretic currents. In (B) are shown the responses to NMDA pulses during the administration of SP. (C–F) show the responses to NMDA pulses at the times indicated following termination of the SP current. (From Dougherty and Willis, 1991)

experiments showed that both EAAs and SP are likely to be involved in the sensitization of STT cells that follows intradermal injection of capsaicin.

When non-NMDA receptors are blocked by CNQX, the responses of STT cells to BRUSH, PRESS, and PINCH are reduced to a minimum, as are the responses to capsaicin itself (Dougherty et al., 1992a). Although sensitization is prevented by CNQX, this is presumably not by a specific effect on sensitization but rather due to interference with synaptic transmission in the afferent pathways leading to the STT cells. On the other hand, AP7, which had no effect on the responses of STT cells

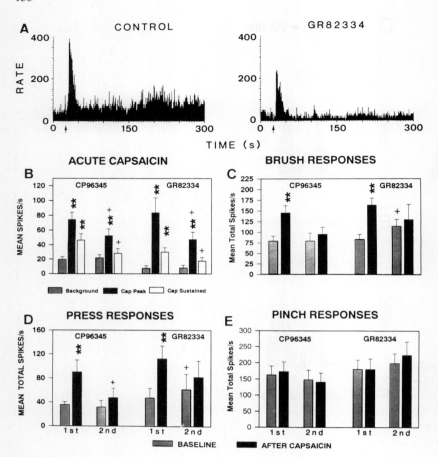

Fig. 4. Effects of NK1 receptor antagonists on the sensitization of primate STT cells. (A) shows the responses of an STT cell to intradermal capsaicin injections before and after administration of the NK1 antagonist, GR82334, by microdialysis. The bar graphs in B-E show the mean responses of populations of STT cells (±S.E.) before and after capsaicin and before and after treatment with an NK1 antagonist. Two NK1 antagonists were used, CP96345 and GR82334. (B) shows the response to capsaicin and the sustained discharge that follows intradermal injection. The NK1 antagonists reduced the responses, especially the sustained component of the discharge, to capsaicin. (C) and (D) show increased responses to brush and press stimuli following the first dose of capsaicin. However, during administration of NK1 antagonists, a second dose of capsaicin failed to increase the responses; i.e., sensitization was blocked. The pinch responses were unaffected by capsaicin injection. (From Dougherty et al., 1994)

to BRUSH or PRESS before capsaicin, prevents the enhancement of these responses following capsaicin (Dougherty et al., 1992a).

Similarly, the NK1 antagonists, CP96345 and GR82334, had no effect on the responses of STT cell to BRUSH, PRESS or PINCH stimuli, but prevented sensitization of these responses by capsaicin (Fig. 4; Dougherty et al., 1994; cf., Radhakrishnan and Henry, 1995). As a control, we also examined the effect of CP96344, which has a similar action to that of CP96345 on Ca^{2+} channels but none on NK1 receptors. CP96344 did not affect sensitization. On the other hand, the NK2 antagonists, MEN10376 and GR94800, did not block sensitization, but instead appeared to enhance the responses of STT cells to BRUSH and PRESS stimuli before injection of capsaicin. We speculated that this may have resulted from a mixed agonist effect of these peptide antagonists (Dougherty et al., 1994).

These studies strongly suggest that sensitization of STT cells by intradermal injection of capsaicin depends on the release of EAAs and SP from the terminals of nociceptors in the dorsal horn. Others have shown that capsaicin causes the release of EAAs and SP in the dorsal horn (Gamse et al., 1979; Sorkin and McAdoo, 1993; Ueda et al., 1994).

Acute experimental arthritis

Acute joint inflammation is manifested in many arthritic conditions, including rheumatoid arthritis and osteoarthritis. Arthritis is most commonly accompanied by pain, which limits the range of motion and limb function. Animal models of inflammation are designed to mimic these human arthritic conditions. We have employed a model of acute joint inflammation, induced by intraarticular injection of kaolin and carrageenan, to study the role of dorsal horn neurons in the integration and transmission of arthritic pain in monkeys (Dougherty et al., 1992b; Sluka et al., 1992; Sorkin et al., 1992; Westlund et al., 1992). A similar model was developed in rats to explore the behavioral changes, as well as dorsal horn changes in neurotransmitters produced by acute arthritis (Sluka and Westlund, 1992, 1993a–d).

Several neurotransmitters and neuromodulators have been shown to be involved in the transmission of nociceptive information related to acute inflammation. For example, there is evidence that the EAAs, GLU and ASP, are released during the development of acute inflammation and mediate the accompanying heat hyperalgesia through both non-NMDA and NMDA EAA receptors (Sluka and Westlund, 1992, 1993d; Sorkin et al., 1992). SP and neurokinin A (NKA) and their receptors, the neurokinin 1 (NK1) and neurokinin 2 (NK2) receptors, are also involved in the integration of arthritic pain (Neugebauer et al., 1994b; Sluka and Westlund, 1993b,d; Sluka et al., 1995a).

After induction of acute inflammation, the primary afferent fibers become sensitized to mechanical stimulation applied to the peripheral receptive field and to joint movement (Schaible and Schmidt, 1985). In addition there is activation of previously silent neurons (Schaible and Schmidt, 1985). All joint afferent fiber types (Groups II, III and IV) become sensitized and develop an increase in background discharges and increased responses to innocuous and noxious joint movement. This increased activity in primary afferents results in central sensitization of dorsal horn neurons (Dougherty et al., 1992b; Neugebauer and Schaible, 1990; Schaible et al., 1987). Pretreatment with capsaicin to eliminate Group IV (unmyelinated) fibers decreases the severity of the acute inflammatory response (Colpaert et al., 1983; Lam and Ferrell, 1989).

Spinothalamic cells and other dorsal horn neurons become sensitized to mechanical stimuli following induction of acute joint inflammation (Fig. 5; Schaible et al., 1987, 1991; Neugebauer et al., 1990; Dougherty et al., 1992b). Additionally, primate STT cells are more responsive to iontophoretically applied GLU and QUIS, a non-NMDA receptor agonist (Dougherty et al., 1992b). Schaible et al. (1990) demonstrated that rat dorsal horn neurons became more sensitive to iontophoretically applied NMDA, although Dougherty et al. (1992b) found a decrease in NMDA responses in primate STT cells. The differences between the two studies may be related to either the time following induction of inflammation or the type of cell tested. Dougherty et al. (1992b) tested the responses of identified spinothalamic cells within the first 3 h after induction of inflammation, whereas Schaible et al. (1991) tested the responses of unidentified dorsal horn neurons 4–8 h after development of inflammation. The sensitization of dorsal horn neurons to mechanical stimuli that occurs in rats with acutely inflamed knee joints is reversed by intravenous administration of either a non-NMDA (CNQX) or an NMDA (ketamine, AP5) receptor antagonist (Schaible et al., 1991; Neugebauer et al., 1993). Recently, we demonstrated that microdialysis administration of a NK1 receptor antagonist, CP99,994, can also reverse the sensitization of STT cells to brushing the peripheral receptive field and flexion of the knee joint (Rees et al., 1995a). Therefore, blockade of any of these three receptors can fully reverse the sensitization of dorsal horn neurons, including STT cells, to peripheral stimuli, indicating that multiple receptor activation is necessary for the sensitization of dorsal horn neurons.

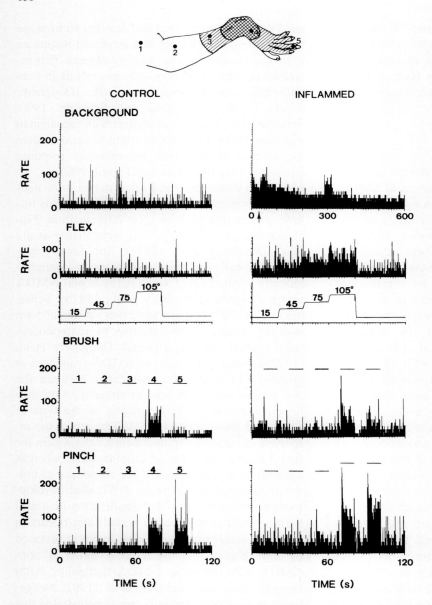

Fig. 5. Background activity and responses to flexion (FLEX), brushing (BRUSH) and pinching (PINCH) of a wide dynamic range spinothalamic tract cell before (left column) and 3 h after (right column) induction of arthritic inflammation. Mechanical stimuli were given at the points labeled on the leg. (From Dougherty et al., 1992)

Behaviorally, rats with inflamed knee joints guard the limb and are hyperalgesic to heat applied to the paw. The heat hyperalgesia, tested by applying a radiant heat source to the paw (see Hargreaves et al., 1988), becomes maximal at 4 h and lasts through 24 h (Sluka and Westlund, 1993b). The heat hyperalgesia can be prevented or reduced by pretreatment or posttreatment with the non-NMDA EAA receptor antagonist, CNQX, or the NMDA antagonist, AP7 (Sluka and Westlund, 1993d; Sluka et al., 1994a,b). On the other hand, pain-related behaviors, such as guarding of the limb, are reversed only by CNQX administered by microdialysis before or after induction of arthritis; AP7 had no effect on pain-related behaviors (Sluka and Westlund, 1993c; Sluka et al., 1994b). Neurokinin receptors in the spinal cord are also involved in the induction and maintenance of heat hyperalgesia (Sluka et al., 1995a). The NK2 receptor antagonist, SR48968, delivered by microdialysis prior to induction of inflammation can prevent the development of the heat hyperalgesia associated with acute inflammation; posttreatment with the NK2 antagonist had no effect on the hyperalgesia (Sluka et al., 1995a). In contrast, posttreatment with the NK1 receptor antagonist, CP99,994-1, by spinal cord microdialysis reversed the arthritis-induced heat hyperalgesia in a dose-dependent manner while pretreatment had no effect (Sluka et al., 1995a). Therefore, several receptors are involved in the development and maintenance of heat hyperalgesia induced by acute joint inflammation. including non-NMDA EAA, NMDA EAA, NK1 and NK2 receptors.

The release of neuropeptides has been monitored by introducing antibody microprobes into the spinal cord before and during acute joint inflammation (Schaible et al., 1990, 1992). The dorsal horn content of neurotransmitters and neuropeptides can also be monitored with computer quantification of immunohistochemically stained tissue (Sluka et al., 1992; Sluka and Westlund, 1993b). The antibody microprobe technique has been used to demonstrate the release of the neuropeptides, SP and NKA, into the dorsal horn during the development of acute inflammation and during flexion of the knee joint (Hope et al., 1990; Schaible et al., 1990). Immunohistochemical staining revealed that there is an initial decrease in staining density for SP in the superficial dorsal horn (Sluka et al., 1992; Sluka and Westlund, 1993b) that is followed by an increase in SP content through 1 week following induction of inflammation (Sluka and Westlund, 1993b). In addition, the content of GLU is increased through the first 24 h of arthritis (Sluka et al., 1992; Sluka and Westlund, 1993b) and is significantly correlated with the decrease in Paw Withdrawal Latency to radiant heat (Sluka and Westlund, 1993b).

In studies using microdialysis and HPLC to measure substances in the extracellular fluid of the dorsal horn, there is a transient release of the EAAs, GLU and ASP, and of the IAAs, GLY and SER, at the time of injection (Sluka and Westlund, 1992; Sorkin et al., 1992). This is followed by a second, more prolonged phase of amino acid release, including ASP, GLU, and GLY in the anesthetized primate and ASP and GLU in the awake rat (Sluka and Westlund. 1992; Sorkin et al., 1992) (Fig. 6). In the awake rat, microdialysis can be used to look at longer time periods and release during the behavioral testing for paw withdrawal latency to radiant heat. An increased release of the IAAs, GLY and SER, is also observed during the PWL test for radiant heat (Sluka et al., 1994b). Thus, a multiplicity of neurochemical changes occur in the dorsal horn in response to the peripheral activation including an increase in both the content and release of glutamate and substance P.

Spinal cord administration of the non-NMDA receptor antagonist, CNQX, by microdialysis prior to the induction of arthritis prevented the release of ASP and GLU at the time of injection and during the prolonged release phase (Sluka and Westlund, 1993c), as well as the increased content of GLU and SP in the superficial dorsal horn (Sluka and Westlund, 1993d). Similarly, the increased release of ASP and GLU in the dorsal horn is prevented by pretreatment with an NMDA receptor antagonist (AP7), and the increased release of GLU is reversed by posttreatment with AP7 (Sluka et al., 1994b) (Fig. 7). The release of GLY and SER during the PWL test is blocked by posttreatment with either a non-NMDA (CNQX) or an NMDA (AP7) receptor antagonist (Sluka et al., 1994b). Further supporting a role of NK1 receptors in the development of acute inflammation, the increased release of the EAAs in the late phase is prevented by pretreatment with CP96,345 (NK1 antagonist) (Sluka and Westlund, 1993d).

Surprisingly, administration of the $GABA_A$ receptor antagonist, bicuculline, into the dorsal horn by microdialysis prior to the induction of inflammation prevented the development of heat hyperalgesia (Sluka et al., 1993a). Furthermore, when administered prior to the induction of inflammation, bicuculline prevented the prolonged release

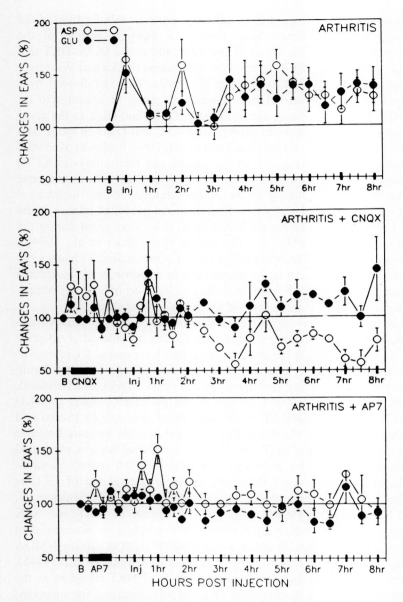

Fig. 6. EAA concentration (percent change ± SEM) was measured using fluorescent high performance liquid chromatography in (A) (control arthritic) and (B) (arthritic animals) pretreated with CNQX and in (C) (arthritic animals pretreated with AP7). Antagonists were delivered for 1 h after baseline and followed by a 1 h (CNQX) or 30 min (AP7) washout The knee joint was injected with 3% kaolin and carrageenan (arrow, Inj). (From Sluka and Westlund, 1993a)

of the EAAs and the increased release of the IAAs during the PWL test to radiant heat (Sluka et al., 1994d). Thus, blockade of non-NMDA EAA, NMDA EAA, NK1 and GABA$_A$ receptors specifically alters the release of EAAs and IAAs induced by acute inflammation.

The most interesting finding in our investigation of the role of neurotransmitters in the spinal cord during the induction of acute arthritis is that either a non-NMDA EAA receptor antagonist (CNQX) or a GABA$_A$ receptor antagonist (bicuculline) administered spinally prior to induction of inflam-

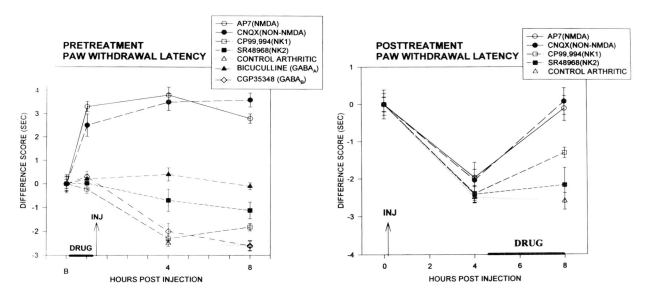

Fig. 7. Differences in paw withdrawal latency before and after induction of inflammation in the paw ipsilateral to the inflamed knee. Animals were either pretreated (top graph) or posttreated (bottom graph) with the drugs (thick bar). The knee joint was injected after control testing (arrow). The drugs were delivered through the microdialysis fiber, and the doses were as follows: CNQX pretreatment (1.2 mM), CNQX posttreatment (1.0 mM), AP7 pretreatment (2 mM), AP7 posttreament (2 mM), CP99994 pre- and posttreament (10 mM), SR48968 pre- and posttreatment (1 mM), bicuculline pretreatment (0.01 mM), CGP35348 pretreatment (3.0 mM).

mation can reduce the degree of joint swelling and the increase in skin temperature typical of the inflammation (Sluka and Westlund, 1993c; Sluka et al., 1993) (Fig. 8). In contrast, an NMDA receptor antagonist, AP7, or a $GABA_B$ receptor antagonist, CGP35348, had no effect on the joint swelling and temperature (Fig. 8). Chemical and surgical sympathectomy also had no effect on the joint swelling or thermographic readings (Sluka et al., 1994c). On the other hand, sectioning the dorsal roots prevented the swelling and thermographic changes (Sluka et al., 1994c) (Fig. 8). These results led us to hypothesize that the spinal cord can influence the degree of joint inflammation, i.e. the neurogenic component of inflammation is made worse by the spinal dorsal horn through activation of the primary afferents.

The best explanation for the spinal control of inflammation is that depolarization of the primary afferents results in antidromic action potentials or dorsal root reflexes (Rees et al., 1994, 1995b; Sluka et al., 1995b). These dorsal root reflexes would then release neuropeptides into the knee joint, resulting in increased joint swelling and temperature. To test this hypothesis we have recorded from the central end of a cut medial articular nerve in rats, cats and monkeys and from the central end of cut dorsal root filaments in cats and monkeys. Since the portion of the nerve recorded was disconnected from the peripheral receptive field, any activity recorded would have to be generated centrally. Sympathetic discharges were ruled out by surgical and chemical sympathetomy (Rees et al., 1994, 1995b). Therefore, antidromic action potentials or dorsal root reflexes could be recorded in both articular afferents and in dorsal roots following induction of inflammation (Rees et al., 1994, 1995b; Sluka et al., 1995b). Most of the dorsal root reflex activity was evoked by mechanical stimulation of the peripheral receptive field (Rees et al., 1994). The dorsal root reflexes were found to occur in C-fibers, Aδ fibers and larger Aβ fibers in

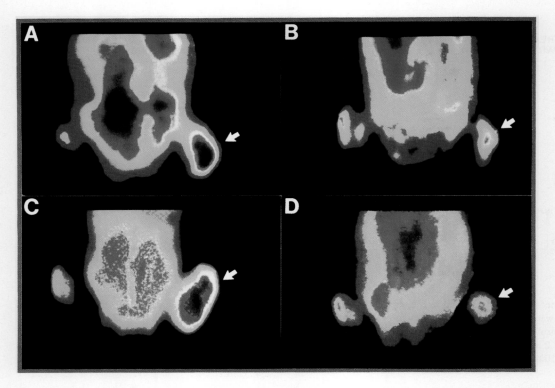

Fig. 8. Lower body thermographic readings (abdomen and both knees) with the inflamed (right) and the contralateral (left) knee visualized 8 h after injection of kaolin and carrageenan into the knee joint cavity for control arthritic animals (A), bicuculline treated arthritic animals (B), sympathectomized arthritic animals (C) and dorsal rhizotomized animals (D). Bar represents a continuous heat scale from 29.5 (red) to 33.4°C (dark blue).

the joint afferents and dorsal roots (Sluka et al., 1995b). Similar to our previous behavioral studies demonstrating a spinal involvement of non-NMDA EAA and $GABA_A$ receptors, dorsal root reflexes could be eliminated with either CNQX or bicuculline (Rees et al., 1995b).

Future directions in research on inflammatory pain and central sensitization

Second messenger systems

As already mentioned, the sensitization produced by intradermal capsaicin injections or by acute experimental arthritis depends on the activation of a number of neurotransmitter receptors (e.g., non-NMDA and NMDA EAA receptors, NK1 and NK2 receptors, $GABA_A$ receptors). Activation of a number of neurotransmitter receptors triggers a cascade of intracellular events that involve second messengers, protein kinases and protein phosphatases. Evidence is beginning to accumulate that implicates such signal transduction events in the sensitization of dorsal horn neurons. For example, microdialysis administration of the metabotropic EAA receptor agonist, trans-ACPD, increases the responsiveness of primate STT cells to innocuous mechanical stimulation (Palecek et al., 1993b), suggesting that metabotropic EAA receptors play a role in central sensitization. In support of this, blockade of metabotropic GLU receptors by L-AP3 reduces the arthritis-induced sensitization of dorsal horn neurons (Neugebauer et al., 1994a). Furthermore, activation of metabotropic GLU receptors by trans-ACPD also enhances the non-NMDA and NMDA responses of dorsal horn neurons (Bleakman et al. 1992; Cerne and Randic, 1992), indicating an intricate interaction between receptors that could contribute to central sensitization. In addition to these observations concerning metabotropic EAA receptors, there is evidence that the sensitization process depends on the release of NO in the spinal cord (Palecek et al., 1993c; cf., Meller and Gebhart, 1993; Coderre and Yashpal, 1994; Meller et al., 1994).

To explore the possibility that protein kinases, such as protein kinase C (PKC), are involved, we have introduced phorbol esters into the dorsal horn by microdialysis and have found that an active phorbol ester can increase the responses of primate STT cells to innocuous mechanical stimuli, whereas an inactive phorbol ester has no effect (Palecek et al., 1993a).

Activation of PKC by phorbol esters or intracellular injection of PKC enhances both non-NMDA and NMDA currents in dorsal horn neurons (Gerber et al., 1989; Chen and Huang, 1992) and increases the release of EAAs from dorsal horn slices (Gerber et al., 1989).

Immediate, early genes

The activation of second messenger systems can result in long-term changes in the central nervous system by activation of immediate early genes, which in turn causes increases or decreases in the expression of other genes (Menétrey et al., 1989; Herdegen et al., 1991; Abbadie and Besson, 1992). The ultimate changes in gene products have the potential to result in morphological and consequent functional alterations in the circuitry of the dorsal horn. Inflammatory pain is associated with the expression of immediate early genes (Menétrey et al., 1989; Abbadie and Besson, 1992). Possibly prolonged inflammatory pain states would result in a rewiring of dorsal horn circuitry.

Therapeutic potential

It is too early to forecast what new therapies for pain might arise from experimental work of the sort discussed here. However, it does seem likely that antagonists of neurotransmitter receptors will continue to receive close attention as candidate therapeutic agents. Whether interventions at the level of second messenger systems will prove feasible is more speculative (Zimmerman and Herdegen, this volume) since second messenger systems have many and diverse effects, and therefore blocking second messengers is likely to result in unexpected side effects. A thorough understanding of the cascade of events involving immediate early genes will be important, since it may be possible to intervene at a molecular level to prevent long-term alterations that may underlie chronic pain states.

Acknowledgements

The authors thank Griselda Gonzales for her help with the illustrations. The work was supported by NIH grants NS 09743, NS 28064, NS 01445, and NS 11255.

References

Abbadie, C. and Besson, J.M. (1992) *C-fos* expression in rat lumbar spinal cord during the development of adjuvant-induced arthritis. *Neuroscience*, 48: 985–993.

Baumann, T.K., Simone, D.A., Shain, C.N. and LaMotte, R.H. (1991) Neurogenic hyperalgesia: the search for the primary cutaneous afferent fibers that contribute to capsaicin-induced pain and hyperalgesia. *J. Neurophysiol.*, 66: 212–227.

Bleakman, D., Rusin, K.I., Chard, P.S., Glaum, S.R. and Miller, R.J. (1992) Metabotropic glutamate receptors potentiate ionotropic glutamate responses in rat dorsal horn. *Mol. Pharmacol.*, 42: 192–196.

Cerne, R. and Randic, M. (1992) Modulation of AMPA and NMDA responses in rat spinal dorsal horn neurons by trans-l-aminocyclopentane-1-3-dicarboxylic acid. *Neurosci. Lett.*, 144: 180–184.

Chen, L. and Huang, L.Y.M. (1992) Protein-kinase-C reduces Mg^{2+} block of NMDA-receptor channels as a mechanism of modulation. *Nature*, 356: 521–523.

Coderre, T.J. and Yashpal, K. (1994) Intracellular messengers contributing to persistent nociception and hyperalgesia induced by persistent nociception and hyperalgesia induced by L-glutamate and substance P in the rat formalin pain model. *Eur. J. Neurosci.*, 6: 1328–1334.

Colpaert, F.C., Donnerer, J. and Lembeck, F. (1983) Effects of capsaicin on inflammation and on substance P content of nervous tissues in rats with adjuvant arthritis. *Life Sci.*, 32: 1827–1834.

Dougherty, P.M. and Willis, W.D. (1991) Enhancement of spinothalamic neuron responses to chemical and mechanical stimuli following combined micro-iontophoretic application of N-methyl-D-aspartic acid and substance P. *Pain*, 47: 85–93.

Dougherty, P.M. and Willis, W.D. (1992) Enhanced responses of spinothalamic tract neurons to excitatory amino acids accompany capsaicin-induced sensitization in the monkey. *J. Neurosci.*, 12: 883–894.

Dougherty, P.M., Palecek, J., Palecková, V., Sorkin, L.S. and Willis, W.D. (1992a) The role of NMDA and non-NMDA excitatory amino acid receptors in the excitation of primate spinothalamic tract cells by mechanical, chemical, thermal and electrical stimuli. *J. Neurosci.*, 12: 3025–3041.

Dougherty, P.M., Sluka, K.A., Sorkin, L.S., Westlund, K.N. and Willis, W.D. (1992b) Neural changes in acute arthritis in monkeys. I. Parallel enhancement of responses of spinothalamic tract neurons to mechanical stimulation and excitatory amino acids. *Brain Res. Rev.*, 17: 1–13.

Dougherty, P.M., Palecek, J., Zorn, S. and Willis, W.D. (1993) Combined application of excitatory amino acids and substance P produced long-lasting changes in responses of primate spinothalamic tract neurons. *Brain Res. Rev.*, 18: 227–246.

Dougherty, P.M., Palecek, J., Palecková, V. and Willis, W.D. (1994) Neurokinin 1 and 2 antagonists attenuate the responses and NK1 antagonists prevent the sensitization of primate spinothalamic tract neurons after intradermal capsaicin. *J. Neurophysiol.*, 72: 1464–1475.

Dougherty, P.M., Palecek, J., Palecková, V. and Willis, W.D. (1995) Infusion of substance P or neurokinin A by microdialysis alters responses of primate spinothalamic tract neurons to cutaneous stimuli and to iontophoretically released excitatory amino acids. *Pain*, in press.

Gamse, R., Molnar, A. and Lembeck, F. (1979) Substance P release from spinal cord slices by capsaicin. *Life Sci.*, 25: 629–636.

Gerber, G., Kangrga, I., Ryu, P.D., Larew, J.S.A. and Randic, M. (1989) Multiple effects of phorbol esters in the rat spinal dorsal horn. *J. Neurosci.*, 9: 3606–3617.

Hargreaves, K., Dubner, R., Brown, F., Flores, C. and Joris, J. (1988) A new and sensitive method for measuring thermal nociception in cutaneous hyperalgesia. *Pain*, 32: 77–88.

Herdegen, T., T'lle, T.R., Bravo, R., Zieglgänsberger, W. and Zimmermann, M. (1991) Sequential expression of JUN B. JUN D and FOS B proteins in rat spinal neurons: cascade of transcriptional operations during nociception. *Neurosci. Lett.*, 129: 221–224.

Hope, P.J., Jarrott, B., Schaible, H.-G. Clarke, R.W. and Duggan, A.W. (1990) Release and spread of immunoreactive neurokinin A in the cat spinal cord in a model of acute arthritis. *Brain Res.*, 533: 292–299.

Kilo, S., Schmelz, M., Koltzenburg, M. and Handwerker, H.O. (1994) Different patterns of hyperalgesia induced by experimental inflammation in human skin. *Brain*, 117: 385–396.

Koltzenburg, M., Torebjörk, H.E. and Wahren, L.K. (1994) Nociceptor mediated central sensitization causes mechanical hyperalgesia in acute chemogenic and chronic neuropathic pain. *Brain*, 117: 579–591.

Lam, F.Y. and Ferrell, W.R. (1989) Capsaicin suppresses substance P-induced joint inflammation in the rat. *Neurosci. Lett.*, 105: 155–158.

LaMotte, R.H., Lundberg, L.E.R. and Torebjörk, H.E. (1992) Pain, hyperalgesia and activity in nociceptive C units in humans after intradermal injection of capsaicin. *J. Physiol.*, 448: 749–764.

LaMotte, R.H., Shain, C.N., Simone, D.A. and Tsai, E.F.P. (1991) Neurogenic hyperalgesia: psychophysical studies of underlying mechanisms. *J. Neurophysiol.*, 66: 190–211.

Meller, S.T. and Gebhart, G.F. (1993) Nitric oxide (NO) and nociceptive processing in the spinal cord. *Pain*, 52: 127–136.

Meller, S.T., Cummings, C.P., Traub, R.J. and Gebhart, G.F. (1994) The role of nitric oxide in the development and maintenance of the hyperalgesia produced by intraplantar injection of carrageenan in the rat. *Neuroscience*, 60: 367–374.

Menétrey, D., Gannon, A., Levine, J.D. and Basbaum, A.I. (1989) Expression of *c-fos* protein in interneurons and projection neurons of the rat spinal cord in response to noxious somatic, articular, and visceral stimulation. *J. Comp. Neurol.*, 285: 177–195.

Neugebauer, V. and Schaible, H.G. (1990) Evidence for a central component in the sensitization of spinal neurons with joint input during development of acute arthritis in cat's knee. *J. Neurophysiol.*, 64: 299–311.

Neugebauer, V., Lücke, T. and Schaible, H.G. (1993) N-methyl-D-aspartate (NMDA) and non-NMDA receptor antagonists block the hyperexcitability of dorsal horn neurons during development of acute arthritis in rat knee joint. *J. Neurophysiol.*, 70: 1365–1377.

Neugebauer, V., Lucke, T. and Schaible, H.-G. (1994a) Requirement of metabotropic glutamate receptors for the generation of inflammation-evoked hyperexcitability in rat spinal cord neurons. *Eur. J. Neurosci.*, 6: 1179–1186.

Neugebauer, V., Schaible, H.G., Weiretter, F. and Freudenberger, U. (1994b) The involvement of substance P and neurokinin-1 receptors in the responses of rat dorsal horn neurons to noxious but not to innocuous mechanical stimuli applied to the knee joint. *Brain Res.*, 666: 207–215.

Palecek, J., Palecková, V., Dougherty, P.M. and Willis, W.D. (1993a) The effect of phorbol esters on the responses of primate spinothalamic neurons to mechanical and thermal stimuli. *J. Neurophysiol.*, 71: 529–537.

Palecek, J., Palecková, V., Dougherty, P.M. and Willis, W.D. (1993b) The effect of trans-ACPD, a metabotropic excitatory amino acid receptor agonist, on the responses of primate spinothalamic tract neurons. *Pain*, 56: 261–269.

Palecek, J., Palecková, V. and Willis, W.D. (1993c) The role of nitric oxide in hyperalgesia induced by capsaicin injection in primates. *Abstracts, 7th World Congress on Pain*, IASP Publications, Seattle, WA, pp. 226–227.

Radhakrishnan, V. and Henry, J.L. (1995) Antagonism of nociceptive responses of cat spinal dorsal horn neurons in vivo by the NK-1 receptor antagonists CP-96,345 and CP-99,994, but not by CP-96,344. *Neuroscience*, 64: 943–958.

Randic, M., Hecimovic, H. and Ryu, P.D. (1990) Substance P modulates glutamate-induced currents in acutely isolated rat dorsal horn neurones. *Neurosci. Lett.*, 117: 74–80.

Rees, H., Sluka, K.A., Westlund, K.N. and Willis, W.D. (1994) Do dorsal root reflexes augment peripheral inflammation? *NeuroReport*, 5: 821–824.

Rees, H., Sluka, K.A., Tsuruoka, M., Chen, P.S. and Willis, W.D. (1995a) The effects of NK1 and NK2 receptor antagonists on the sensitization of STT cells following acute inflammation in the anaesthetized primate. *J. Physiol.*, 483.P: 152P.

Rees, H., Sluka, K.A., Westlund, K.N. and Willis, W.D. (1995b) The role of glutamate and GABA receptors in the generation of dorsal root reflexes by acute arthritis in the anaesthetized rat. *J. Physiol.*, 484: 437–445.

Schaible, H.G. and Schmidt, R.F. (1985) Effects of an experimental arthritis on the sensory properties of fine articular afferent units. *J. Neurophysiol.*, 54: 1109–1122.

Schaible, H., Schmidt, R.F. and Willis, W.D. (1987) Enhancement of the responses of ascending tract cells in the cat spinal cord by acute inflammation of the knee joint. *Exp. Brain Res.*, 66: 489–499.

Schaible, H.G., Jarrott, B., Hope, P.J. and Duggan, A.W. (1990) Release of immunoreactive substance P in the cat spinal cord during development of acute arthritis in cat's knee: a study with antibody bearing microprobes. *Brain Res.*, 529: 214–223.

Schaible, H.G., Grubb, B.D., Neugebauer, V. and Oppman, M. (1991) The effects of NMDA antagonists on neuronal activity in cat spinal cord evoked by acute inflammation in the knee joint. *J. Neurosci.*, 3: 981–991.

Schaible, H.G., Hope, P.J., Lang, C.W. and Duggan, A.W. (1992) Calcitonin gene-related peptide causes intraspinal spreading of substance P released by peripheral stimulation. *Eur. J. Neurosci.*, 4: 750–757.

Simone, D.A., Baumann, T.K. and LaMotte, R.H. (1989) Dose-dependent pain and mechanical hyperalgesia in humans after intradermal injection of capsaicin. *Pain*, 38: 99–107.

Simone, D.A., Sorkin, L.S., Oh, U., Chung, J.M., Owens, C., LaMotte, R.H. and Willis, W.D. (1991) Neurogenic hyperalgesia: central neural correlates in responses of spinothalamic tract neurons. *J. Neurophysiol.*, 66: 228–246.

Sluka, K.A. and Westlund, K.N. (1992) An experimental arthritis in rat: dorsal horn aspartate and glutamate increases. *Neurosci. Lett.*, 145: 141–144.

Sluka, K.A. and Westlund, K.N. (1993a) An experimental arthritis model in rats- the effects of NMDA and non-NMDA antagonists on aspartate and glutamate release in the dorsal horn. *Neurosci. Lett.*, 149: 99–102.

Sluka, K.A. and Westlund, K.N. (1993b) Behavioral and immunohistochemical changes in an experimental arthritis model in rats. *Pain*, 55: 367–377.

Sluka, K.A. and Westlund, K.N. (1993c) Centrally administered non-NMDA but not NMDA receptor antagonists block peripheral knee joint inflammation. *Pain*, 55: 217–225.

Sluka, K.A. and Westlund, K.N. (1993d) Spinal cord amino acid release and content in an arthritis model- the effects of pretreatment with non-NMDA, NMDA and NK1 receptor antagonists. *Brain Res.*, 627: 89–103.

Sluka, K.A., Dougherty, P.M., Sorkin, L.S., Willis, W.D. and Westlund, K.N. (1992) Neural changes in acute arthritis in monkeys. III. Changes in substance P, calcitonin gene-related peptide and glutamate in the dorsal horn of the spinal cord. *Brain Res. Rev.*, 17: 29–38.

Sluka, K.A., Willis, W.D. and Westlund, K.N. (1993) Joint inflammation and hyperalgesia are reduced by spinal bicuculline. *NeuroReport*, 5: 109–112.

Sluka, K.A., Jordan, H.H. and Westlund, K.N. (1994a) Reduction in joint swelling and hyperalgesia following post-treatment with a non-NMDA glutamate receptor antagonist. *Pain*, 59: 95–100.

Sluka, K.A., Jordan, H.H., Willis, W.D. and Westlund, K.N. (1994b) Differential effects of N-methyl-D-aspartate (NMDA) and non-NMDA receptor antagonists on spinal release of amino acids after development of acute arthritis in rats. *Brain Res.*, 664: 77–84.

Sluka, K.A., Lawand, N.B. and Westlund, K.N. (1994c) Joint inflammation is reduced by dorsal rhizotomy and not by sympathectomy or spinal cord transection. *Ann. Rheum. Dis.*, 53: 309–314.

Sluka, K.A., Willis, W.D. and Westlund, K.N. (1994d) Arthritis induced release of excitatory amino acids is prevented by spinal administration of a $GABA_A$ and not by a $GABA_B$ receptor antagonist in rats. *J. Pharmacol. Exp. Ther.*, 271: 76–82.

Sluka, K.A., Milton, M.A., Westlund, K.N. and Willis, W.D. (1995a) Involvement of neurokinin receptors in the joint inflammation and heat hyperalgesia following acute inflammation in unanaesthetized rats. *J. Physiol.*, 483.P: 152–153P.

Sluka, K.A., Rees, H., Westlund, K.N. and Willis, W.D. (1995b) Fiber types contributing to dorsal root reflexes induced by joint inflammation in cats and monkeys. *J. Neurophysiol.*, 74: 981–989.

Sluka, K.A., Willis, W.D. and Westlund, K.N. (1995c) The role of dorsal root reflexes in neurogenic inflammation. *Pain Forum*, 4: 141–149.

Sorkin, L.S. and McAdoo, D.J. (1993) Amino acids and serotonin are released into the lumbar spinal cord of the anesthetized cat following intradermal capsaicin injections. *Brain Res.*, 607: 89–98.

Sorkin, L.S., Westlund, K.N., Sluka, K.A., Dougherty, P.M. and Willis, W.D. (1992) Neural changes in acute arthritis in monkeys. IV. Time course of amino acid release into the lumbar dorsal horn. *Brain Res. Rev.*, 17: 39–50.

Torebjörk, H.E., Lundberg, L.E.R. and LaMotte, R.H. (1992) Central changes in processing of mechanoreceptive input in capsaicin-induced secondary hyperalgesia in humans. *J. Physiol.*, 448: 765–780.

Ueda, M., Kuraishi, Y., Sugimoto, K. and Satoh, M. (1994) Evidence that glutamate is released from capsaicin-sensitive primary afferent fibers in rats: study with on-line continuous monitoring of glutamate. *Neurosci. Res.*, 20: 231–237.

Urban, L., Thompson, S.W.N. and Dray, A. (1994) Modulation of spinal excitability: co-operation between neurokinin and excitatory amino acid neurotransmitters. *Trends Neurosci.*, 17: 432–438.

Westlund, K.N., Sun, Y.C., Sluka, K.A., Dougherty, P.M., Sorkin, L.S. and Willis, W.D. (1992) Neural changes in acute arthritis in monkeys. II. Increased glutamate immunoreactivity in the medial articular nerve. *Brain Res. Rev.*, 17: 15–27.

Willis, W.D. (1994) Central plastic responses to pain. In: G.F. Gebhart, D.L. Hammond and T.S. Jensen (Eds.), *Proc. 7th World Congress on Pain.* IASP Press, Seattle, WA, pp. 301–324.

CHAPTER 13

Neurophysiology of chronic inflammatory pain: electrophysiological recordings from spinal cord neurons in rats with prolonged acute and chronic unilateral inflammation at the ankle

Hans-Georg Schaible and Robert F. Schmidt

Physiologisches Institut der Universität Würzburg, Röntgenring 9, D-97070 Würzburg, Germany

Introduction

The nature of chronic pain is still not fully elucidated. Clinical evidence suggests that the development and persistence of chronic pain may depend on several factors. These include the presence of a pathophysiological condition such as inflammatory, degenerative or other types of tissue lesions, an affliction of neurons in the peripheral or central nervous system, and, last not least, the involvement of psychological and social factors. While in many cases chronic pain is obviously correlated with the presence of somatic and/or visceral or nerval disease, chronic pain may be poorly correlated with the presence and severity of disease in other cases. Occasionally, clinical diagnosis may not reveal any organic dysfunction as a possible cause of chronic pain. Hence the term "chronic pain" per se is just a description of a painful condition that is entirely based on the duration of pain. It does not offer any clue as to the causes involved unless it is more specified.

Recently, we have performed experiments on the spinal events related to the neurophysiology of long-lasting inflammatory pain. The experiments aimed at finding out differences and similarities in the nociceptive processing in the spinal cord under normal conditions, during prolonged acute inflammation (2 days), and during chronic inflammation (lasting up to 3 weeks). In order to study nociceptive mechanism of the spinal cord during prolonged acute and chronic inflammation, Freund`s complete adjuvant (FCA) was injected subdermally around the ankle. These injections produced a local inflammation in the ankle region within hours which persisted over 20 days. There was no evidence for secondary lesions. Throughout 20 days, the rats were hypersensitive to pressure applied to the afflicted ankle suggesting the presence of hyperalgesia. No other gross behavioral changes were noted, except that some rats showed disturbances of the gait within the first week. Histological examination of the tissue around the ankle revealed the presence of inflammatory cells. These were mainly located in the periarticular tissue (Grubb et al., 1993; Hanesch et al., 1993).

Discharge properties of spinal cord neurons in control rats and in rats with prolonged acute and chronic inflammation

The aim of this study was to determine the discharge and receptive field properties of spinal cord neurons with ankle input in spinal segments L4-L6, both in control rats and in rats with adjuvant-induced unilateral inflammation at the ankle. Recordings were made 2, 6, 13, or 20 days after inoculation of FCA in the ankle region of the left hindlimb. The recording sites (segments, depths) of the neurons in control rats and in rats with inflammation were similar (for details of sampling see Grubb et al., 1993).

Inflammation-dependent changes were seen in neurons in the superficial and deep dorsal horn and in the ventral horn. The following alterations have been identified: (1) Throughout the three weeks after induction of the unilateral inflammation, there was a progressive reduction in the proportion of nociceptive specific neurons, and a concomitant increase in the proportion of neurons showing a wide dynamic range response profile. In the superficial and deep dorsal horn the reduction in the proportion of NS neurons was significant (chi-squared test, $P < 0.01$). (2) At all times after induction of inflammation, the mechanical thresholds at the ankle joint for driving the neurons with ankle input were significantly lower in rats with inflammation than in control rats. This is shown in Fig. 1A for the neurons located in the deep dorsal horn. The figure shows the mean pressure (mediolateral compression of the ankle) which was necessary to activate the neurons, at the time points indicated. A similar drop in threshold during inflammation was found in neurons in the superficial dorsal horn and in the ventral horn. Using a subjective pressure scale (palpation of the tissue), a decrease in threshold was also found for the stimulation of areas remote from the ankle. Low intensity stimuli applied to inflamed and non-inflamed areas in arthritic rats activated a higher proportion of neurons than stimuli applied to the same regions in control rats. (3) The receptive fields of neurons with ankle input were markedly larger in rats with inflammation. At 2, 13, and 20 days post inoculation, there was a large increase in the proportion of neurons with ankle input which were also activated by pressure applied to the thigh (Fig. 1B). In the later stages of inflammation, an increasing number of neurons with ankle joint input were found whose receptive fields included either the abdomen, the back or the tail, indicating even further expansion of the receptive fields. There was also an increase in the number of neurons with contralateral excitatory inputs. At 20 days post inoculation, 13 out of 52 neurons (25%) had contralateral excitatory receptive fields

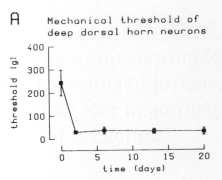

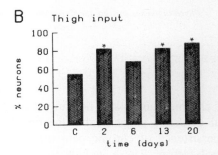

Fig. 1. (A) Average firing threshold of spinal cord neurons in the deep dorsal horn for pressure applied to the ankle with calibrated forceps. The data are from neurons with ankle input in control rats (at 0 days), and from neurons in rats with a unilateral inflammation at the ankle at several time points after the start of inflammation. (B) Proportion of neurons with ankle input whose receptive fields included the thigh. The samples of neurons are from control rats (C), and from rats with unilateral inflammation at the ankle at different days post inoculation of Freund`s complete adjuvant.

which was significantly different from 8 out of 82 neurons (10%) in the control rats. By contrast, the proportion of neurons with contralateral inhibitory receptive fields remained stable. (4) The proportion of spontaneously active neurons was higher in rats during the initial and later stages of inflammation. The mean spontaneous discharge frequency was not increased.

Collectively, the changes of the discharge properties suggest that the neurons with input from the ankle were in a state of hyperexcitability during inflammation at the ankle joint. This statement is made because previous experiments in an acute model of inflammation, the acute kaolin/carrageenan-induced arthritis, had allowed us to determine the features of inflammation-evoked hyperexcitability in a more direct way (see Schaible and Grubb, 1993). Using the latter model, single neurons were continuously monitored during the development of inflammation, and the resulting changes of discharges could be documented in long-term recordings in one and the same neuron. The comparison between the response properties of spinal cord neurons with input from the normal and inflamed ankle (the study described here) suggests that the changes of the discharge properties in rats with prolonged acute and chronic inflammation are quite similar to those observed during development of acute kaolin/carrageenan-evoked inflammation in spinalized cats (Schaible et al., 1987; Neugebauer and Schaible 1990) and in non-spinalized rats (Neugebauer et al., 1993, 1994a,b, 1995). In spinalized polyarthritic rats, spinal cord neurons with input from inflamed areas show in addition ongoing bursting discharges (Menetrey and Besson, 1982).

The present findings and the earlier work of Menetrey and Besson (1982) in polyarthritic rats show that spinal cord neurons can remain in a state of hyperexcitability over weeks. This is remarkable in view of convincing evidence that the development of inflammation may lead to an increase in the effectiveness of descending inhibition which counteracts the expression of hyperexcitability of neurons to some extent (Schaible et al., 1991). In addition, heterotopic inhibitory influences are more active in polyarthritic rats (Calvino et al., 1987). Local mechanisms in the spinal cord also seem to build up inhibitory influences (see below).

The hyperexcitability represents a functional plasticity of the nociceptive system. It is thought to contribute significantly to the neuronal basis of hyperalgesia and allodynia in awake humans and animals suffering from inflammatory lesions of peripheral tissue (Dubner and Ruda, 1992; Coderre et al., 1993; McMahon et al., 1993; Schaible and Grubb, 1993; Urban et al., 1994; Willis, 1994). There is now plenty of evidence that the inflammation-evoked hyperexcitability of spinal cord neurons is due to increased and altered afferent input from the inflamed region as well as to changes in spinal cord neurons. In acute knee joint inflammation, non-nociceptive and nociceptive articular primary afferent units including initially mechano-insensitive (silent) nociceptors become sensitized (Schaible and Schmidt, 1985, 1988; Grigg et al., 1986). Similar conclusions have been drawn from recordings of afferent fibres from the normal and chronically inflamed ankle (Guilbaud et al., 1985; Grubb et al., 1991). The increased responses of spinal cord neurons to mechanical stimuli applied to non-inflamed tissue, and the expansion of receptive fields, are usually taken as evidence for changes in the central nervous system itself (Woolf, 1983; Guilbaud et al. 1986; Cook et al. 1987; Hylden et al. 1989; Neugebauer and Schaible 1990; Simone et al. 1991).

The involvement of excitatory amino acids and their receptors in the responses of spinal cord neurons to sensory outflow from the normal and the chronically inflamed ankle

In regard to the mode of operation of the spinal nociceptive system, and the mechanisms underlying chronic pain, an important question is whether the sensory transmission in the spinal cord involves the same transmitter/receptor systems in situations of acute and chronic nociception. Although the phenomena of hyperexcitability are similar in acute and chronic inflammation (see above), this does not necessarily imply that the same mechanisms are at work. Recently we investigated the involvement of N-methyl-D-aspartate (NMDA) and non-NMDA receptors in the synaptic transmission of afferent inflow induced by innocuous and noxious pressure applied to the normal and acutely inflamed knee joint in the rat and cat (Schaible et al., 1991; Neugebauer et al., 1993a,b, 1994). Comparable recordings in rats with prolonged unilateral inflammation at the ankle aimed to identify the importance of this transmitter/receptor system under the conditions of chronic inflammation (Neugebauer et al., 1994b).

Unilateral adjuvant inflammation was induced at the rat ankle, 2 or 20 days prior to the evaluation of the contribution of NMDA and non-NMDA receptors to the nociceptive processing in wide dynamic range neurons with ankle input. The ankle was compressed across the mediolateral axis at quantitatively controlled innocuous and noxious intensities using a motor-driven mechanical device. Either the non-NMDA antagonist 6-cyano-7-nitroquinoxaline-2,3-dione (CNQX) or the NMDA antagonists ketamine or D,L-2-amino-5-phosphonovalerate (AP5) were ionophoretically administered to the spinal recording site to test for their effects on the responses to mechanical stimuli applied to the ankle.

Twenty neurons with ankle input were identified in the superficial (240–401 μm; $n = 4$) and deep (585–1212 μm; $n = 16$) dorsal horn of the segments L3-L4. Ten were recorded in the prolonged acute stage (2 days post inoculation), and ten in the chronic stage (20 days post inoculation). The responses to (normally) innocuous and to noxious compression of the inflamed ankle were reduced by both NMDA and non-NMDA antagonists. As an example, Fig. 2 shows a WDR neuron recorded 20 days after induction of inflammation (depth 953 μm in L3). The receptive field was located in the ankle and the deep structures of thigh, knee, lower leg and paw. The cell had no cutaneous input and no ongoing discharges but was readily excited by ionophoretically administered AMPA and NMDA. CNQX selectively suppressed the responses to AMPA, and AP5 suppressed selectively the responses to NMDA (not shown). As seen in Fig. 2, each of these antagonists reduced the responses to innocuous and noxious pressure to 30–35% of the predrug values, and the simultaneous application of both antagonists caused a further reduction of the responses to noxious pressure (8% of control) and abolished the responses to light compression.

Marked reductions of the effects of the non-NMDA and NMDA receptor antagonists were seen in 9/10 neurons recorded at 2 days post inoculation, and in 10/10 neurons recorded at 20 days post inoculation. Ongoing activity, present in 12 neurons, was decreased by CNQX (9 neurons) and by ketamine or AP5 (9 neurons). In 4 out of 10 neurons recorded at 2 days, and in 5 out of 10 neurons recorded at 20 days after induction of in-

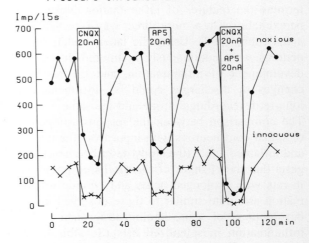

Fig. 2. Involvement of N-methyl-D-aspartate (NMDA) and non-NMDA receptors in the responses of a WDR spinal cord neuron to local mechanical stimulation of the ipsilateral inflamed ankle. The recording was made 20 days post inoculation of Freund's complete adjuvant into the ankle. Dots show the responses to high intensity (noxious) pressure applied to the ankle, crosses show the responses to low intensity (normally innocuous) pressure. The boxes indicate the focal intraspinal administration of the non-NMDA receptor antagonist CNQX (ejected at 20 nA) and of the NMDA receptor antagonist AP5 (ejected at 20 nA) and the co-administration of both antagonists. More details in the text.

flammation, both the NMDA and non-NMDA antagonists led to a reversible reduction in the receptive field size. In 15 out of 20 neurons the mechanical response thresholds increased during application of antagonists.

An involvement of excitatory amino acid receptors in the development of hyperexcitability of spinal cord neurons has also been described in several other models of acute inflammation and tissue injury (see Coderre et al., 1993; McMahon et al., 1993; Neugebauer et al., 1993b; Urban et al., 1994). The present experiments showed that prolonged acute and chronic inflammation-induced activity of dorsal horn neurons can be reduced by non-NMDA and NMDA receptor antagonists, similar to that evoked by acute inflammation (hours after induction). These data reveal the importance of ionotropic excitatory amino acid receptors in spinal cord nociception under chronic pathophysiological conditions. Even at a chronic stage of inflammation, the enhanced responses to mechanical stimuli and the expanded receptive fields are largely dependent on the continuous activation of non-NMDA and NMDA receptors. The pattern of the effect of the antagonists in prolonged acute and chronic inflammation is different from the pattern seen in control rats without inflammation. In normal rats, non-NMDA receptors are involved in both the responses to innocuous and noxious stimuli, whereas NMDA receptors are only activated during application of noxious stimuli to the knee and the ankle (Neugebauer et al., 1993a).

Evidence for increased activity of inhibitory kappa opioid ligand(s)

The sensory outflow from inflammatory lesions such as arthritis seems to induce an enhancement of tonic opioidergic inhibition of spinal cord neurons (Lombard and Besson, 1989). Furthermore, there are marked alterations in the amount of endogenous opioids contained in the spinal cord. A pronounced long-term activation has been described in particular for the prodynorphin synthesis in spinal cord neurons (Millan et al., 1985, 1987, 1988; Höllt et al., 1987; Iadarola et al., 1988a,b; Ruda et al., 1988; Nahin et al., 1989; Weihe et al., 1989; Przewlocka et al., 1992). There is also a small but significant increase in the concentration of opioid peptides derived from pro-enkephalin A, e.g. met-enkephalin (Cesselin et al., 1980, 1984; Noguchi et al., 1992; Przewlocka et al., 1992). Behavioral data suggest that the upregulation of the dynorphin synthesis under inflammatory conditions may exert a hypoalgesic effect (Millan et al., 1985, 1987, 1988; Millan and Colpaert, 1991).

In order to get more insight into the site and mode of action of endogenous kappa opioid ligands under normal versus inflammatory conditions, we have ionophoretically administered the kappa opioid agonist U50,488H and the kappa antagonist nor-Binaltorphamine (nor-BNI) to the vicinity of neurons with sensory input from the normal ankle (experiments on control rats) and of neurons with sensory input from the inflamed ankle (experiments on rats with unilateral inflammation). Again, we recorded in the prolonged acute (2 days post inoculation) and in the chronic phase (around 20 days post inoculation). Most neurons were located in the deep dorsal horn (for details see Stiller et al., 1993).

The majority of the neurons with ankle input seem to possess kappa receptors since they were influenced by U50,488H. The effects were, however, not homogeneous. In most neurons in control animals and in rats with inflammation, U50,488H reduced the responses to noxious pressure. In a smaller number of cells, the responses to pressure were either increased or remained unaffected. Ongoing discharges were either reduced or increased or remained unaffected during application of U50,488H. The effects of U50,488H were noted within a few minutes after starting the ejection. The various effects of U50,488H could be reduced by preceeding and simultaneous coadministration of nor-BNI.

The discharges of neurons with inputs from the normal and inflamed ankle were influenced by nor-BNI, but the proportion of neurons affected by this kappa opioid antagonist was significantly higher in neurons with input from the inflamed ankle. Ongoing activity was increased in 7 out of 19 (37%) neurons in control rats, in 16 out of 24 (67%) neurons in the prolonged phase of acute inflammation, and in 15 of 23 (65%) in the chronic phase of inflammation. An example of the latter action is displayed in Fig. 3A (recording from a neuron at 20 days post inoculation). During administration of nor-BNI the ongoing activity showed a marked increase outlasting the administration of the antagonist.

The responses to pressure were also influenced by the administration of nor-BNI. In control rats, the responses to pressure were increased in 9 cells (36%), reduced in 7 cells (28%), and unaffected in 9 cells (36%). In the prolonged phase of acute inflammation, significantly more neurons (11 of 15, 73%) showed enhanced responses to pressure, but not in the chronic phase. Fig. 3B shows the responses of a neuron recorded at 2 days post inoculation. The peristimulus time histograms display the responses to stimuli preceding the ejection of

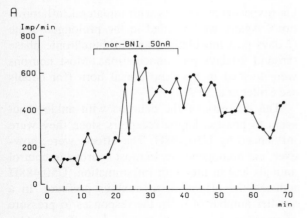

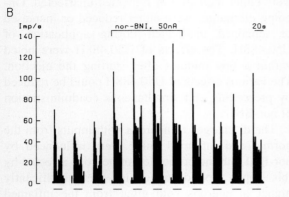

Fig. 3. (A) Effects of the kappa opioid antagonist nor-binaltorphamine (nor-BNI) on the ongoing discharges of a spinal cord neuron with input from the inflamed ankle. The deep dorsal horn neuron (depth 749 μm) was recorded 20 days post inoculation of Freund`s complete adjuvant (FCA). As indicated, nor-BNI was ejected at a current of 50 nA. (B) Effects of nor-BNI (ejected at 50 nA) on the responses of a neuron to pressure applied to the inflamed ankle. This neuron had no resting discharge. It was recorded 2 days post inoculation of FCA (depth 690 μm).

nor-BNI, the responses obtained during ejection of nor-BNI and the responses after termination of the ejection of the antagonist. During the application of nor-BNI, the responses to pressure across the ankle were enhanced. This effect of nor-BNI was reversible although the reversal had a slow time course.

These data suggest that the majority of deep dorsal horn neurons with ankle input are equipped with kappa opioid receptors. In most cases the effect of the antagonist consisted of an increase in the ongoing and/or evoked activity suggesting that the endogenous agonist had continuously reduced the neuron`s excitability. Since dynorphin synthesis is upregulated under inflammatory conditions, the more pronounced effect of nor-BNI in rats with inflammation most likely is a consequence of the enhanced background inhibition due to the increased release of kappa ligands. So far there is no evidence that the density of μ, δ and κ binding sites in the lumbar cord undergoes changes in the presence of a unilateral inflammation (Iadarola et al. 1988b; Millan et al. 1988). The results suggest that the enhanced activity of kappa opioids counteracts to some extent inflammation-induced activity in the majority of spinal neurons with joint input. It should be noted, however, that a small number of neurons in our study also showed excitatory and/or facilitatory effects of U50,488H, and in these neurons nor-BNI caused a reduction of the responses to pressure and of the ongoing discharges (Stiller et al., 1993).

Upregulation of the synthesis of calcitonin gene-related peptide during inflammation

While dynorphin and other opioid peptides are mainly or exclusively inhibitory, other peptides such as the tachykinins and calcitonin gene-related peptide (CGRP) have excitatory and/or facilitatory effects. CGRP is produced in the dorsal root ganglia and transported to the peripheral and spinal terminations. In the joint, CGRP is involved in neurogenic inflammation (Green et al., 1992; Karimian and Ferrell, 1994). In the spinal cord, CGRP has facilitatory effects on the nociceptive processing (Wiesenfeld-Hallin et al., 1984; Oku et al., 1987; Cridland and Henry, 1988; Miletic and Tan, 1988; Biella et al., 1991; Neugebauer et al., 1994). On this background it is of interest whether the synthesis of CGRP is upregulated under the conditions of prolonged acute and chronic inflammation.

The proportion of CGRP-immunoreactive perikarya was determined in the dorsal root ganglia L4-L6 in 4 control rats, and in 10 rats with a unilateral inflammation in the ankle region of the left hindlimb (Hanesch et al., 1993). In control rats, about 24% of 20 419 perikarya showed CGRP-like immunoreactivity. In rats with unilateral inflammation, the proportion of CGRP-positive neurons was increased on the inflamed side to about 32% of 11 454 cells at 2 days post inoculation (significantly different from ganglia in normal rats with $P < 0.001$), and to about 29% of 10 739 perikarya at 20 days post inoculation (different from ganglia of control rats with $P < 0.01$). By contrast, no significant changes were found between the control ganglia and those on the non-injected side of the rats with inflammation (about 25% at day 2 and about 24% at day 20).

Several studies have shown an overall upregulation of the synthesis of CGRP in the dorsal root ganglia in acute and chronic phases of inflammation (Kuraishi et al., 1989; Marlier et al., 1991; Donnerer et al., 1992; Smith et al., 1992; Collin et al., 1993). Furthermore, the present results demonstrate that the enhanced synthesis of CGRP in acute and chronic phases of peripheral inflammation may result at least in part from the synthesis of CGRP in neurons which do not produce CGRP in detectable amounts under normal conditions.

Conclusions

The experimental results briefly summarized in this review provide evidence that the nociceptive processing in the spinal cord under chronic inflammatory conditions exhibits numerous phenomena which are also seen during acute inflammation of the knee joint. Excitatory amino acids and their NMDA and non-NMDA receptors are similarly involved in the nociceptive processing under the conditions of acute, prolonged and chronic inflammation. An upregulation of the inhibitory kappa opioid system has been found in both prolonged acute and chronic inflammation. From these experiments we conclude that at least the transmitter/receptor systems studied are similarly involved in the processing of afferent inputs from acutely and chronically inflamed regions. These findings do not exclude that more pronounced differences exist in other transmitter/receptor systems between acute and chronic inflammation.

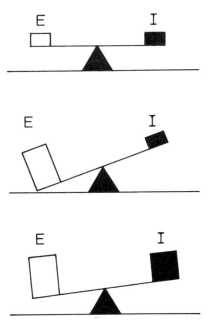

Fig. 4. Balance between excitatory and inhibitory influences on spinal cord neurons with input from inflamed tissue. Top: equipotent influences of excitatory and inhibitory processes under control conditions. Middle: selective increase of excitatory influences e.g. during development of inflammation. Bottom: increase of inhibitory influences which counteract excitatory factors to some extent.

An important difference between acute and chronic inflammation is that the synthesis of peptidergic transmitters is upregulated under chronic inflammatory conditions. This has been convincingly demonstrated for the kappa ligand dynorphin and for the neuropeptide CGRP. Thus the relative importance of transmitter/receptor systems may change in the presence of inflammation.

When excitatory as well as inhibitory drives are changing it may be important how the new balance is set. This is displayed in Fig. 4. The top shows the normal situation in which excitatory and inhibitory systems keep the neurons in a balance. The middle graph shows the increase of excitatory drives such as during development of inflammation. In order to keep the neuronal activity under control, inhibitory systems may show an upregulation (bottom). It may be speculated that in patients the severity of chronic pain is dependent on the relative capacities of the excitatory and inhibitory systems to develop enhanced levels of function.

References

Biella, G., Panara, C., Pecile, A. and Sotgiu, M.L. (1991) Facilitatory role of calcitonin gene-related peptide (CGRP) on excitation induced by substance P (SP) and noxious stimuli in rat spinal dorsal horn neurons. An iontophoretic study in vivo. *Brain Res.*, 559: 352–356.

Calvino, B., Villanueva, L. and LeBars, D. (1987) Dorsal horn (convergent) neurones in the intact anaesthetised arthritic rat. II. Heterotopic inhibitory influences. *Pain*, 31: 359–379.

Cesselin, F., Montastruc, J.L., Gros, C., Bourgoin, S. and Hamon, M. (1980) Met-enkephalin levels and opiate receptors in the spinal cord of chronic suffering rats. *Brain Res.*, 191: 289–293.

Cesselin, F., Bourgoin, S., Artaud, F. and Hamon, M. (1984) Basic and regulatory mechanisms of in vitro release of Met-enkephalin from the dorsal zone of the rat spinal cord. *J. Neurochem.*, 43: 763–774.

Coderre, T.J., Katz, J., Vaccarino, A.L. and Melzack, R. (1993) Contribution of central neuroplasticity to pathological pain: review of clinical and experimental evidence. *Pain*, 52: 259–285.

Collin, E., Mantelet, S., Frechilla, D., Pohl, M., Bourgoin, S., Hamon, M. and Cesselin, F. (1993) Increased in vivo release of calcitonin gene-related peptide-like material from the spinal cord in arthritic rats. *Pain*, 54: 203–211.

Cook, A.J., Woolf, C.J., Wall, P.D. and McMahon, S.B. (1987) Dynamic receptive field plasticity in rat spinal cord dorsal horn following C-primary afferent input. *Nature*, 325: 151–153.

Cridland, R.A. and Henry, J.L. (1988) Effects of intrathecal administration of neuropeptides on a spinal nociceptive reflex in the rat: VIP, galanin, CGRP, TRH, somatostatin and angiotensin. *Neuropeptides*, 11: 23–32.

Donnerer, J., Schuligoi, R. and Stein, C. (1992) Increased content and transport of substance P and calcitonin gene-related peptide in sensory nerves innervating inflamed tissue: evidence for a regulatory function of nerve growth factor in vivo. *Neuroscience*, 49: 693–698.

Dubner, R. and Ruda, M.A. (1992) Activity-dependent neuronal plasticity following tissue injury and inflammation. *Trends Neurosci.*, 15: 96–103.

Green, P.G., Basbaum, A.I. and Levine, J.D. (1992) Sensory neuropeptide interactions in the production of plasma extravasation in the rat. *Neuroscience*, 50: 745–749.

Grigg, P., Schaible, H.-G. and Schmidt, R.F. (1986) Mechanical sensitivity of group III and IV afferents from posterior articular nerve in normal and inflamed cat knee. *J. Neurophysiol.*, 55: 635–643.

Grubb, B.D., Birrell, G.J., McQueen, D.S. and Iggo, A. (1991) The role of PGE2 in the sensitization of mechanoreceptors in normal and inflamed ankle joints of the rat. *Exp. Brain Res.*, 84: 383–392.

Grubb, B.D., Stiller, R.U. and Schaible, H.-G. (1993) Dynamic changes in the receptive field properties of spinal cord neurons with ankle input in rats with chronic unilateral inflammation in the ankle region. *Exp. Brain Res.*, 92: 441–452.

Guilbaud, G., Iggo, A. and Tegner, R. (1985) Sensory receptors in ankle joint capsules of normal and arthritic rats. *Exp. Brain Res.*, 58: 29–40.

Guilbaud, G., Kayser, V., Benoist, J.R. and Gautron, M. (1986) Modifications in the responsiveness of rat ventrobasal thalamic neurons at different stages of carrageenan-produced inflammation. *Brain Res.*, 385: 86–98.

Hanesch, U., Pfrommer, U., Grubb, B.D. and Schaible, H.-G. (1993) Acute and chronic phases of unilateral inflammation in rat's ankle are associated with an increase in the proportion of calcitonin gene-related peptide-immunoreactive dorsal root ganglion cells. *Eur. J. Neurosci.*, 5: 154–161.

Höllt, V., Haarmann, I., Millan, M.J. and Herz, A. (1987) Prodynorphin gene expression is enhanced in the spinal cord of chronic arthritic rats. *Neurosci. Lett.*, 73: 90–94.

Hylden, J.L.K., Nahin, R.L., Traub, R.J. and Dubner, R. (1989) Expansion of receptive fields of spinal lamina I projection neurons in rats with unilateral adjuvant-induced inflammation: the contribution of dorsal horn mechanisms. *Pain*, 37: 229–243.

Iadarola, M.J., Brady, L.S., Draisci, G. and Dubner, R. (1988a) Enhancement of dynorphin gene expression in spinal cord following experimental inflammation: stimulus specificity, behavioural parameters and opioid receptor binding. *Pain*, 35: 313–326.

Iadarola, M.J., Douglass, J., Civelli, O. and Naranjo, J.R. (1988b) Differential activation of spinal cord dynorphin and enkephalin neurons during hyperalgesia: evidence using cDNA hybridization. *Brain Res.*, 455: 205–212.

Karimian, M. and Ferrell, W.R. (1994) Plasma protein extravasation into the rat knee joint induced by calcitonin gene-related peptide. *Neurosci. Lett.*, 166: 39–42.

Kuraishi, Y., Nanayama, T., Ohno, H., Fujii, N., Ataka, A., Yajima, H. and Satoh, M. (1989) Calcitonin gene-related peptide increases in the dorsal root ganglia of adjuvant arthritic rat. *Peptides*, 10: 447–452.

Lombard, M.-C. and Besson, J.-M. (1989) Electrophysiological evidence for a tonic activity of spinal cord intrinsic opioid systems in a chronic pain model. *Brain Res.*, 477: 48–56.

Marlier, L., Poulat, P., Rajaofetra, N. and Privat, A. (1991) Modifications of serotonin-, substance P- and calcitonin gene-related peptide-like immunoreactivities in the dorsal horn of the spinal cord of arthritic rats: a quantitative immunocytochemical study. *Exp. Brain Res.*, 85: 482–490.

McMahon, S.B., Lewin, G.R. and Wall, P.D. (1993) Central hyperexcitability triggered by noxious inputs. *Curr. Opin. Neurobiol.*, 3: 602–610.

Menetrey, D. and Besson, J.-M. (1982) Electrophysiological characteristics of dorsal horn cells in rats with cutaneous inflammation resulting from chronic arthritis. *Pain*, 13: 243–364.

Miletic, V. and Tan, H. (1988) Iontophoretic application of calcitonin gene-related peptide produces a slow and prolonged excitation of neurons in the cat lumbar dorsal horn. *Brain Res.*, 446: 169–172.

Millan, M.J. and Colpaert, F.C. (1991) Opioid systems in the response to inflammatory pain: sustained blockade suggests role of κ- but not μ-opioid receptors in the modulation of nociception, behaviour and pathology. *Neuroscience*, 42: 541–553.

Millan, M.J., Millan, M.H., Pilcher, C.W.T., Czlonkowski, A., Herz, A. and Colpaert, F.C. (1985) Spinal cord dynorphin may modulate nociception via kappa-opioid receptor in chronic arthritic rats. *Brain Res.*, 340: 156–159.

Millan, M.J., Czlonkowski, A., Pilcher, C.W.T., Almeida, O.F.X., Millan, M.H., Colpaert, F.C. and Herz, A. (1987) A model of chronic pain in the rat: functional correlates of alterations in the activity of opioid systems. *J. Neurosci.*, 7: 77–87.

Millan, M.J., Czlonkowski, A., Morris, B., Stein, C., Arendt, R., Huber, H., Höllt, V. and Herz, A. (1988) Inflammation of the hind limb as a model of unilateral, localized pain: influence on multiple opioid systems in the spinal cord of the rat. *Pain*, 35: 299–312.

Nahin, R.L., Hylden, J.L.K., Iadarola, M.J. and Dubner, R. (1989) Peripheral inflammation is associated with increased dynophin immunoreactivity in both projection and local circuit neurons in the superficial dorsal horn of the rat spinal cord. *Neurosci. Lett.*, 96: 247–252.

Neugebauer, V. and Schaible, H.-G. (1990) Evidence for a central component in the sensitization of spinal neurons with joint input during development of acute arthritis in cat's knee. *J. Neurophysiol.*, 64: 299–311.

Neugebauer, V., Lücke, T. and Schaible, H.-G. (1993a) Differential effects of N-methyl-D-aspartate (NMDA) and non-NMDA receptor antagonists on the responses of rat spinal neurons with joint input. *Neurosci. Lett.*, 155: 29–32.

Neugebauer, V., Lücke, T. and Schaible, H.-G. (1993b) N-methyl-D-aspartate (NMDA) and non-NMDA receptor antagonists block the hyperexcitability of dorsal horn neurons during development of acute arthritis in rat's knee joint. *J. Neurophysiol.*, 70: 1365–1377.

Neugebauer, V., Lücke, T., Grubb, B.D. and Schaible, H.-G. (1994a) The involvement of N-methyl-D-aspartate (NMDA) and non-NMDA receptors in the responsiveness of rat spinal neurons with input from the chronically inflamed ankle. *Neurosci. Lett.*, 170: 237–240.

Neugebauer, V., Lücke, T. and Schaible, H.-G. (1994b) Requirement of metabotropic glutamate receptors for the generation of inflammation-evoked hyperexcitability in rat spinal cord neurons. *Eur. J. Neurosci.*, 6: 1179–1186.

Neugebauer, V., Weiretter, F., Rümenapp, P. and Schaible, H.-G. (1994c) Substance P (NK-1) and CGRP receptors are involved in inflammation-induced hyperexcitability of rat dorsal horn neurons. *Soc. Neurosci. Abstr.*, 20: 15.

Neugebauer, V., Weiretter, F. and Schaible, H.-G. (1995) The involvement of substance P and neurokinin-1 receptors in the hyperexcitability of dorsal horn neurons during development of acute arthritis in rat's knee joint. *J. Neurophysiol.*, 73: 1574–1583.

Oku, R., Satoh, M., Fujii, N., Otaka, A., Yajima, H. and Tagaki, H. (1987) Calcitonin gene-related peptide promotes mechanical nociception by potentiating release of substance P from the spinal dorsal horn in rats. *Brain Res.*, 403: 350–354.

Przewlocka, B., Lason, W. and Pzewlocki, R. (1992) Time dependent changes in the activity of opioid systems in the spinal cord of monoarthritic rats - a release and in situ hybridization study. *Neuroscience*, 46: 209–216.

Ruda, M.A., Iadarola, M.J., Cohen, L.V. and Young, W.S. (1988) In situ hybridization histochemistry and immunocytochemistry reveal an increase in spinal dynorphin biosynthesis in a rat model of peripheral inflammation and hyperalgesia. *Proc. Natl. Acad. Sci. USA*, 85: 622–626.

Schaible, H.-G. and Grubb, B.D (1993) Afferent and spinal mechanisms of joint pain. *Pain*, 55: 5–54.

Schaible, H.-G. and Schmidt, R.F. (1985) Effects of an experimental arthritis on the sensory properties of fine articular afferent units. *J. Neurophysiol.*, 54: 1109–1122.

Schaible, H.-G. and Schmidt, R.F. (1988) The time course of mechanosensitivity changes in articular afferents during a developing experimental arthritis. *J. Neurophysiol.*, 60: 2180–2195.

Schaible, H.-G., Schmidt, R.F. and Willis, W.D. (1987) Enhancement of the responses of ascending tract cells in the cat spinal cord by acute inflammation of the knee joint. *Exp. Brain Res.*, 66: 489–499.

Schaible, H.-G., Grubb, B.D., Neugebauer, V. and Oppmann, M. (1991a) The effects of NMDA antagonists on neuronal activity in cat spinal cord evoked by acute inflammation in the knee joint. *Eur. J. Neurosci.*, 3: 981–991.

Schaible, H.-G., Neugebauer, V., Cervero, F. and Schmidt, R.F. (1991b) Changes in tonic descending inhibition of spinal neurons with articular input during the development of acute arthritis in the cat. *J. Neurophysiol.*, 66: 1021–1032.

Simone, D.A., Sorkin, L.S., Oh, U., Chung, J.M., Owens, C., LaMotte, R.H. and Willis, W.D. (1991) Neurogenic hyperalgesia: central neural correlates in responses of spinothalamic tract neurons. *J. Neurophysiol.*, 66: 228–246.

Smith, G.D., Harmar, A.J., McQueen, D.S. and Seckl, J.R. (1992) Increase in substance P and CGRP, but not somatostatin content of innervating dorsal root ganglia in adjuvant monoarthritis in the rat. *Neurosci. Lett.*, 137: 257–260.

Stiller, R.U., Grubb, B.D. and Schaible, H.-G. (1993) Neurophysiological evidence for increased kappa opioidergic control of spinal neurons in rats with unilateral inflammation at the ankle. *Eur. J. Neurosci.*, 5: 1520–1527.

Urban, L., Thompson, S.W.N. and Dray, A. (1994) Modulation of spinal excitability: cooperation between neurokinin and excitatory amino acid transmitters. *Trends Neurosci.*, 17: 432–438.

Weihe, E., Millan, M.J., Höllt, V., Nohr, D. and Herz, A. (1989) A model of chronic pain in the rat: induction of the gene encoding pro-dynorphin by arthritis results in a lightning-up of dynorphin neurons in lumbosacral spinal cord. *Neuroscience*, 31: 77–95.

Wiesenfeld-Hallin, Z., Hökfelt, T., Lundberg, J.M., Forssmann, W.G., Reinecke, M., Tschopp, F.A. and Fischer, J.A. (1984) Immunoreactive calcitonin gene-related peptide and substance P coexist in sensory neurons to the spinal cord and interact in spinal behavioral responses of the rat. *Neurosci. Lett.*, 52: 199–204.

Willis, W.D. (1994) Central plastic responses to pain. In: G.F. Gebhart, D.L. Hammond and T.S. Jensen (Eds.), *Proc. 7th World Congress on Pain, Progress in Pain Research and Management*, Vol. 2, IASP Press, Seattle, WA, pp. 301–324.

Woolf, C.J. (1983) Evidence for a central component of post-injury pain hypersensitivity. *Nature (London)*, 306: 686–688.

CHAPTER 14

Acute mechanical hyperalgesia in the rat can be produced by coactivation of spinal ionotropic AMPA and metabotropic glutamate receptors, activation of phospholipase A_2 and generation of cyclooxygenase products

S.T. Meller, C. Dykstra and G.F. Gebhart

Department of Pharmacology, University of Iowa, Iowa City, IA 52242, USA

Introduction

Persistent pain is often associated with altered sensitivity to cutaneous stimuli which is manifest as hyperalgesia (increased sensitivity to noxious stimuli) and allodynia (non-noxious stimuli perceived as noxious) (Willis, 1992). As a consequence of tissue injury, inflammatory mediators released at the site of injury contribute to sensitization of nociceptors (Handwerker and Reeh, 1992; Levine et al., 1992). The increase in neuronal activity leads to a neuronal "plasticity" in the dorsal horn of the spinal cord, which contributes to the development of hyperalgesia and allodynia (Dubner and Ruda, 1992). Although recent efforts to describe and define hyperalgesia indicate that excitatory amino acids (EAAs) are intimately involved in mechanisms that result in long-term, use-dependent changes in neuronal excitability in the spinal cord (Wilcox, 1991; Dubner and Ruda, 1992; Meller and Gebhart, 1993), EAA receptor subtypes and signal transduction mechanisms subsequent to receptor activation that result in hyperalgesia are incompletely defined.

Recent investigations have focused on mechanisms of thermal hyperalgesia and it is now clear that activation of *N*-methyl-D-aspartate (NMDA) receptors in the spinal cord play a significant role in thermal hyperalgesia. Further, this NMDA-receptor produced and maintained thermal hyperalgesia is mediated through production of nitric oxide (NO) (Meller et al., 1992a,b, 1994, 1995; Kitto et al., 1992; Malmberg and Yaksh, 1993) and activation and translocation of protein kinase C (PKC) (Mao et al., 1992a,b, 1993; Meller et al., 1995).

In contrast, the receptor subtypes that are involved in acute or persistent mechanical hyperalgesia are not well defined; in preliminary studies we found that coactivation of AMPA and metabotropic glutamate receptor subtypes are able to produce an acute mechanical hyperalgesia (Meller et al., 1993a). However, there is little known about the signal transduction mechanisms that underlie mechanical hyperalgesia (although see Palacek et al., 1994). Activation of non-NMDA receptors is able to trigger a number of signal transduction pathways in other systems (Farooqui and Horrocks, 1991; Baskys, 1992; Pin et al., 1992; Schoepp, 1993; Schoepp and Conn, 1993). For example, activation of non-NMDA ionotropic receptors has been shown to produce NO (Garthwaite et al., 1989; Wood et al., 1990; Marin et al., 1993) while activation of the metabotropic receptor has been linked to phospholipase C (PLC), arachidonic acid (AA) and cAMP (Nakanishi, 1992; Schoepp, 1993). Recent evidence also suggests that coactivation of AMPA and metabotropic glutamate receptors is linked to production of AA through activation of phospholipase A_2 (PLA_2) (Dumuis et al., 1990).

Given the role of AA in synaptic plasticity mechanisms such as long-term potentiation (Lynch et al., 1989; Lynch, 1991), it is possible that AA, like NO, might act either directly or through one of its lipoxygenase (LOX) or cyclooxygenase (COX) products to produce hyperalgesia. However, there has been limited investigation into the possible role of AA or its metabolites in nociception in the spinal cord. Although recent reports indicate that intrathecal (i.t.) administration of various prostanoids are able to produce both thermal and mechanical hyperalgesia (Taiwo and Levine, 1988; Uda et al., 1990; Minami et al., 1992), it is not at all clear whether there is some modality-specificity for different metabolites of AA. Therefore, the aim of the present study was to characterize the glutamate receptor subtypes and the signal transduction mechanisms that are involved in acute mechanical hyperalgesia.

Experimental procedures

Animals and surgery

A total of seventy male Sprague–Dawley rats (Harlan Sprague-Dawley, Indianapolis, IN; 400–525 g), housed in a room at constant temperature (22°C) on a 12/12 h light/dark cycle with food and water available ad libitum, were used in the present study (Table 1).

All rats were deeply anesthetized with an intraperitoneal injection of sodium pentobarbital (45 mg/kg, Nembutal®, Abbott Labs., North Chicago, IL) and a catheter (8.25 cm; PE-10) was permanently placed into the lumbar i.t. space through an incision in the dura over the atlantooccipital joint. All rats were allowed 3–7 days to recover from surgery before testing; there was no difference in the responses of rats when tested at 3 days and 7 days (Meller et al., 1992a).

Nociceptive testing

On the day of the experiment, rats were allowed to freely crawl inside a canvas garden glove.

TABLE 1

Summary of experimental groups, drugs tested, number of doses and number of rats in each

Manipulation	Number of doses	n
NMDA dose-response curve	9	5
AMPA and trans-ACPD dose-response curves	11	5
QA dose-response curve	7	5
AMPA + trans-ACPD dose-response curve	6	5
QA and AMPA + trans-ACPD with DNQX	11	5
QA and AMPA + trans-ACPD with AP3	11	5
QA and AMPA + trans-ACPD with APV, L-NAME, MB	13	5
QA and AMPA + trans-ACPD with chelerythine	7	5
QA and AMPA + trans-ACPD with neomycin	7	5
QA and AMPA + trans-ACPD with baicalein	7	5
QA and AMPA + trans-ACPD with indomethacin	21	5
QA with NDGA	12	5
AMPA + trans-ACPD with NDGA	11	5
QA and AMPA + trans-ACPD with mepacrine	23	5
Total	156	70

Withdrawal of the tail to mechanical pressure was measured using a set of calibrated nylon monofilaments (Semmes-Weinstein Anesthesiometer, Stoelting Co., Chicago, IL) requiring different pressures to bow the filament. The tip of each filament produced a punctate stimulus which did not damage the skin; filaments varied in diameter from the smallest filament used (0.25 mm) to the largest (1.05 mm) (Table 2). Mechanical stimulation was increased in a graded manner on the dorsal surface of the tail at one of two sites (either 8 or 15 cm from the tip) using successively greater diameter filaments until the tail was reflexively withdrawn. Thresholds for both sites were averaged. This method generates reliable and reproducible nociceptive withdrawal reflexes (Zhuo and Gebhart, 1991; Zhuo et al., 1993). Thresholds were checked at least twice for each trial at each site. The overall baseline withdrawal threshold for

TABLE 2

The diameter and force (g) for each filament used in the present experiments

Filament	Diameter	Force (g)
4.17	0.25	1.2
4.31	0.29	1.8
4.56	0.37	4.9
4.74	0.37	5.5
4.93	0.42	7.6
5.07	0.49	13.4
5.18	0.51	15.2
5.46	0.55	21.0
5.88	0.71	55.0
6.10	0.82	92.0
6.45	1.05	156.0

untreated rats was 98.0 ± 2.1 g (range 55–156 g); baseline withdrawal thresholds for the different experimental groups are provided in the figure legends. These intensities of mechanical stimulation were considered to be noxious as the response thresholds for high threshold mechanoreceptors in rat hindlimb skin has been reported to be in the range of 0.5–2.6 g (Lynn and Carpenter, 1982). Studies of persistent mechanical hyperalgesia are typically performed on the hindlimb (so that one side can be compared with the other) and we have made the assumption that mechanical stimulation of the tail activates neural mechanisms analogous to those activated by mechanical stimulation of a hindlimb.

Experimental protocol

In the present experiments, receptor selective EAA agonists were administered into the lumbar i.t. space to awake rats to determine which EAA receptors could produce an acute mechanical hyperalgesia. Here, mechanical hyperalgesia is defined as a reduction in mechanical withdrawal threshold. Rats were divided into 14 groups (Table 1). On average, rats received between 6 and 21 i.t. doses of drugs during the course of a 2–3 h experiment (see Table 1); all rats received a vehicle control before drug administration. For the dose-response curves, subsequent administration of agonist was not made until mechanical withdrawal thresholds returned to baseline. For studies with antagonists or enzyme blockers, agonists were administered at 15 min intervals until full recovery of the agonist-produced effect. Subsequent administration of antagonists or enzyme blockers were not made until baseline mechanical withdrawal thresholds and agonist-produced mechanical-hyperalgesia had returned to pre-treatment values.

Rats were injected with NMDA (100 amol-1 nmol), α-amino-3-hydroxy-5-methylisoxazole-4-proprionate (AMPA; 10 fmol to 100 pmol), *trans*-1-amino-1,3-cyclopentanedicarboxylate (*trans*-ACPD; 10 fmol to 100 pmol), quisqualate (QA; 5 pmol to 1 nmol) or a 1:1 combination of AMPA and *trans*-ACPD (total dose = 20 fmol to 20 pmol) in a dose-dependent manner and mechanical thresholds were tested at 0.5, 1, 2, 5, 10 and 15 min post-drug. For all of the EAA receptor agonists, in the dose ranges studied, mechanical withdrawal thresholds returned to pre-drug values within 15 min. For all analyses, the maximum effect after drug administration was used.

After determining which EAA agonists produced an acute mechanical hyperalgesia, their effects were examined before and 10 min after i.t. administration of either the NMDA receptor selective antagonist 2-amino-5-phosphonopentanoate (APV; 100 pmol), the AMPA receptor selective antagonist 6,7-dinitroquinoxaline-2,3-dione (DNQX; 10 pmol to 1 nmol) or the metabotropic receptor antagonist 2-amino-3-phosphonopropionate (AP3; 100 pmol to 0 nmol). If an antagonist attenuated agonist-produced mechanical hyperalgesia, testing continued every 15 min until complete recovery of the agonist-produced effect. A separate, but similarly structured series of experiments examined the effect of pretreatment with the protein kinase C (PKC) inhibitor chelerythrine (100 pmol), the NO synthase (NOS) inhibitor N^G-nitro-L-arginine methyl ester (L-NAME, 10 nmol), the soluble guanylate cyclase (GC-S) inhibitor methylene blue (1 nmol), the PLA_2 inhibitor mepacrine (100 pmol to 10 nmol), the PLC inhibitor neomycin (10 nmol), the non-selective eicosanoid inhibitor nordihydroguaiarate (NDGA; 100 pmol to 10 nmol), the COX inhibitor indomethacin (100 pmol to 10 nmol) or the LOX inhibitor baicalein (1 nmol).

Drugs

NMDA, AMPA, QA, APV, AP3, mepacrine, neomycin, NDGA, indomethacin and baicalein were purchased from Sigma Chemical Co. (St. Louis, MO). Chelerythrine was purchased from LC Laboratories (Woburn, MA), trans-ACPD was purchased from Tocris Neuramin (Essex, UK), DNQX was purchased from Research Biochemical Int. (Natick, MA), L-NAME was purchased from Bachem (Torrance, CA) and methylene blue was purchased from Fisher Scientific Co. (Fair Lawn, NJ). Stock solutions of all drugs were made up fresh in saline and diluted according to the concentration needed except for NDGA, baicalein and indomethacin, which were dissolved in 10%, 10% or 100% DMSO, respectively, and diluted accordingly. All drugs were given in a volume of 1 μl followed by a flush of 10 μl preservative-free saline. In all rats the effect of vehicle alone (saline, 10% and 100% DMSO) was examined with corresponding pH adjustments (pH 2.0–9.0) for the drugs that were administered.

Statistical analyses

Withdrawal thresholds are reported in grams (g) or expressed as the maximum percentage difference from time 0 (pre-drug) according to the formula: (baseline threshold − post-drug withdrawal threshold)/baseline threshold × 100. Data are presented as mean ± SEM. For rats which demonstrated a dose-dependent mechanical hyperalgesia, the ED_{50} (50% of maximum effect) was determined using sigmoid curve fitting. All data were analyzed using the Kruskall–Wallis k-sample test. In all cases $P < 0.05$ was considered significant.

Results

Receptor subtypes

I.t. administration of a wide range of doses of NMDA (100 amol to 1 nmol) did not produce any change in mechanical withdrawal threshold of the tail (Fig. 1); greater doses produced a caudally directed biting and scratching behavior that precluded further testing in this paradigm. In contrast, i.t. administration of quisqualate (QA) produced a rapid (within 30 s), transient (lasting less than 5 min), dose-dependent (5 pmol to 1 nmol) mechanical hyperalgesia ($ED_{50} = 27.4 \pm 7.4$ pmol; maximum observed effect, $86.1 \pm 0.2\%$) (Fig. 1). The effect was both reproducible and reliable. Desensitization was avoided by maintaining a 15 min interval between repeated doses.

As QA activates both ionotropic, AMPA-sensitive and metabotropic, trans-ACPD-sensitive glutamate receptors (Schoepp et al., 1990), the

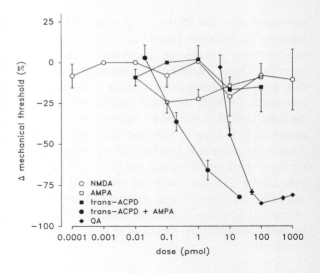

Fig. 1. Effect of NMDA (1 fmol to 1 pmol, $n = 5$), AMPA (10 fmol to 100 pmol, $n = 5$), trans-ACPD (10 fmol to 100 pmol, $n = 5$), QA (10 fmol to 10 nmol, $n = 5$) or AMPA + trans-ACPD (20 fmol to 20 pmol, $n = 5$) on mechanical withdrawal thresholds. The dose administered is represented on the x-axis in picomoles (pmol) and changes in mechanical withdrawal thresholds are represented on the y-axis as % change calculated by the formula: (trial threshold − baseline threshold)/(baseline threshold) × 100 expressed as a percentage. All data points are expressed as mean ± SEM. Baseline mechanical thresholds between groups were not different: NMDA, 82.6 ± 4.1 g; AMPA, 101.1 ± 8.9 g; trans-ACPD, 104.8 ± 7.0 g; QA, 84.6 ± 4.1 g; AMPA + trans-ACPD, 92.0 ± 0.1 g. A total of 20 rats were used; the same 5 rats were used, on the same day, for the AMPA alone and trans-ACPD alone dose-response curves.

receptor subtype(s) involved in QA-produced acute mechanical hyperalgesia was further investigated using two different approaches. First, we examined the effect of receptor selective antagonists on QA-produced mechanical hyperalegsia and, second, we examined the effect of agonists selective for AMPA and metabotropic glutamate receptors administered alone and in combination.

The i.t. administration of the AMPA receptor selective antagonist DNQX (10 pmol to 1 nmol) had no effect on baseline mechanical withdrawal thresholds, but did produce a dose-dependent attenuation of QA-produced mechanical hyperalgesia ($ED_{50} = 149.2 \pm 28.4$ pmol) (Fig. 2A), suggesting that activation of AMPA receptors is involved in QA-produced mechanical hyperalgesia. Maximum observed effect was produced by 1 nmol DNQX (75.3 ± 7.4% attenuation of QA-produced mechanical hyperalgesia) (Fig. 2A). In addition to the effect of DNQX, QA-produced mechanical hyperalgesia was also attenuated in a dose-dependent manner by pre-treatment with AP3 (100 pmol to 10 nmol, $ED_{50} = 3.4 \pm 0.8$ nmol) (Fig. 2B), an antagonist at the metabotropic glutamate receptor. Maximum observed effect was produced by 10 nmol AP3 (70.4 ± 4.3% attenuation of QA-produced mechanical hyperalgesia) (Fig. 2B). AP3, like DNQX, had no effect on baseline mechanical withdrawal thresholds (Fig. 2B). In contrast, i.t. pretreatment with a dose of the NMDA receptor selective antagonist APV (100 pmol) that attenuates thermal hyperalgesia (Meller et al., 1995) affected neither baseline mechanical withdrawal thresholds nor the mechanical hyperalgesia produced by QA (Fig. 3A). These results suggest that both AMPA and metabotropic glutamate receptors, but not NMDA receptors, are involved in QA-produced hyperalgesia. However, when tested alone, neither i.t. administration of AMPA (10 fmol to 100 pmol), an AMPA receptor agonist, nor trans-ACPD (10 fmol to 100 pmol), a metabotropic receptor selective agonist, produced any evidence of mechanical hyperlagesia over the dose ranges tested (Fig. 1). Greater doses could not be administered because the agonists produced caudally-directed biting and scratching behavior that prevented testing in this paradigm. Although

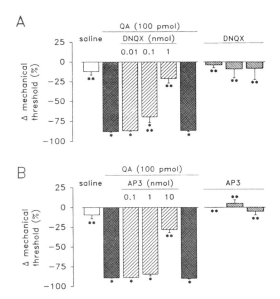

Fig. 2. Effect of either DNQX (10 pmol to 1 nmol) or AP3 (100 pmol to 10 nmol) on mechanical hyperalgesia produced by QA (100 pmol). Changes in mechanical withdrawal threshold are represented on the y-axis as % change calculated by the formula: (trial threshold − baseline threshold)/(baseline threshold) × 100 expressed as a percentage. All data points are expressed as mean ± SEM. Baseline mechanical thresholds between groups were not different: QA pre-DNQX, 104.8 ± 11.4 g; QA pre-AP3, 117.6 ± 10.7 g. A total of 10 rats were used for the data in this figure and Fig. 4; QA and AMPA + trans-ACPD were tested in the same rat after either DNQX ($n = 5$) or AP3 ($n = 5$) pretreatments. *Significantly different from saline, **significantly different from agonist alone.

activation of AMPA or metabotropic glutamate receptors alone do not produce an acute mechanical hyperalgesia, coadministration of AMPA and trans-ACPD (1:1; total combined dose 20 fmol to 20 pmol) did produce a dose-dependent mechanical hyperalgesia ($ED_{50} = 329.8 \pm 135.4$ fmol; maximum observed effect, 82.2 ± 1.1%) with a magnitude and time-course similar to the mechanical hyperalgesia produced by QA (Fig. 1). As with QA, this effect was both reproducible and reliable upon repeated administration. In addition, the maximum mechanical hyperalgesia produced by AMPA + trans-ACPD (20 pmol) was attenuated in a dose-dependent manner by i.t. administration of either DNQX (10 pmol to 1 nmol) or AP3 (100 pmol to 10 nmol) ($ED_{50} = 132.0 \pm 8.6$ pmol

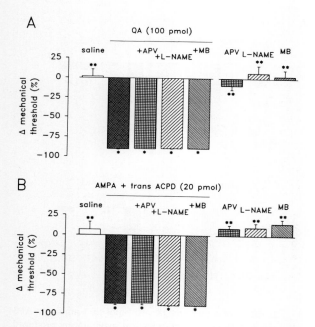

Fig. 3. Effect of APV (100 pmol), L-NAME (10 nmol) or methylene blue (1 nmol) on the mechanical hyperalgesia produced by QA (100 pmol) or AMPA + *trans*-ACPD (20 pmol). Changes in mechanical withdrawal threshold are represented on the y-axis as % change calculated by the formula: (trial threshold − baseline threshold)/(baseline threshold) × 100 expressed as a percentage. All data points are expressed as mean ± SEM. Baseline mechanical thresholds between groups were not different: QA pre-APV, pre-L-NAME, pre-methylene blue; 111.2 ± 7.0 g; AMPA + *trans*-ACPD pre-APV pre-L-NAME, pre-methylene blue, 124.0 ± 0.1 g. A total of 5 rats were used. *Significantly different from saline; **significantly different from agonist alone.

and 2.2 ± 0.6 nmol, respectively) (Fig. 4), but not APV (100 pmol) (Fig. 3B). Maximum observed effect was produced by 1 nmol DNQX (89.5 ± 6.5% attenuation of the AMPA + *trans*-ACPD-produced mechanical hyperalgesia) or 10 nmol AP3 (86.2 ± 9.2% attenuation of the AMPA + *trans*-ACPD-produced mechanical hyperalgesia) (Fig. 4). The i.t. administration of saline, at pH values ranging from 2.0 to 9.0, 10% DMSO in saline or 100% DMSO did not alter mechanical withdrawal thresholds.

Signal transduction mechanisms

We next examined the role of different signal transduction pathways in QA- (or AMPA + *trans*-ACPD)-produced acute mechanical hyperalgesia. Neither the QA- nor AMPA + *trans*-ACPD-produced mechanical hyperalgesia or baseline mechanical withdrawal thresholds were affected by doses of L-NAME (10 nmol), methylene blue (1 nmol) or chelerythrine (100 pmol) (Figs. 3 and 5) that completely block acute thermal hyperalgesia (Meller et al., 1995). In addition, the QA- and AMPA + *trans*-ACPD-produced mechanical hyperalgesia was unaffected by the PLC inhibitor neomycin (10 nmol) (Fig.6).

In contrast, both the QA- and AMPA + *trans*-ACPD-produced mechanical hyperalgesia were attenuated in a dose-dependent manner by administration of the PLA_2 inhibitor mepacrine (100 pmol to 10 nmol, ED_{50} = 2.6 ± 0.6 nmol and 0.9 ±

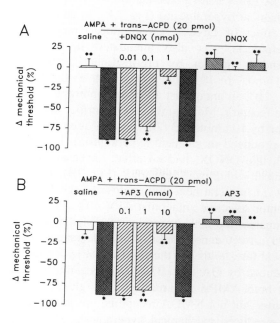

Fig. 4. Effect of either DNQX (10 pmol to 1 nmol) or AP3 (100 pmol to 10 nmol) on mechanical hyperalgesia produced by AMPA + *trans*-ACPD (20 pmol). Changes in mechanical withdrawal threshold are represented on the y-axis as % change calculated by the formula: (trial threshold − baseline threshold)/(baseline threshold) × 100 expressed as a percentage. All data points are expressed as mean ± SEM. Baseline mechanical thresholds between groups were not different: AMPA + *trans*-ACPD pre-DNQX, 101.1 ± 8.9 g; AMPA + *trans*-ACPD pre-AP3, 117.6 ± 10.7 g. *Significantly different from saline, **significantly different from agonist alone.

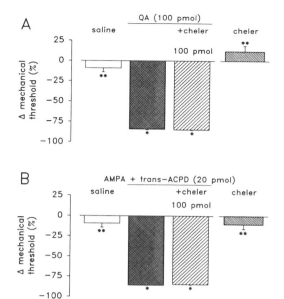

Fig. 5. Effect of chelerythrine (100 pmol) on the mechanical hyperalgesia produced by QA (100 pmol) or AMPA + *trans*-ACPD (20 pmol). Changes in mechanical withdrawal threshold are represented on the *y*-axis as % change calculated by the formula: (trial threshold − baseline threshold)/(baseline threshold) × 100 expressed as a percentage. All data points are expressed as mean ± SEM. Baseline mechanical thresholds between groups were not different: QA pre-chelerythrine; 80.9 ± 4.1 and AMPA + *trans*-ACPD pre-chelerythrine; 88.3 ± 3.3. A total of 5 rats were used. *Significantly different from saline; **significantly different from agonist alone.

0.2 nmol, respectively) (Fig. 7). Maximum observed effect was produced by 10 nmol mepacrine (78.8 ± 9.2% and 73.7 ± 7.3% attenuation of QA- and AMPA + *trans*-ACPD-produced mechanical hyperalgesia, respectively) (Fig. 7). Mepacrine had no effect on baseline mechanical withdrawal thresholds (Fig. 7).

Activation of PLA_2 produces AA (Glaser et al., 1993) which is then rapidly converted primarily to prostaglandin and leukotriene products by COX and LOX enzymes, respectively. The non-selective COX and LOX inhibitor NDGA (100 pmol to 10 nmol) produced a dose-dependent inhibition of acute QA or AMPA + *trans*-ACPD-produced mechanical hyperalgesia (ED_{50} = 2.1 ± 0.5 nmol and 2.3 ± 0.7 nmol, respectively) (Fig. 8). Maximum observed effect was produced by 10 nmol NDGA (90.9 ± 5.6% and 101.3 ± 8.4% attenuation of QA- and AMPA + *trans*-ACPD-produced mechanical hyperalgesia, respectively) (Fig. 8). As this inhibitor is non-selective, we tested selective inhibitors of the COX (indomethacin) and LOX (baicalein) pathways. Both the QA- and AMPA + *trans*-ACPD-produced mechanical hyperalgesia were attenuated in a dose-dependent manner by administration of indomethacin (100 pmol to 10 nmol, ED_{50} = 1.3 ± 0.1 nmol and 1.8 ± 0.3 nmol, respectively) (Fig. 9). Maximum observed effect was produced by 10 nmol indomethacin (86.2 ± 9.2% and 100.3 ± 16.0% attenuation of QA- and AMPA + *trans*-ACPD-produced mechanical hyperalgesia, respectively) (Fig. 9). Baicalein (1 nmol) was without effect on the acute mechanical hyperalgesia produced by QA or

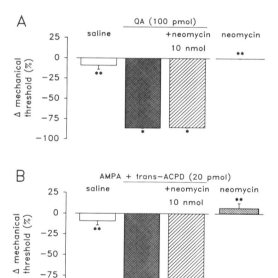

Fig. 6. Effect of neomycin (10 nmol) on the mechanical hyperalgesia produced by QA (100 pmol) or AMPA + *trans*-ACPD (20 pmol). Changes in mechanical withdrawal threshold are represented on the *y*-axis as % change calculated by the formula: (trial threshold − baseline threshold)/(baseline threshold) × 100 expressed as a percentage. All data points are expressed as mean ± SEM. Baseline mechanical thresholds in each group were not different: QA pre-neomycin, 80.9 ± 4.1 and AMPA + *trans*-ACPD pre-neomycin, 80.9 ± 6.6 g. A total of 5 rats were used. *Significantly different from saline; **significantly different from agonist alone.

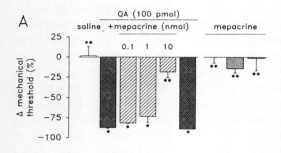

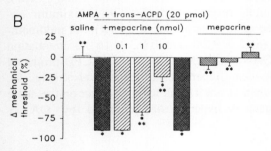

Fig. 7. Effect of mepacrine (100 pmol to 10 nmol) on the mechanical hyperalgesia produced by QA (100 pmol) or AMPA + *trans*-ACPD (20 pmol). Changes in mechanical withdrawal threshold are represented on the y-axis as % change calculated by the formula: (trial threshold − baseline threshold)/(baseline threshold) × 100 expressed as a percentage. All data points are expressed as mean ± SEM. Baseline mechanical thresholds between groups were not different: QA pre-mepacrine; 117.6 ± 5.7 g and AMPA + *trans*-ACPD pre-mepacrine; 117.6 ± 5.7 g. A total of 5 rats were used. *Significantly different from saline, **significantly different from agonist alone.

AMPA + *trans*-ACPD (Fig. 10). Neither baicalein, NDGA nor indomethacin had any effect on baseline mechanical withdrawal thresholds (Figs. 8–10).

Discussion

The results of the present study support the hypothesis that coactivation of spinal AMPA and metabotropic glutamate receptors produce an acute mechanical hyperalgesia in the rat. Further, this agonist-produced acute mechanical hyperalgesia appears to be mediated through activation of PLA_2 and production of COX products, but likely not through activation of NMDA receptors, PKC, PLC or generation of NO, cGMP or LOX products.

Hyperalgesia

The International Association for the Study of Pain has defined hyperalgesia as "an increased response to a stimulus that is normally painful." Acute mechanical hyperalgesia is defined here as a reduction in mechanical withdrawal threshold (provided the pressure at threshold is in the noxious range). The minimum tail withdrawal thresholds after i.t. administration of QA or AMPA + *trans*-ACPD in the present study were in the range of 9–12 g, which is greater than the reported mechanical thresholds for activation of Aδ- and C-fibers (Lynn and Carpenter, 1982).

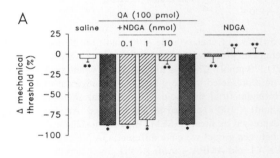

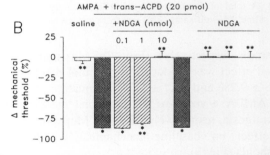

Fig. 8. Effect of NDGA (100 pmol to 10 nmol) on the mechanical hyperalgesia produced by QA (100 pmol) or AMPA + *trans*-ACPD (20 pmol). Changes in mechanical withdrawal threshold are represented on the y-axis as % change calculated by the formula: (trial threshold − baseline threshold)/(baseline threshold) × 100 expressed as a percentage. All data points are expressed as mean ± SEM. Baseline mechanical thresholds between groups were not different: QA pre-NDGA; 104.8 ± 7.0 g; AMPA + *trans*-ACPD pre-NDGA; 88.3 ± 3.3 g. A total of 10 rats were used; QA (± NDGA; $n = 5$) and AMPA + *trans*-ACPD (± NDGA; $n = 5$). *Significantly different from saline, **significantly different from agonist alone.

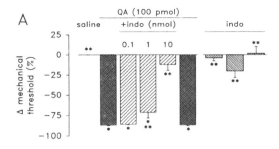

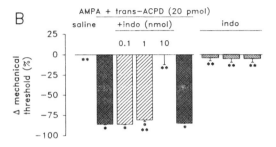

Fig. 9. Effect of indomethacin (100 pmol to 10 nmol) on the mechanical hyperalgesia produced by QA (100 pmol) or AMPA + *trans*-ACPD (20 pmol). Changes in mechanical withdrawal threshold are represented on the y-axis as % change calculated by the formula: (trial threshold − baseline threshold)/(baseline threshold) × 100 expressed as a percentage. All data points are expressed as mean ± SEM. Baseline mechanical thresholds between groups were not different: QA pre-indomethacin; 92.0 ± 0.1 g and AMPA + *trans*-ACPD pre-indomethacin; 88.3 ± 3.3 g. A total of 5 rats were used; a 30 min interval was given between QA (± indomethacin) and AMPA + *trans*-ACPD (± indomethacin). *Significantly different from saline, **significantly different from agonist alone.

Receptor subtypes

At present, relatively little is known about the EAA receptor subtypes involved in mechanical hyperalgesia. This is surprising considering the large and growing number of recent reports on mechanisms of hyperalgesia; these reports, however, have centered on thermal hyperalgesia. While activation of NMDA receptors is clearly linked to thermal hyperalgesia (see Meller and Gebhart, 1993 for review), one cannot assume that NMDA receptors are also involved in mechanical hyperalgesia in the same manner. In fact, the present report demonstrates that NMDA receptor activation alone does not produce an acute hyperalgesia to a punctate mechanical stimulus; coactivation of AMPA and metabotropic glutamate receptors is required. In addition, the present findings show that neither AMPA nor *trans*-ACPD alone altered nociceptive mechanical withdrawal reflexes, but coactivation of AMPA and metabotropic glutamate receptors produced an acute mechanical hyperalgesia.

Although these results relate directly to acute, short-term mechanical hyperalgesia, we believe that similar mechanisms are important to persistent mechanical hyperalgesia. In support, Dougherty et al. (1992b) reported that the sensitization of spinothalamic tract neurons to mechanical stimulation after intra-articular injection of carrageenan was enhanced by non-NMDA receptor agonists such as

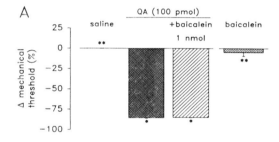

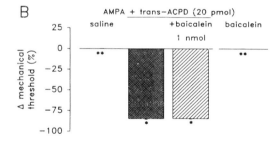

Fig. 10. Effect of baicalein (1 nmol) on the mechanical hyperalgesia produced by QA (100 pmol) or AMPA + *trans*-ACPD (20 pmol). Changes in mechanical withdrawal threshold are represented on the y-axis as % change calculated by the formula: (trial threshold − baseline threshold)/(baseline threshold) × 100 expressed as a percentage. All data points are expressed as mean ± SEM. Baseline mechanical thresholds between groups were not different: QA pre-baicalein and AMPA + *trans*-ACPD pre-baicalein; 80.9 ± 6.6 g. A total of 5 rats were used. *Significantly different from saline; **significantly different from agonist alone.

QA; NMDA was without effect. Preliminary findings by us (Meller et al., 1993b) using intraplantar injection of the inflammatory agent zymosan suggest that either AMPA or metabotropic glutamate receptor antagonists attenuate persistent mechanical hyperalgesia (but do not affect thermal hyperalgesia). Further, Tal and Bennett (1993) have reported that the mechanical hyperalgesia produced in a model of neuropathic pain is unaffected by the NMDA receptor antagonist dextrorphan (thermal hyperalgesia was blocked). Consistent with this, Rueff et al. (1993) reported that the thermal hyperalgesia, but not the mechanical hyperalgesia, produced following nerve growth factor treatment is blocked by the NMDA receptor channel blocker MK-801. In further support of a role for AMPA and metabotropic receptors in mechanical hyperalgesia (and a lack of effect of NMDA receptors), mechanical hypersensitivity following spinal cord ischemia is abolished by AMPA but not NMDA receptor antagonists (Yu et al., 1992). In addition, this same group (Hao et al., 1992) reported that systemic tocainamide relieves the mechanical hypersensitivity produced by spinal cord ischemia. While this is consistent with the clinical use and efficacy of local anesthetics in relieving chronic neuropathic pain, Biella et al. (1993) has shown that systemic lidocaine actually enhances the effects of NMDA, but reduces the effects of QA on hyperexcitable dorsal horn neurons. This is entirely consistent with the reported reduction in mechanical hypersensitivity found in the clinical literature being due to actions of local anesthetics on the QA-sensitive mechanisms that we propose are important to mechanical hyperalgesia.

Early evidence that suggested a prominent role for NMDA receptors in hyperalgesia included reports that NMDA receptor antagonists blocked the wind-up produced by mustard oil (Woolf and Thompson, 1991) and that NMDA receptor antagonists reduced the expansion of spinal cord neuron receptive fields produced by inflammation (see Dubner and Ruda, 1992); such data suggested an obligatory role for NMDA receptors in hyperalgesia. However, it has subsequently been reported that the wind-up produced by mustard oil can be blocked by metabotropic glutamate receptor antagonists (Young et al., 1994) and the expansion of receptive fields produced by inflammation is reduced by CNQX (Neugebauer et al., 1993b). These electrophysiological results suggest that AMPA and metabotropic glutamate receptors may play a prominent role in hyperalgesia, as also suggested in the present report.

There is also evidence of a role for NMDA receptors in mechanical hyperalgesia. For example, we found that the mechanical hyperalgesia produced by intraplantar zymosan is attenuated by MK-801 (Meller et al., 1993b). Similarly, the mechanical hyperalgesia produced by intraplantar carrageenan is attenuated by either MK-801 or APV (Ren and Dubner, 1993). In addition, NMDA receptor activation has been shown to enhance the responses of dorsal horn neurons to noxious and non-noxious mechanical stimulation (Sher and Mitchell, 1990; Dougherty and Willis, 1992; Dougherty et al., 1992a) and NMDA receptor antagonists have been shown to attenuate the hyper-responsiveness of dorsal horn neurons to mechanical stimulation following inflammation produced by carrageenan and kaolin (Neugebauer et al., 1993a,b) or capsaicin (Dougherty and Willis, 1992; Dougherty et al., 1992a). Further, a recent clinical report (Kristensen et al., 1992) has shown that mechanical hypersensitivity in a neuropathic pain patient was abolished by i.t. administration of an NMDA receptor antagonist. These results appear to contradict the present results that acute NMDA receptor activation, at the doses tested, does not produce an acute mechanical hyperalgesia. However, it is possible that NMDA receptor activation is able to enhance the effects of AMPA and metabotropic glutamate receptor coactivation to produce persistent mechanical hyperalgesia, or possibly that NMDA receptor activation allows for the expression of the events involved in mechanical hyperalgesia (i.e., AMPA and metabotropic glutamate receptor activation). That is, NMDA receptor activation would allow for a cascade of events (AMPA and metabotropic glutamate receptor activation) leading to the expression of mechanical hyperalgesia. The reduction in expanded receptive fields of the same neuron by MK-801 or

DNQX supports this hypothesis (Neugebauer et al., 1993b).

It is generally thought that activation of KA/AMPA receptors is involved in fast synaptic transmission. Therefore, one might expect that an antagonist at AMPA/KA receptors like DNQX might block sensory transmission when administered into the i.t. space. At the doses tested, we observed no effect on baseline mechanical withdrawal thresholds (or thermal withdrawal latencies; Meller, Dykstra and Gebhart, unpublished observations). However, when greater doses of DNQX were given (data not shown), we observed that the hindlimbs of the rats were unresponsive to noxious thermal or mechanical stimuli, suggesting a blockade of sensory and/or motor transmission. This fits with the hypothesis that AMPA/KA receptors are involved in fast synaptic transmission. One explanation for the attenuation of mechanical hyperalgesia by DNQX in the absence of an effect on baseline withdrawal thresholds is that different AMPA/KA receptor subtypes may be involved. Recent evidence suggests that some AMPA-sensitive receptors are able to gate calcium (Sommer and Seeburg, 1992); we have previously suggested that the fast excitatory AMPA/KA receptors that are responsible for spinal cord transmission are different from those AMPA receptors that are involved in the development of mechanical hyperalgesia (sodium versus calcium permeability?) (Meller et al., 1993a). This hypothesis has not been examined directly in electrophysiological studies in the spinal cord, although immunocytochemical experiments by Tolle et al. (1993) suggest that sodium-gated AMPA receptors might be more numerous in lamina II while calcium-gated AMPA receptors might be more predominant in lamina IV.

An interaction between AMPA and metabotropic glutamate receptors (and possibly NMDA) is further supported by several recent electrophysiological studies that have revealed a cooperativity between these two receptor subtypes on spinal cord neurons (Bleakman et al., 1992; Cerne and Randic, 1992). In the study by Bleakman et al. (1992), it was reported that the metabotropic glutamate receptor agonist 1S,3R-ACPD enhanced responses to AMPA, KA and NMDA. They suggested that these effects may play a role in the regulation of ionotropic responses mediated by glutamate in the spinal cord. Cerne and Randic (1992) reported that *trans*-ACPD potentiated the responses of both NMDA and AMPA and suggested that metabotropic receptors may contribute to the strength of primary afferent nociceptive transmission. However, in a previous study we found that while AMPA + *trans*-ACPD produced an acute mechanical hyperalgesia, NMDA + *trans*-ACPD did not (Meller et al., 1993a); *trans*-ACPD did potentiate an NMDA-produced acute thermal hyperalgesia.

Signal transduction mechanisms

There are a number of intracellular enzymes and second messenger systems that have been linked to glutamate receptor activation (Mayer and Miller, 1990; Farooqui and Horrocks, 1991; Schoepp, 1993). For example, NMDA receptors have been linked to the activation of NOS (Bredt and Snyder, 1989; Garthwaite et al., 1989), PKC (Vaccarino et al., 1987; Etoh et al., 1991) and PLA$_2$ (Dumuis et al., 1988; Tapia-Aranciba et al., 1992) and other ionotropic non-NMDA glutamate receptors have been shown to be able to activate NOS (Wood et al., 1990; Marin et el., 1993); the metabotropic receptor has been linked to activation of adenylate cyclase and PLC (Nakanishi, 1992; Schoepp, 1993). The intracellular events activated by calcium-permeable AMPA receptors have not been reported, but coactivation of AMPA and metabotropic glutamate receptors has been reported to produce AA through activation of PLA$_2$ (Dumuis et al., 1990). We tested the role of each in the mechanisms underlying QA and AMPA + *trans*-ACPD-produced mechanical hyperalgesia.

There have been several reports that activation of non-NMDA receptors is able to produce NO (Wood et al., 1990; Marin et al., 1993). While it might be reasonable to suggest that NO may be a final, labile intracellular mediator of both thermal and mechanical hyperalgesia, this appears not to be the case. Acute mechanical hyperalgesia produced by coactivation of AMPA and metabotropic glutamate receptors is unaffected by i.t. administration of doses of L-NAME or methylene blue that we have found to completely block thermal hyperalgesia (Meller et al., 1992a, 1995). In addition, we have shown that the persistent mechanical hyperalgesia produced by the intraplantar injection of zymosan (Meller et al., 1993b) or carageenan (Meller et al., 1994) is also not mediated through production of NO as it is unaffected by i.t. treatment with L-NAME, hemoglobin or methylene blue (at doses that attenuate thermal hyperalgesia in the same models; Meller et al., 1992a,b, 1995).

Activation of the metabotropic receptor is able to activate PKC through production of phosphotidyl inositol (PI) (Schoepp, 1993). While activation and translocation of PKC is involved in thermal hyperalgesia (Mao et al., 1992b; Meller et al., 1993b, 1995), administration of selective PKC inhibitors such as chelerythrine do not alter the acute mechanical hyperalgesia produced by coactivation of AMPA and metabotropic glutamate receptors, indicating that activation of PKC is not necessary for acute mechanical hyperalgesia (this dose of chelerythrine does block agonist-produced thermal hyperalgesia and persistent thermal hyperalgesia; Meller et al., 1993b, 1995).

We also examined the role of PLC, which has been linked to activation of the metabotropic receptor (Farooqui and Horrocks, 1991; Schoepp, 1993). However, as with NO and PKC inhibitors, we found that the PLC inhibitor neomycin did not affect the acute mechanical hyperalgesia produced by coactivation of AMPA and metabotropic glutamate receptors. Neomycin was also found to be without effect on the persistent mechanical hyperalgesia produced by the intraplantar administration of zymosan (Meller et al., 1993b). While neomycin has been reported to be an agonist at the polyamine-sensitive site on the NMDA receptor (Pullman et al., 1992), given that NMDA appears not to contribute to acute mechanical hyperalgesia, it is unlikely that an action of neomycin at the NMDA receptor would confound the results reported here.

It is possible that the metabotropic receptor may activate an AMPA sensitive calcium channel either directly or by production of an intracellular messenger. If an AMPA-sensitive calcium channel (inactive at resting membrane potential) is activated (by phosphorylation?), then an increase in intracellular calcium could activate PLA_2. Until recently, there was no evidence to support this hypothesis. However, Dumuis et al. (1990) reported that coactivation of AMPA and metabotropic glutamate receptors is able to produce AA through activation of PLA_2; the present study documents that the PLA_2 inhibitor mepacrine can block the acute mechanical hyperalgesia produced by coactivation of AMPA and metabotropic glutamate receptors. While NMDA receptor activation is also able to activate PLA_2 (Dumuis et al., 1988; Tapia-Aranciba et al., 1992) and produce AA (Lazarewicz et al., 1990, 1992), it is unlikely that this pathway is involved in mechanical hyperalgesia as NMDA receptor activation does not produce an acute mechanical hyperalgesia. Activation of the NMDA receptor may, however, contribute to the potentiation of AMPA and metabotropic glutamate receptor-mediated mechanical hyperalgesia; NMDA receptor antagonists do reduce the persistent mechanical hyperalgesia produced by the intraplantar injection of zymosan (Meller et al., 1993b).

One of the major actions of PLA_2 is to produce AA (Glaser et al., 1993) and it has been recently suggested that AA may be involved in mechanisms of synaptic plasticity such as LTP (Lynch et al., 1989; Williams et al., 1989; Lynch, 1991). In addition, induction of LTP has been reported to be accompanied by an increase in the extra- and intracellular content of AA (Bliss et al., 1990; Lynch, 1991) and AA has been found to potentiate the release of glutamate (Dorman et al., 1992). Although we did not use inhibitors of AA synthesis, we examined the role of AA metabolites such as prostaglandins and leukotrienes. Previous evidence suggests that both may play a role in hyperalgesia. For example, the i.t. administration of PGE_2 produces a dose-dependent hyperalgesia in both thermal (tail-flick) and mechanical (Randall-Selitto) nociceptive tests (Taiwo and Levine, 1988) while PGD_2 produces thermal hyperalgesia (mechanical was not tested) (Uda et al., 1990). In addition, a recent report suggests that i.t. administration of $PGF_{2\alpha}$ produces spontaneous agitation and touch-evoked agitation indicative of mechanical allodynia and hyperalgesia (Minami et al., 1992). The results of the present study suggests that spinal prostaglandins, but not leukotrienes, are important to the mechanisms that underlie mechanical hyperalgesia. If prostanoids are important, however, it is not at all clear whether they act as intra- or intercellular messengers. If they act as intracellular signalling molecules then, once produced, they would activate prostanoid receptors on nearby cells. How this fits mechanistically remains to be determined. However, it is clear that NO likely acts as an intracellular signalling molecule in mechanisms of thermal hyperalgesia. This may be important as preliminary studies suggest that generation of prostaglandins are not involved in the acute thermal hyperalgesia produced by low doses of NMDA (Meller et al., 1994) or in the persistent thermal hyperalgesia produced by the intraplantar injection of zymosan (Meller et al., 1993b). This, however, is in contrast to the effect of indomethacin on the prolonged thermal hyperalgesia produced by i.t. administration of AMPA, NMDA or substance P that was reported by Malmberg and Yaksh (1992). They did not examine the acute thermal hyperalgesia produced by NMDA, but reported an attenuation by indomethacin of a late onset hyperalgesia (30 min) that develops following resolution of an intense biting and scratching behavior produced by i.t. administration of NMDA, AMPA or substance P; the behavior and hyperalgesia observed was the same with all three agonists. Obviously, more work needs to be done to examine the precise role of prostanoids in hyperalgesia.

The present results suggest that coactivation of AMPA and metabotropic glutamate receptors, activation of PLA_2 and production of COX products are important to mechanisms of acute mechanical hyperalgesia. Preliminary evidence indicates that the same mediators are associated with the persistent mechanical hyperalgesia produced by intraplantar administration of zymosan (Meller et al., 1993b). The recent reports that both AMPA (Massicotte and Baudry, 1990; Massicotte et al., 1991; Catania et al., 1993) and metabotropic glutamate receptor binding (Catania et al., 1993) are selectively increased by PLA_2 activation supports these findings. In addition, PLA_2 activation has been shown to enhance excitatory amino acid release and AMPA-stimulated calcium influx (Aronica et al., 1992). This feedforward mechanism is important in mechanisms of synaptic plasticity as an increased sensitivity to AMPA contributes to LTP (Kauer et al., 1988; Davies et al., 1989). This is supported by evidence in the spinal cord that there is a preferential increase in the sensitivity of spinothalamic tract neurons to non-NMDA agonists such as AMPA that parallels the increase in mechanical sensitivity following kaolin and carrageenan-produced inflammation (Dougherty et al., 1992b).

In summary, the present results document that acute mechanical hyperalgesia can be produced by coactivation of spinal ionotropic AMPA and metabotropic glutamate receptor subtypes, activation of PLA_2 and production of COX products. Given previous reports on thermal hyperalgesia, the present results are consistent with a role for spinal NMDA receptors, NO and PKC in thermal hyperalgesia and for coactivation of spinal AMPA and metabotropic glutamate receptors, activation of PLA_2 and production of COX products in mechanical hyperalgesia. Further, these data suggest that acute thermal and mechanical hyperalgesia rely principally on activation of different signal transduction mechanisms in the spinal cord.

Acknowledgements

The authors wish to thank Mike Burcham for preparation of graphics. Supported by DHSS grants DA 02879 and NS 29844.

References

Aronica, E., Casabona, G., Genazzani, A.A., Catania, M.V., Contestabile, A., Virgili, M. and Nicoletti, F. (1992) Mellitin enhances excitatory amino acid release and AMPA-stimulated $^{45}Ca^{2+}$ influx in cultured neurons. *Brain Res.*, 586: 72–77.

Baskys, A. (1992) Metabotropic glutamate receptors and "slow" excitatory actions of glutamate agonists in the hippocampus. *Trends Neurosci.*, 15: 92–96.

Biella, G., Lacerenza, M., Marchettini, P. and Sotgiu, L. (1993) Diverse modulation by systemic lidocaine of iontophoretic NMDA and quisqualic acid induced excitations of rat dorsal horn neurons. *Neurosci. Lett.*, 157: 207–210.

Bleakman, D., Rusin, K.I., Chard, P.S., Glaum, S.R. and Miller, R.J. (1992) Metabotropic glutamate receptors potentiate ionotropic glutamate responses in the rat dorsal horn. *Mol. Pharmacol.*, 42: 192–196.

Bliss, T.V., Errington, M.L., Lynch, M.A. and Williams, J.H. (1990) Presynaptic mechanisms in hippocampal long-term potentiation. *Cold Spring Harbor Symp. Quant. Biol.*, 55: 119–129.

Bredt, D.S. and Snyder, S.H. (1989) Nitric oxide mediates the glutamate-linked enhancement of cGMP levels in the cerebellum. *Proc. Natl. Acad. Sci. USA*, 86: 9030–9033.

Catania, M.V., Hollingsworth, Z., Penney, J.B. and Young, A.B. (1993) Phospholipase A_2 modulates different subtypes of excitatory amino acid receptors: autoradiographic evidence. *J. Neurochem.*, 60: 236–245.

Cerne, R. and Randic, M., (1992) Modulation of AMPA and NMDA responses in rat spinal dorsal horn neurons by trans-1-aminocyclopentane-1,3-dicarboxyclic acid. *Neurosci. Lett.*, 144: 180–184.

Davies, S.N., Lester, R.A.J., Reyman, K.G. and Collingridge, G.L. (1989) Temporally distinct pre- and post-synaptic mechanisms maintain long-term potentiation. *Nature*, 338: 500–503.

Dorman, R.V., Hamm, T.F., Damron, D.S. and Freeman, E.J. (1992) Modulation of glutamate release from hippocampal mossy fiber nerve endings by arachidonic acid and ecosanoids. *Adv. Exp. Med. Biol.*, 318: 121–136.

Dougherty, P.M. and Willis, W.D. (1992) Enhanced responses of spinothalamic tract neurons to excitatory amino acids accompany capsaicin-induced sensitization in the monkey. *J. Neurosci.*, 12: 883–894.

Dougherty, P.M., Palacek, J., Paleckova, V., Sorkin, L.S. and Willis, W.D. (1992a) The role of NMDA and non-NMDA excitatory aminom acid receptors in the excitation of primate spinothalamic tract neurons by mechanical, chemical, thermal and electrical stimuli. *J. Neurosci.*, 12: 3025–3041.

Dougherty, P.M., Sluka, K.A., Sorkin, L.S., Westlund, K.N. and Willis, W.D. (1992b) Neural changes in acute arthritis in monkeys. I. Parallel enhancement of responses of spinothalamic tract neurons to mechanical stimulation and excitatory amino acids. *Brain Res. Rev.*, 17: 1–13.

Dubner, R. and Ruda, M.A. (1992) Activity-dependent neuronal plasticity following tissue injury and inflammation. *Trends Neurosci.*, 15: 96–103.

Dumuis, A., Sebben, M., Haynes, L., Pin, J.P. and Bockaert, J. (1988) NMDA receptors activate the arachidonic acid cascade in striatal neurons. *Nature*, 336: 68–70.

Dumuis, A., Pin, J.P., Oomagari, K., Sebben, M. and Bockaert, J. (1990) Arachidonic acid released from striatal neurons by joint stimulation of ionotropic and metabotropic glutamate receptors. *Nature*, 347: 182–184.

Etoh, S., Baba, A. and Iwata, H. (1991) NMDA induces protein kinase C translocation in hippocampal slices of immature rat brain. *Neurosci. Lett.*, 126: 119–122.

Farooqui, A.A. and Horrocks, L.A. (1991) Excitatory amino acids receptors, neural membrane phospholipid metabolism and neurological disorders. *Brain Res. Rev.*, 16: 171–191.

Garthwaite, J., Garthwaite, G., Palmer, R.M.J. and Moncada, S. (1989) NMDA receptor activation induces nitric oxide synthesis from arginine in rat brain slices. *Eur. J. Pharmacol.*, 172: 413–416.

Glaser, K.B., Mobilio, D., Chang, J.Y. and Senko, N. (1993) Phospholipase A_2 enzymes: regulation and inhibition. *Trends Pharmacol. Sci.*, 14: 92–98.

Handwerker, H.O. and Reeh, P.W. (1992) Nociceptors. chemosensitivity and sensitization by chemical agents. In: W.D. Willis, Jr. (Ed.), *Hyperalgesia and Allodynia*, Raven Press, New York, pp. 107–115.

Hao, J.-X., Yu, Y.-X., Sieger, A. and Wiesenfeld-Hallin, Z. (1992) Systemic tocainamide relieves mechanical hypersensitivity and normalizes the responses of hyperexcitable dorsal horn wide-dynamic-range neurons after transient spinal cord ischemia in rats. *Exp. Brain Res.*, 91: 229–235.

Kauer, J.A., Malenka, R.C. and Nicholl, R.A. (1988) A persistent postsynaptic modification mediates long-term potentiation in the hippocampus. *Neuron*, 1: 911–917.

Kitto, K.F., Haley, J.E. and Wilcox, G.L. (1992) Involvement of nitric oxide in spinally mediated hyperalgesia in the mouse. *Neurosci. Lett.*, 148: 1–5.

Kristensen, J.D., Svensson, B. and Gordh, T. (1992) The NMDA-receptor antagonist CPP abolishes neurogenic "wind-up pain" after intrathecal administration in humans. *Pain*, 51: 249–253.

Lazarewicz, J.W., Wroblewski, J.T. and Costa, E. (1990) N-methyl-D-aspartate-sensitive glutamate receptors induce calcium-mediated arachidonic acid release in primary cultures of cerebellar granule cells. *J. Neurochem.*, 55: 1875–1876.

Lazarewicz, J.W., Salinska, E. and Wroblewski, J.T. (1992) NMDA receptor-mediated arachidonic acid release in neurons: role in signal transduction and pathological aspects. In: N.G. Bazan (Ed.), *Neurobiology of Fatty Acids*. Plenum Press, New York, pp. 73–89.

Levine, J.D., Taiwo, Y.O. and Heller, P.H. (1992) Hyperalgesic pain: inflammatory and neuropathic. In: W.D. Willis, Jr. (Ed.), *Hyperalgesia and Allodynia*. Raven Press, New York, pp. 117–123.

Lynch, M.A. (1991) Presynaptic mechanisms in the maintenance of long-term potentiation: the role of arachidonic acid. In: H. Wheal and A. Thompson (Eds.), *Excitatory Amino Acids and Synaptic Function*. Academic Press, New York, pp. 355–374.

Lynch, M.A., Errington, M.L. and Bliss, T.V.P. (1989) Nordihydroguaiaretic acid blocks the synaptic component of long-term potentiation and the associated increases in release of glutamate and arachidonate: an in vivo study in the dentate gyrus in the rat. *Neuroscience*, 30: 693–701.

Lynn, B. and Carpenter, S.E. (1982) Primary afferent units

from the hairy skin of the rat hind limb. *Brain Res.*, 238: 29–43.

Malmberg, A.B. and Yaksh, T.L. (1992) Hyperalgesia mediated by spinal glutamate or substance P receptor blocked by spinal cyclooxygenase inhibition. *Science*, 257: 1276–1279.

Malmberg, A.B. and Yaksh, T.L. (1993) Spinal nitric oxide synthesis inhibition blocks NMDA induced thermal hyperalgesia and produces antinociception in the formalin test in rats. *Pain*, 54: 291–300.

Mao, J., Hayes, R.L., Price, D.D., Coghill, R.C., Lu, J. and Mayer, D.J. (1992a) Post-injury treatment with GM1 ganglioside reduces nociceptive behaviors and spinal cord metabolic activity in rats with experimental peripheral mononeuropathy. *Brain Res.*, 584: 18–27.

Mao, J., Price, D.D., Mayer, D.J. and Hayes, R.L. (1992b) Pain-related increases in spinal cord membrane-bound protein kinase C following peripheral nerve injury. *Brain Res.*, 588: 144–149.

Mao, J., Mayer, D.J., Hayes, R.L. and Price, D.D. (1993) Spatial patterns of increased spinal cord membrane-bound protein kinase C and their relation to increases in ^{14}C-2-deoxyglucose metabolic activity in rats with painful peripheral mononeuropathy. *J. Neurophysiol.*, 70: 470–481.

Marin, P., Quignard, J.F., Lafon-Cazal, M. and Bockaert, J. (1993) Non-classical glutamate receptors, blocked by both NMDA and non-NMDA antagonists, stimulate nitric oxide production in neurons. *Neuropharmacology*, 32: 29–36.

Massicotte, G. and Baudry, M. (1990) Modulation of L-Â-amino-3-hydroxy-5-methylisoxazole-4-proprionate (AMPA)/quisqualate receptors by phospholipase A_2 treatment. *Neurosci. Lett.*, 118: 245–248.

Massicotte, G., Vanderklish, P., Lynch, G. and Baudry, M. (1991) Modulation of DL-Â-amino-3-hydroxy-5-methyl-4-isoxazoleroprionic acid/quisqualate receptors by phospholipase A_2: a necessary step in long-term potentiation. *Proc. Natl. Acad. Sci. USA*, 88: 1893–1897.

Mayer, M.L. and Miller, R.J. (1990) Excitatory amino acid receptors, second messengers and regulation of intracellular Ca^{2+} in mammalian neurons. *Trends Pharmacol. Sci.*, 11: 254–260.

Meller, S.T. and Gebhart, G.F. (1993) NO and nociceptive processing in the spinal cord. *Pain*, 52: 127–136.

Meller, S.T., Dykstra, C. and Gebhart, G.F. (1992a) Production of endogenous nitric oxide and activation of soluble guanylate cyclase are required for N-methyl-D-aspartate-produced facilitation of the nociceptive tail-flick reflex. *Eur. J. Pharmacol.*, 214: 93–96.

Meller, S.T., Pechman, P.S., Gebhart, G.F. and Maves, T.J. (1992b) Nitric oxide mediates the thermal hyperalgesia produced in a model of neuropathic pain in the rat. *Neuroscience*, 50: 7–10.

Meller, S.T., Dykstra, C. and Gebhart, G.F. (1993a) Acute mechanical hyperalgesia in the rat is produced by coactivation of ionotropic AMPA and metabotropic glutamate receptors. *NeuroReport*, 4: 879–882.

Meller, S.T., Dykstra, C. and Gebhart, G.F. (1993b) Characterization of the spinal mechanisms of thermal and mechanical hyperalgesia following intraplantar zymosan. *Soc. Neurosci. Abstr.*, 19: 967.

Meller, S.T., Cummings, C.J., Traub, R.J. and Gebhart, G.F. (1994) The role of nitric oxide in the development and maintenance of hyperalgesia and edema produced by the intraplantar carrageenan in the rat. *Neuroscience*, 60: 367–374.

Meller, S.T., Dykstra, C. and Gebhart, G.F. (1995) Acute thermal hyperalgesia is produced by activation of NMDA receptors, activation of protein kinase C and production of nitric oxide (NO) and cGMP. *Neuroscience*, in press.

Minami, T., Uda, R., Horiguchi, S., Ito, S., Hyodo, M. and Hayaishi, O. (1992) Allodynia evoked by intrathecal administration of prostaglandin $F_{2Â}$ to conscious mice. *Pain*, 50: 223–229.

Nakanishi, S. (1992) Molecular diversity of glutamate receptors and implications for brain function. *Science*, 258: 597–603.

Neugebauer, V., Kornhuber, J., Lucke, T. and Schaible, H.-G. (1993a) The clinically available NMDA receptor antagonist memantine is antinociceptive on rat spinal neurons. *NeuroReport*, 4: 1259–1262.

Neugebauer, V., Lucke, T. and Schaible, H.-G. (1993b) N-methyl-D-aspartate (NMDA) and non-NMDA receptor antagonists block the hyperexcitability of dorsal horn neurons during development of acute arthritis in rat's knee joint. *J. Neurophysiol.*, 70: 1365–1377.

Palacek, J., Paleckova, V., Dougherty, P.M. and Willis, W.D. (1994) The effect of phorbol esters on the responses of primate spinothalamic neurons to mechanical and thermal stimuli. *J. Neurophysiol.*, 71: 529–537.

Pin, J.P., Waeber, C., Prezeau, L., Bockaert, J. and Heinemann, S.F. (1992) Alternative splicing generates metabotropic glutamate receptors inducing different patterns of calcium release in *Xenopus* oocyctes. *Proc. Natl. Acad. Sci. USA*, 89: 10331–10335.

Pullman, L.M., Stumpo, R.J., Powel, R.J., Paschetto, K.A. and Britt, M. (1992) Neomycin is an agonist at a polyamine site on the N-methyl-D-aspartate receptor. *J. Neurochem.*, 59: 2087–2093.

Ren, K. and Dubner, R. (1993) NMDA receptor antagonists attenuate mechanical hyperalgesia in rats with unilateral inflammation of the hindpaw. *Neurosci. Lett.*, 163: 19–21.

Rueff, A., Lewin, G.R. and Mendell, L.M. (1993) Peripheral and central mechanisms of NGF-induced hyperalgesia in adult rats. *Soc. Neurosci. Abstr.*, 18: 1563.

Schoepp, D. (1993) The biochemical pharmacology of metabotropic glutamate receptors. *Biochem. Soc. Trans.*, 21: 97–102.

Schoepp, D. and Conn, P.J. (1993) Metabotropic glutamate receptors in brain function and pathology. *Trends Pharmacol. Sci.*, 14: 13–20.

Schoepp, D., Bockaert, J. and Sladeczek, F. (1990) Pharmacological and functional characteristics of metabotropic excitatory amino acid receptors. *Trends Pharmacol. Sci.,* 11: 508–515.

Sher, G.D. and Mitchell, D. (1990) Intrathecal N-methyl-D-aspartate induces hyperexcitability in rat dorsal horn convergent neurons. *Neurosci. Lett.,* 119: 199–202.

Sommer, B. and Seeburg, P.H. (1992) Glutamate receptor channels: novel properties and new clones. *Trends Pharmacol. Sci.,* 13: 291–296.

Taiwo, Y.O. and Levine, J.D. (1988) Prostaglandins inhibit endogenous pain control mechanisms by blocking transmission at spinal noradrenergic synapses. *J. Neurosci.,* 8: 1346–1349.

Tal, M. and Bennett, G.J. (1993) Dextrorphan relieves neuropathic heat-evoked hyperalgesia in the rat. *Neurosci. Lett.,* 151: 107–110.

Tapia-Arancibia, L., Rage, F., Recasens, M. and Pin, J.P. (1992) NMDA receptor activation stimulates phospholipase A_2. *Eur. J. Pharmacol.,* 225: 253–262.

Tolle, T.R., Berthele, A., Zieglgansberger, W., Seeburg, P.H. and Wisden, W. (1993) The differential expression of 16 NMDA and non-NMDA receptor subunits in the rat spinal cord and in periaqueductal gray. *J. Neurosci.,* 13: 5009–5028.

Uda, R., Horiguchi, S., Ito, S., Hyodo, M. and Hayaishi, O. (1990) Nociceptive effects induced by intrathecal administration of prostaglin D_2, E_2, or F_2 to conscious mice. *Brain Res.,* 510: 26–32.

Vaccarino, F., Guidotti, A. and Costa, E. (1987) Ganglioside inhibition of glutamate-mediated protein kinase C translocation in primary cultures of cerebellar neurons. *Proc. Natl. Acad. Sci. USA,* 84: 8707–8711.

Wilcox, G.L. (1991) Excitatory neurotransmitters and pain. In: M.R. Bond, J.E. Charlton and C.J. Woolf (Eds.), *Pain, Research and Clinical Management,* Vol. 4. Elsevier, Amsterdam, pp. 97–117.

Williams, J.H., Errington, M.L., Lynch, M.A. and Bliss, T.V.P. (1989) Arachidonic acid induces a long-term activity-dependent enhancement of synaptic transmission in the hippocampus. *Nature,* 341: 739–742.

Willis, Jr., W.D. (1992) *Hyperalgesia and Allodynia.* Raven Press, New York.

Wood, P.L., Emmett, M.R., Rao, T.S., Cler, J., Mick, S. and Iyengar, S. (1990) Inhibition of nitric oxide synthase blocks N-methyl-D-aspartate-, quisqualate-, kainate-, harmaline-, and pentylenetetrazole-dependent increases in cerebellar cyclic GMP in vivo. *J. Neurochem.,* 55: 346–348.

Woolf, C.J. and Thompson, S.W.N. (1991) The induction and maintenance of central sensitization is dependent on N-methyl-D-aspartic acid receptor activation: implications for the treatment of post-injury pain hypersensitivity states. *Pain,* 44: 293–300.

Young, M.R., Fleetwood-Walker, S.M., Mitchell, R. and Munro, F.E. (1994) Evidence for a role of metabotropic glutamate receptors in sustained nociceptive inputs to rat dorsal horn neurons. *Neuropharmacology,* 33: 141–144.

Yu, Y.-X., Hao, J.-X., Sieger, A. and Wiesenfeld-Hallin, Z. (1992) Systemic excitatory amino acid receptor antagonists of the Â-amino-3-hydroxy-5-methyl-4isoazoleproprionic acid (AMPA) receptor and the N-methyl-D-aspartate (NMDA) receptor relieve mechanical hypersensitivity after transient spinal cord ischemia in rats. *J. Pharmacol. Exp. Ther.,* 267: 140–144.

Zhuo, M. and Gebhart, G.F. (1991) Tonic cholinergic inhibition of spinal mechanical transmission. *Pain,* 46: 211–222.

Zhuo, M., Meller, S.T. and Gebhart, G.F. (1993) Endogenous nitric oxide is required for tonic spinal cholinergic inhibition of nociceptive mechanical transmission. *Pain,* 54: 71–78.

CHAPTER 15

Involvement of glutamatergic neurotransmission and protein kinase C in spinal plasticity and the development of chronic pain

T.R. Tölle, A. Berthele, J. Schadrack and W. Zieglgänsberger

Max-Planck-Institute of Psychiatry, Clinical Institute, Clinical Neuropharmacology, Kraepelinstrasse 2, 80804 Munich, Germany

Introduction

Primary afferent fibers transmitting somatosensory stimuli into the dorsal horn of the spinal cord use either glutamate or aspartate as principal excitatory transmitters. Noxious stimuli evoke excitatory synaptic responses in spinal cord neurons mediated via ligand-gated, cation-selective ion channels, so-called ionotropic glutamate receptors (NMDA, AMPA, and kainate subtype) and/or G-protein coupled metabotropic receptors (mGluR) (Headley and Grillner, 1990; Dougherty et al., 1992; Meller et al., 1993; Randic et al., 1993; Neugebauer et al., 1994).

Molecular cloning studies have demonstrated that ionotropic glutamate receptor constituents belong to the same gene family (for review see: Seeburg, 1993; Hollmann and Heinemann, 1994). Recombinant expression studies showed that native glutamate receptors are supposed to be heterooligomeric assemblies of five subunits (Seeburg, 1993; Wisden and Seeburg, 1993). Electrophysiological experiments utilizing patch-clamp techniques revealed that the subunit composition in these heterooligomeric assemblies is crucial for selectivity and kinetics of ion fluxes and activation of intracellular second messenger systems (for review see: Hollmann and Heinemann, 1994).

Protein phosphorylation of ligand-gated ion channels appears to be another major mechanism to regulate the properties of the diverse types of glutamate activated receptor channels (for review see: Raymond et al., 1993). NMDA and some AMPA receptor proteins carry consensus sequences for phosphorylation by protein kinase C (PKC) or cAMP dependent protein kinase (Raymond et al., 1993). Activation of PKC, a calcium-dependent protein kinase that phosphorylates a wide variety of substrates, including ion channels (Nishizuka, 1986; Tingley et al., 1993), increases the responses of dorsal horn neurons to excitatory amino acids and enhances NMDA-activated currents (Chen and Huang, 1992). Recent behavioral studies indicate that PKC may act as a cellular link in neuronal plasticity suggesting a role for protein phosphorylation in secondary hyperalgesia and the development of chronic pain (Coderre, 1992; Tölle et al., 1995a; Yashpal et al., 1995; for review see Coderre et al., 1993).

Since the functional properties of neurons are critically determined by their individual re-ceptor subunit composition we investigated the distribution of ionotropic glutamate receptor subunit mRNA in spinal cord neurons. It may be assumed that a differential distribution of glutamate receptor subunits may be tailored to fit the neurons individual response characteristics that are required to participate in particular spinal neuronal circuits.

Inflammatory or other persisting noxious stimuli change the functional properties of neurons. One key feature of central sensitization and the development of chronic pain is an increase in neuronal excitability. Alterations in glutamate receptor subunit composition, subunit phosphorylation or both may be the underlying basic molecular mechanisms. In the present study, we induced monoarthritis in rats and investigated the expression patterns of ionotropic glutamate receptor subunit mRNA as well as the binding intensity and the laminar distribution of PKC in the spinal cord dorsal horn (Tölle et al., 1995a).

Ionotropic glutamate receptors

Three glutamate ionotropic receptor subtypes are classified on the basis of their selectivity to synthetic agonists as (1) N-methyl-D-aspartate (NMDA), (2) α-amino-3-hydroxy-5-methyl-isoxazole-4-propionate (AMPA), and (3) kainate receptors (Wisden and Seeburg, 1993; Hollmann and Heinemann, 1994). Autoradiographic studies indicate the presence of all three binding sites in the mammalian spinal cord (Monaghan and Cotman, 1985; Westbrook, 1994).

Molecular cloning studies have identified families of receptor subunits which can assemble in homomeric or heteromeric receptor configurations in vitro (Wisden and Seeburg, 1993; Hollmann and Heinemann, 1994). Immunochemical studies suggest a pentameric, heterooligomeric assembly of at least two different subunits in native channels of each family (Wenthold et al., 1992; Brose et al., 1993; see Fig. 1). Receptor diversity is further increased by alternative RNA splicing of subunit transcripts (Wisden and Seeburg, 1993; Hollmann and Heinemann, 1994; see Table 1 for overview of subunits/splice variants investigated in the present study). However, the exact subunit composition of any channel type is not known.

AMPA receptors, consisting of the GluR-A to -D subunits, often colocalize with NMDA receptors in the same synapse and mediate fast excitatory transmission (Wisden and Seeburg, 1993). Each of the GluR-A to -D subunits exists as two variants, which differ in the amino acid sequence of a small C-terminal segment. Variants are generated by alternative splicing of two adjacent exons (see Table 1). Since these modules impart differing pharmacological and kinetic properties to the receptor, the respective splice variants are designated "flip" and "flop" (Sommer et al., 1990). Expressed in in vitro systems, heterooligomeric AMPA receptor configurations lacking the GluR-B subunit are permeable to Ca^{2+} (Burnashev et al., 1992).

Kainate receptors can be constructed as heterooligomeric and/or homomeric combinations from two classes of subunits: KA-1 and -2 and GluR-5, -6, and -7 (Hollmann and Heinemann, 1994). In contrast to AMPA receptors, kainate receptors mediate currents which desensitize rapidly in the presence of kainate. The physiological function of kainate receptors is not yet understood, and it is likely that members of this family are mainly located on presynaptic terminals and dendrites (Good et al., 1992). Kainate receptors, particularly those containing the GluR-5 subunit, are synthesized in dorsal root ganglion cells and may function as presynaptic autoreceptors on primary afferent terminals (Headley and Grillner, 1990; Wisden und Seeburg, 1993).

mRNA expression of NMDA receptor subunits vary in a brain region specific manner (Buller et al., 1994). NMDA receptors are believed to exist in vivo as heterooligomeric assemblies of the NMDAR1 subunit and at least one of the NMDAR2A to -D subunits and have distinct electrophysiological and pharmacological profiles (Moriyoshi et al., 1991; Meguro et al., 1992; Monyer et al., 1992). They are characterized by their voltage dependent Mg^{2+} block, slow gating kinetics and permeability to Ca^{2+} (Moriyoshi et al., 1991; Monyer et al., 1992). Eight possible mRNA splice variants can be generated from the principal NMDAR1 subunit by the alternative splicing of one cassette in the N-terminal region and the individual or combined deletion of two consecutive exons in the C-terminal region (Nakanishi et al., 1992; Laurie and Seeburg, 1994; see Table 1). Alternative splicing of the NMDAR1 subunit mRNA results in changes of functional properties of these variants, when expressed as homomeric NMDAR1 receptors in vitro, such as sensitivity to NMDA agonists/antagonists, blockade by Zn^{2+} or potentiation by PKC (Durand et al., 1992; Nakanishi et al., 1992; Hollmann et al., 1993; Tingley et al., 1993).

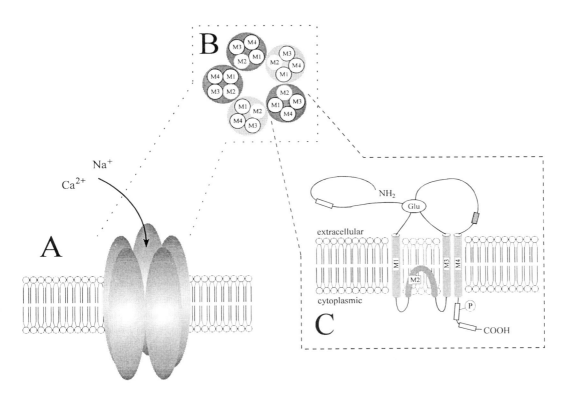

Fig. 1. Schematic structural representation of the ionotropic glutamate receptor (adapted from Raymond et al., 1993; Bettler and Mulle, 1995; Dani and Mayer, 1995). (A) Putative pentameric structure of the ionotropic glutamate receptor. Ionotropic glutamate receptors gate ion fluxes of Na^+ and/or Ca^{2+} depending on subunit composition. In analogy with the nicotinic acetylcholine receptor, native ionotropic glutamate receptors are likely to be composed of five subunit monomers. This model is supported by immunoprecipitation studies of native AMPA receptors (Wenthold et al., 1992). (B) Sectional view of (A). Molecular cloning studies suggest that native ionotropic glutamate receptors consist of heterooligomeric assemblies of subunits (see Table 1), forming the central aqueous pore. The exact composition of any native ionotropic glutamate receptor is not known yet. All ionotropic glutamate receptor subunits show highest sequence similarity in a region comprising four hydrophobic domains (M1 to M4), which are involved in the membrane folding of subunit protein. M2, possibly in conjunction with M1, is very likely to line the channel pore (Wo and Oswald, 1995). (C) The three transmembrane domain topology model of ionotropic glutamate receptor subunits, based on recent experimental data (Hollmann et al., 1994; Wo and Oswald, 1994; Bennett and Dingledine, 1995), suggests that the M1, M3 and M4 domains span the membrane, whereas the M2 domain forms a reentrant membrane segment with both ends facing the cytoplasm. The N-terminal domain and the M3-M4 loop of each of the five subunit proteins are supposed to be located extracellularly and to be related to the glutamate binding site (Glu) (Wo and Oswald, 1995). Finally, the current model suggests that the C-terminal domain, which carries multiple consensus sequences for phosphorylation, is located intracellularly (Tingley et al., 1993; for review see Hollmann et al., 1994; Wo and Oswald, 1995). Boxes indicate sequences of AMPA (grey box) and NMDA subunits (white boxes) affected by mRNA splicing (Seeburg, 1993).

Distribution of ionotropic glutamate receptor subunit mRNA in the spinal cord

Segments of adult rat lumbar spinal cords were dissected from non-perfusion-fixed animals and frozen on dry ice. Sections cut on a cryostat were processed for radioactive in situ hybridization according to Wisden and Morris (1994). Synthetic oligonucleotides were 3'-labelled with $[\alpha\text{-}^{35}S]$-dATP. Probe sequences complementary to rat subunit and splice variant mRNAs were used as previously described (Tölle et al., 1993, 1995b,c). In these experiments so-called "pan" probes detect all splice variants of the respective subunit mRNA;

TABLE I

Summary of ionotropic glutamate receptor subunits and splice variants (Hollmann et al., 1993; Köhler et al., 1994) investigated for mRNA expression in the rat lumbar spinal cord

Glutamate receptor subtype	Subunit	Splice variant	Modification
AMPA	GluR-A		
		GluR-A flip	C-terminal alternate exon 15
		GluR-A flop	C-terminal alternate exon 14
	GluR-B		
		GluR-B flip	C-terminal alternate exon 15
		GluR-B flop	C-terminal alternate exon 14
	GluR-C		
		GluR-C flip	C-terminal alternate exon 15
		GluR-C flop	C-terminal alternate exon 14
	GluR-D		
		GluR-D flip	C-terminal alternate exon 15
		GluR-D flop	C-terminal alternate exon 14
Kainate	GluR-5		
	GluR-6		
	GluR-7		
	KA-1		
	KA-2		
NMDA	NMDAR1		
		NMDAR1-1a	-
		NMDAR1-1b	N-terminal insertion of exon 5
		NMDAR1-2a	C-terminal deletion of exon 21
		NMDAR1-2b	N-terminal insertion of exon 5 and C-terminal deletion of exon 21
		NMDAR1-3a	C-terminal partial deletion of exon 22
		NMDAR1-3b	N-terminal insertion of exon 5 and C-terminal partial deletion of exon 22
		NMDAR1-4a	C-terminal deletion of exon 21 and partial deletion of exon 22
		NMDAR1-4b	N-terminal insertion of exon 5 and C-terminal deletion of exon 21 and partial deletion of exon 22
	NMDAR2A		
	NMDAR2B		
	NMDAR2C		
	NMDAR2D		

Native AMPA, kainate or NMDA receptors are believed to consist of hetero-oligomeric assemblies of receptor subunits. Receptor diversity is further increased by post-transcriptional modifications of subunits due to mRNA splicing of the primary transcript. Nomenclature for AMPA and kainate receptor subunits were adopted from Wisden and Seeburg (1993) and for NMDA receptor subunits from Hollmann and Heinemann (1994).

NMDAR1 splice variant probes detect either N-terminal (NMDAR1-a/b probes) or C-terminal variations (NMDAR1-1 to -4 probes) of NMDAR1 subunit mRNA (see Table 1). Following hybridization, sections were either exposed to X-ray film or, for cellular resolution, dipped in emulsion. After development, sections were counterstained.

AMPA and kainate receptor subunits

Pan oligonucleotide probes, which detect both splice variants of the respective subunit, show that all four AMPA receptor subunit mRNAs are present in the spinal cord, although differences in distribution are pronounced (Fig. 2). In laminae I and II of the dorsal horn, GluR-A and -B are the most

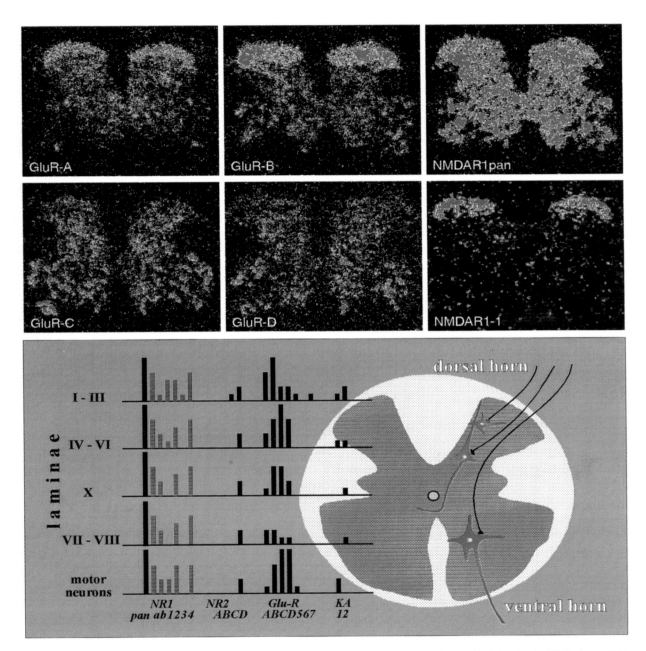

Fig. 2. Expression of glutamate receptor subunits in the rat lumbar spinal cord. Distribution of AMPA subunit (GluR-A, GluR-B, GluR-C and GluR-D), NMDAR1 subunit and NMDAR1-1 splice variants mRNA in rat lumbar spinal cord. Sections were hybridized with synthetic oligonucleotide probes, which recognize both flip- and flop variants of respective AMPA subunits (GluR-A, GluR-B, GluR-C, and GluR-D), all splice variants of the NMDAR1 subunit (NMDAR1pan), or C-terminally unspliced variants of NMDAR1 subunit (NMDAR1-1) and exposed to X-ray film. Autoradiographs are depicted in pseudocoloured mode. Red represents high level mRNA expression. In the diagram the relative abundance of mRNAs encoding for NMDAR1, NMDAR2, AMPA and kainate receptor subunits (black columns)/splice variants (grey columns) is illustrated as height of individual columns. Abundance is depicted for each subtype or splice variant investigated (given below columns) with reference to the cytoarchitectural organisation of the rat lumbar spinal cord. Laminae according to Rexed. Abbreviations: NR1, NMDAR1; NR2, NMDAR2.

abundant transcripts (Fig. 2). While the GluR-A gene is preferentially expressed in neurons of laminae I and II-outer (IIo), the GluR-B mRNA is more evenly expressed throughout all superficial laminae. GluR-A mRNA expression in the superficial dorsal horn consists of both flip and flop variant (Tölle et al., 1995c). In contrast, most of the dorsal horn GluR-B mRNA is derived from GluR-B Flip (Tölle et al., 1995c). Expression of mRNAs encoding GluR-C and GluR-D subunits in the dorsal horn is moderate or low (Fig. 2). The GluR-C gene is the only member of the AMPA subunit family to be expressed abundantly in large neurons in laminae IV and V (Fig. 2). Expression of the GluR-B and GluR-D genes in these large cells is weaker (Fig. 2, Tölle et al., 1993).

In the ventral horn, GluR-B, -C and -D mRNAs are present in motor neurons, with the GluR-C and -D subunit genes being the most prominent (Fig. 2).

Levels of the kainate receptor transcripts are much lower than those of the AMPA receptor transcripts. GluR-6 mRNA is not expressed at all in the spinal cord, although as a positive control, the same probe detects GluR-6 transcripts in the hippocampus. Exposure of hybridized sections to X-ray film reveals that KA-1 mRNA expression is low, but nevertheless the most prominent member of this class in motor neurons, whereas KA-2 mRNA is the most abundant in the dorsal horn (diagram Fig. 2). The GluR-5 and -7 probes give very low signals on X-ray film (Tölle et al., 1993). However, emulsion dipping of hybridized sections, which allows the assessment of mRNA expression patterns with cellular resolution, reveals small cells expressing the KA-2 gene and very few cells expressing GluR-5 and -7 mRNA in the substantia gelatinosa (Tölle et al., 1993).

NMDA receptor subunits

In the present *in situ* hybridization study, only NMDAR1, NMDAR2C and NMDAR2D mRNAs are detectable in the rat lumbar spinal cord. The NMDAR1 mRNA is present in virtually every neuron at high levels, but there is a large abundance mismatch in mRNA expression between NMDAR1 and the NMDAR2A to -D subunits (Figs. 2, 3 and diagram Fig. 2). In the spinal cord dorsal horn each NMDAR1 splice variant shows a distinct expression pattern. The mRNAs encoding the N-terminal splice variants NMDAR1-a are ubiquitously expressed (Fig. 3; Tölle et al., 1993). However, although found in virtually every neuron throughout laminae I–VI and X of the dorsal horn (Fig. 3), the NMDAR1-a specific signal is considerably lower than that of the NMDAR1-pan probe. The N-terminal splice variants NMDAR1-b are weakly expressed, having a scattered pattern (Fig. 3). Emulsion autoradiography with the NMDAR1-b probe reveals silver grains located preferentially over neurons in laminae II-inner (IIi) and III, but also specifically over a small number of neurons in laminae I and II-outer (IIo) and occasionally over neurons of laminae IV and V (Fig. 3). Differences in expression patterns of the C-terminal splice variants are even more pronounced. The NMDAR1-1 variants are almost exclusively found in high abundance in neurons of laminae I to III, while labelled neurons in lamina IV are rare (Figs. 2 and 3). NMDAR1-2 and -4 mRNAs are evenly distributed throughout all laminae of the dorsal horn with NMDAR1-4 showing a higher level of expression (Fig. 3). High NMDAR1-4 and moderate NMDAR1-2 expression is particularly noted in large neurons in laminae IV and V (Fig. 3). In the ventral horn of the spinal cord all splice variants, except NMDAR1-3, are detected in laminae

Fig. 3. Expression of NMDAR1 splice variant mRNA in the rat lumbar spinal cord revealed by in situ hybridization. High-power bright-field photomicrographs of emulsion dipped sections showing the expression of the NMDAR1-pan and NMDAR1 splice variant receptor subunit mRNA in the spinal cord dorsal horn (top; laminae I–IV of Rexed) and in motor neurons of the spinal cord ventral horn (bottom; lamina IX). Both sets of photomicrographs represent the in situ hybridizations of the NMDAR1-pan probe and the splice variant probes NMDAR1-a, -b, -1 to -4 to seven consecutive sections (14 μm) allowing the detection of co-expressed splice variant mRNAs in single motor neurons and adjacent neurons of the dorsal horn. Small dots (silver grains) indicate mRNA expression in respective counterstained neurons. Bar upper right, 30 μm; bar lower right, 25 μm.

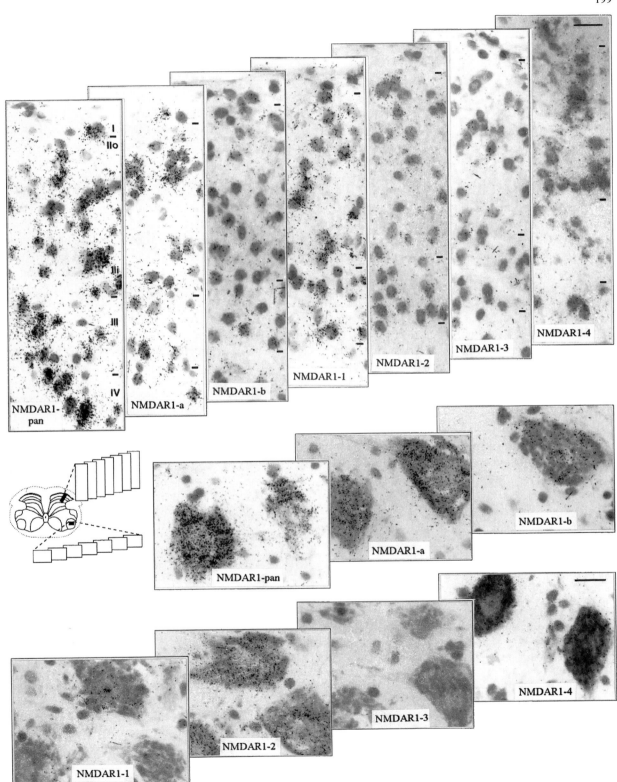

VII–IX (diagram Fig. 2, Fig. 3). NMDAR1-a, NMDAR1-2 and NMDAR1-4 mRNAs are abundantly present in large and small motor neurons (Fig. 3) and in small ventral horn cells of lamina VII which are presumably interneurons. NMDAR1-b mRNAs are only moderately expressed with the same pattern of distribution. At the level of X-ray film, NMDAR1-1 mRNAs are barely detectable, but emulsion autoradiography reveals moderate expression of these splice variants in motor neurons with the majority of silver grains being located over the cell nucleus (Fig. 3, Tölle et al., 1995b). Reconstruction of adjacent sections of the same neurons (up to 3 sections per neuron) shows that many neurons of the dorsal horn and the majority of motor neurons express multiple splice variants of the NMDAR1 gene.

The NMDAR2-C mRNA is detectable only in scattered cells in laminae I and II of the substantia gelatinosa whereas NMDAR2-D mRNA is present at low levels throughout the grey matter and slightly more concentrated in the substantia gelatinosa and lamina X. The NMDAR2-D gene is also expressed at low levels in motor neurons (diagram Fig. 2; Tölle et al., 1993).

Expression of ionotropic glutamate receptor subunit mRNA and binding of [^{3}H]phorbol-12,13-dibutyrate following monoarthritis

Female rats, weighing 150 g at the start of the induction of monoarthritis, were used in this study. The ethical guidelines for investigation of experimental pain in conscious animals were followed (Zimmermann, 1983). Monoarthritis was induced by intra-articular injection of 50 μl of complete Freund's adjuvant (CFA) according to Butler et al. (1992). Intra-articular CFA caused an inflammatory reaction leading to an increase in paw volume which was apparent 4–8 h after the injection, reaching maximum values 2 weeks later and remaining elevated throughout the time period investigated (28 days). Following various survival times (2, 4, 14, 25 days) after induction of the monoarthritis, animals were randomly chosen from a pool of inflamed and non-inflamed control animals and processed for in situ hybridization of glutamate receptor subunit mRNA (GluR-A to -D; GluR-5 to -7; NMDAR1-pan; NMDAR2-A to D) or for receptor-autoradiography with [^{3}H]phorbol-12,13-dibutyrate ([^{3}H]PDBU), a phorbol ester which binds to and activates the diacylglycerol (DAG) binding site on PKC. For [^{3}H]PDBU binding, the lumbar spinal cords were processed with minor modifications as described elsewhere (Worley et al., 1986).

In control animals the highest binding for [^{3}H]-PDBU is found in laminae I–II of the dorsal horn (Fig. 4; 1050 ± 40 pmol/g tissue). Compared to these values, binding in the area around the central canal (lamina X) is moderate (400 ± 30 pmol/g tissue). With the sensitivity of the applied bio-

Fig. 4. Alterations in the expression of glutamate receptor subunit mRNA and in the binding of protein kinase C (PKC) in the rat lumbar spinal cord following experimentally induced chronic monoarthritis. (Top) Column chart depicting expression levels of mRNAs encoding all NMDAR1 splice variants (NMDAR1pan probe) and GluR-A and GluR-B (representative of AMPA family) subunits in the spinal cord dorsal horn (laminae I–III) following 2 and 4 days of monoarthritis induced with injection of complete Freund's adjuvant into the tibio-tarsal joint of one hind limb. Columns represent the relative abundance of each mRNA expressed ipsi- and contralaterally to the monoarthritis compared to levels of control animals, assessed by computer-assisted densitometry of in situ hybridization X-ray autoradiographs. Following monoarthritis, there were no significant alterations in the mRNA expression of all glutamate receptor subunits investigated. The in situ hybridization X-ray autoradiographs illustrate the expression pattern of the NMDAR1 transcript in the spinal cord of control animals and following 4 days of monoarthritis. (Bottom) Column chart depicting binding of protein kinase C (PKC) in the spinal cord dorsal horn (laminae I–II) following 2, 4, 14 and 25 days of monoarthritis induced with complete Freund's adjuvant injected into the tibio-tarsal joint of one hind limb. Data were obtained from binding studies with [^{3}H]PDBU, a phorbol ester which binds to the diacylglycerol binding site on PKC. Columns represent the relative binding affinity of protein kinase C ipsi- and contralateral to the monoarthritis compared to levels of control animals, assessed by computer-assisted densitometry of autoradiographs. Following monoarthritis, specific binding was increased significantly on day 4 and 14 on the ipsilateral and on day 14 on the contralateral side. The X-ray autoradiographs illustrate the level and distribution of [^{3}H]PDBU binding in the spinal cord of control animals and following 4 days of monoarthritis.

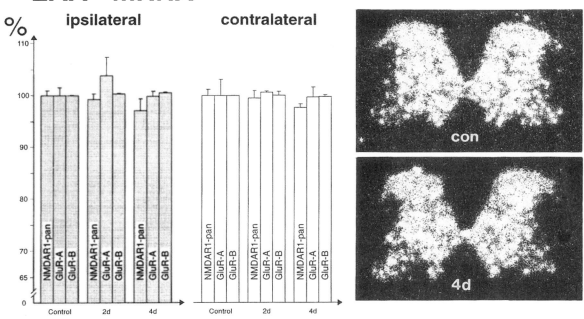

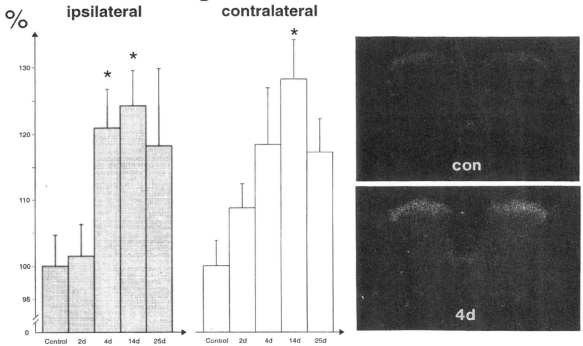

chemical detection system other spinal structures show no quantifiable binding. After CFA, the ipsilateral and the contralateral superficial dorsal horn show enhanced binding of [^{3}H]PDBU (Fig. 4). The observed increase in specific binding of ^{3}H-PDBU is significant on the ipsilateral side at day 4 and on the contralateral side at day 14. The maximal increase of binding is observed at day 14 reaching values of about 1340 ± 50 pmol/g tissue on both sides of the spinal cord. The changes of specific binding around the central canal are not significant. Besides the increase in binding in superficial laminae I–II of the dorsal horn, binding sites are now also observed in deeper laminae III and IV of the dorsal horn (Tölle et al., 1995a).

In situ hybridization of lumbar spinal cord sections of monoarthritic rats was performed as described above. mRNA expression of AMPA, kainate or NMDA receptor subunits was quantified by computer-assisted densitometry of X-ray autoradiographs and emulsion dipped sections.

Compared to controls, no changes are detectable in the abundance or anatomical distribution of any of the ionotropic glutamate receptor subunit mRNAs investigated in the dorsal and ventral horn of monoarthritic rat spinal cords (Fig. 4, Tölle et al., 1995a).

Synaptic plasticity in the spinal cord: glutamate receptor expression and phosphorylation

The NMDA, AMPA and kainate subtypes of the ionotropic glutamate receptors are involved in mono- and polysynaptic noxious and non-noxious transmission in the spinal cord (Headley and Grillner, 1990; Dougherty et al., 1992; Randic et al., 1993). There is mounting evidence that the NMDA receptor plays a particularly important role in development and maintenance of neuronal plasticity in the spinal cord following tissue injury and inflammation (for review see: Dubner and Ruda, 1992; Coderre et al., 1993; Zieglgänsberger and Tölle, 1993).

The properties imparted by the numerous glutamate receptor subunits on native glutamate receptor channels, and the objective of NMDAR1 splice variants on heteromeric NMDAR1-NMDAR2 receptors in particular, have yet to be determined. Conceivably, the molecular diversity of the NMDAR1 subunits and their differential anatomical distribution in the spinal cord dorsal horn are responsible for the heterogeneity in physiological responses, such as the response types of spinal cord neurons to tetanic stimulation, either showing a long-term enhancement or a decrease of EPSP amplitude, or no influence at all (Randic et al., 1993). Also the modulatory effect of NMDA receptor antagonists on C fibre-evoked activity, which apparently depends on the location of the neuron in the dorsal horn, or the differential susceptibility to "wind-up" (Tölle et al., 1989; Dickenson and Sullivan, 1990), may both be attributed to a different subunit assembly of NMDA receptors carried by these neurons. In models of tonic pain, application of NMDA antagonists reduces behavioral hyperalgesia (Ren et al., 1992; Coderre et al., 1993), and in CFA inflammation the enlarged receptive fields of spinal cord dorsal horn neurons are reduced by the non-competitive NMDA antagonist MK-801 (Ren et al., 1992). In this pathological status, regulation of the voltage dependent Mg^{2+} ion block of NMDA receptors is in part responsible for the central sensitization to pain perception (Stevens, 1992).

NMDA and some AMPA receptor proteins carry consensus sequences for phosphorylation by PKC or cAMP dependent protein kinase (Raymond et al., 1993). Protein phosphorylation of ligand-gated channels appears to be a major mechanism to regulate NMDA receptor properties, such as open times and desensitization (Raymond et al., 1993).

Alternative splicing of the C-terminal region of NMDAR1 subunits is suggested to comprise control of NMDA receptor phosphorylation. Recent experimental data suggested a three transmembrane domain topology for ionotropic glutamate receptors (see Fig. 1; reviewed by Hollmann et al., 1994; Wo and Oswald, 1995). A consensus sequence for phosphorylation of NMDAR1 subunits by PKC is located within a single alternatively spliced cassette in the C-terminus following the fourth hydrophobic domain (Tingley et al., 1993), which is, according to the proposed topology model, located intracellularly (Fig. 1; Hollmann et al., 1994). NMDAR1-1a/b and NMDAR1-3a/b are the only splice variants containing this cassette. Their mRNAs are found in superficial laminae of the dorsal horn, together with a high-level binding of [^{3}H]PDBU (Worley et al., 1986; Tölle et al., 1994; Yashpal et al., 1995).

Molecular cloning studies have identified at least seven isoforms of PKC which are the products of three distinct (γ, β and α) genes (Nishizuka, 1986). Autoradiographic as well as immunohistochemical mapping with monoclonal antibodies shows a heterogeneous localization of PKC in the rat brain and a particular abundance in the substantia gelatinosa of the spinal cord (Worley et al., 1986; Mori et al., 1990).

Intracellular injection of PKC increases the responses of dorsal horn neurons to excitatory amino acids and enhances NMDA-activated currents in trigeminal neurons, suggesting a role of protein phosphorylation in central sensitization by noxious stimuli (Chen and Huang, 1992). Recent behavioural studies also indicate that PKC may act as a cellular link in neuronal plasticity evoked by noxious stimuli (Coderre et al., 1993). Nociceptive responses to injury induced by subcutaneous formalin injection are suppressed following intrathecal application of an inhibitor of PKC, and are enhanced after treatment with phorbol esters (Coderre, 1992; Coderre et al., 1993; Yashpal and Coderre, 1993; Yashpal et al., 1995).

In the present study we observed a bilateral increase of [^{3}H]PDBU binding and an expansion of [^{3}H]PDBU binding to deeper laminae of the dorsal horn following unilateral monoarthritis. Electrophysiological studies suggest that in the rat spinal dorsal horn activators of PKC, like PDBU, enhance the basal and evoked release of amino acid neurotransmitters (Gerber et al., 1989) and PKC may also be involved in the regulation of postsynaptic NMDA- and NK_1 receptors (Chen and Huang, 1992; Swartz et al., 1993; Munro et al., 1993) and calcium currents (Rane et al., 1989). The pathophysiological significance of a bilateral increase in [^{3}H]PDBU binding is further suggested by the observation that persistent pain produced by formalin injury and mechanical hyperalgesia in the hindpaw contralateral to a thermal injury are influenced by PKC inhibitors (Coderre et al., 1993). A synopsis of these findings suggests PKC as a crucial mediator of intracellular changes that can induce enhanced excitability of neurons. Whether the increased binding of [^{3}H]PDBU reflects de novo – target gene derived – synthesis of PKC or functional activation of non-active forms of PKC remains to be resolved. Whether the increased binding of [^{3}H]PDBU and/or the detection of [^{3}H]-PDBU in deeper laminae of the dorsal horn is based on an increase of all PKC isoforms, or whether this increase is restricted to certain PKC isoforms with distinct biochemical properties is currently investigated (Tölle et al., in preparation).

The present data on changes in the expression of glutamate receptor subunit genes following persistent pain are still restricted to observations on the expression of AMPA, kainate and NMDAR1-pan/NMDAR2 subunit mRNA. They suggest that the bilateral increase of neuronal excitability following unilateral inflammation is more likely to result from a phosphorylation of glutamate receptors than from alterations in receptor subunit expression and concomitant switches to subunit assemblies with, for example, different Ca^{2+} gating characteristics. However, since alternative splicing of the NMDAR1 subunit gene may provide a mechanism for regulating the sensitivity of glutamate receptors by protein phosphorylation, alterations in the expression of NMDAR1 splice variants are currently under investigation. We suggest that, although the expression of NMDAR1-1 and NMDAR1-3 mRNAs is only moderate in dorsal horn neurons, the enhanced excitability of neurons in this structure following persistent noxious stimulation may be due to NMDA receptor phosphorylation. Ongoing investigations with antibodies that are raised to detect only NMDA receptor complexes in a phosphorylated state will help to elucidate whether this molecular mechanism to rapidly and reversibly change the excitability of neurons is indeed important to understand spinal cord plasticity. Moreover, the integral function of metabotropic glutamate receptors (Anneser et al., 1995; Boxall et al., 1996) and the fundamental aspect of co-activation of receptors for peptidergic neurotransmitters (Urban et al., 1994), such as the tachykinins, has to be taken into consideration, to better understand the activity-dependent long-term changes in the spinal cord following persistent noxious stimulation.

References

Anneser, J., Berthele, A., Laurie, D.J., Tölle, T.R. and Zieglgänsberger, W. (1995) Differential distribution of metabotropic glutamate receptor mRNA in rat spinal cord neurons. *Eur. J. Neurosci.*, Suppl. 8: 14.07 (abstr.).

Bennett, J.A. and Dingledine, R. (1995) Topology profile for a glutamate receptor: three transmembrane domains and a channel-lining reentrant membrane loop. *Neuron*, 14: 373–384.

Bettler, B. and Mulle, C. (1995) Neurotransmitter receptors II: AMPA and kainate receptors. *Neuropharmacology*, 34: 123–139.

Boxall, S.J., Thompson, S.W.N., Dray, A., Dickenson, A.H. and Urban, L. (1996) Metabotropic glutamate receptor activation contributes to nociceptive reflex activity in the rat spinal cord *in vitro*. *Neuroscience*, 74: 13–20.

Brose, N., Gasic, G.P., Vetter, D.E., Sullivan, J.M. and Heinemann, S.F. (1993) Protein chemical characterization of the NMDA receptor subunit NMDA R1. *J. Biol. Chem.*, 268: 22663–22671.

Buller, A.L., Larson, H.C., Schneider, B.E., Beaton, J.A., Morrisett, R.A. and Monaghan, D.T. (1994) The molecular basis of NMDA receptor subtypes: native receptor diversity is predicted by subunit composition. *J. Neurosci.*, 14: 5471–5484.

Burnashev, N., Monyer, H., Seeburg, P.H. and Sakmann, B. (1992) Divalent ion permeability of AMPA receptor channels is dominated by the edited form of a single subunit. *Neuron*, 8: 189–198.

Butler, S.H., Godefroy, F., Besson, J.M. and Weil-Fugazza, J. (1992) A limited arthritic model for chronic pain studies in the rat. *Pain*, 48: 73–81.

Chen, L. and Huang, L.Y. (1992) Protein kinase C reduces Mg^{2+} block of NMDA-receptor channels as a mechanism of modulation. *Nature*, 521–523.

Coderre, T.J. (1992) Contribution of protein kinase C to central sensitization and persistent pain following tissue injury. *Neurosci. Lett.*, 140: 181–184.

Coderre, T.J., Katz, J., Vaccarino, A.L. and Melzack, R. (1993) Contribution of central neuroplasticity to pathological pain: review of clinical and experimental evidence. *Pain*, 52: 259–285.

Dani, J.A. and Mayer, M.L. (1995) Structure and function of glutamate and nicotinic acetylcholine receptors. *Curr. Opin. Neurobiol.*, 5: 310–317.

Dickenson, A.H. and Sullivan, A.F. (1990) Differential effects of excitatory amino acid antagonists on dorsal horn nociceptive neurones in the rat. *Brain Res.*, 506: 31–39.

Dougherty, P.M., Palecek, J., Paleckova, V., Sorkin, L.S. and Willis, W.D. (1992) The role of NMDA and non-NMDA amino acid receptors in the excitation of primate spinothalamic tract neurons by mechanical, chemical, thermal and electrical stimuli. *J. Neurosci.*, 12: 3025–3041.

Dubner, R. and Ruda, M.A. (1992) Activity-dependent neuronal plasticity following tissue injury. *Trends Neurosci.*, 15: 96–103.

Durand, G.M., Gregor, P., Zheng, H., Bennett, M.V.L., Uhl, G. and Zukin, R.S. (1992) Cloning of an apparent splice variant of the rat N-methyl-D-aspartate receptor NMDAR1 with altered sensitivity to polyamines and activators of protein kinase C. *Proc. Natl. Acad. Sci. USA*, 89: 9359–9363.

Gerber, G., Kangrga, I., Ryu, P.D., Larew, J.S.A. and Randic, M. (1989) Multiple effects of phorbol esters in the rat spinal dorsal horn. *J. Neurosci.*, 9: 3606–3617.

Good, P.F., Moran, T., Rogers, S.W., Heinemann, S., Morrison, J.H. (1992) Distribution of the kainate class glutamate receptor subunits (GluR5–GluR7) in the temporal cortex of the macaque: localization to identified projection neurons of the entorhinal cortex. *Soc. Neurosci. Abstr.*, 18: 1466.

Headley, P.M. and Grillner, S. (1990) Excitatory amino acids and synaptic transmission: the evidence for physiological function. *Trends Pharmacol. Sci.*, 11: 205–211.

Hollmann, M. and Heinemann, S. (1994) Cloned glutamate receptors. *Annu. Rev. Neurosci.*, 17: 31–108.

Hollmann, M., Boulter, J., Maron, C., Beasley, L., Sullivan, J., Pecht, G. and Heinemann, S. (1993) Zinc potentiates agonist-induced currents at certain splice variants of the NMDA receptor. *Neuron*, 10: 943–954.

Hollmann, M., Maron, C. and Heinemann, S. (1994) N-glycosylation site tagging suggests a three transmembrane domain topology for the glutamate receptor GluR1. *Neuron*, 13: 1331–1343.

Köhler, M., Kornau, H.C. and Seeburg, P.H. (1994) The organization of the gene for the functionally dominant α-amino-3-hydroxy-5-methyl-isoxazole-4-propionic acid receptor subunit GluR-B. *J. Biol. Chem.*, 269: 17367–17370.

Laurie, D.J. and Seeburg, P.H. (1994) Regional and developmental heterogeneity in splicing of the rat brain NMDAR1 mRNA. *J. Neurosci.*, 14: 3180–3194.

Meguro, H., Mori, H., Araki, K., Kushiya, E., Kutsuwada, T., Yamazaki, M., Kumanishi, T., Arakawa, M., Sakimura, K. and Mishina, M. (1992) Functional characterization of a heteromeric NMDA receptor channel expressed from cloned cDNAs. *Nature*, 357: 70–74.

Meller, S.T., Dykstra, C.L. and Gebhardt, G.F. (1993) Acute mechanical hyperalgesia is produced by coactivation of AMPA and metabotropic glutamate receptors. *NeuroReport*, 4: 879–882.

Monaghan, D.T. and Cotman, C.W. (1985) Distribution of NMDA-sensitive L-[^{3}H]glutamate binding sites in rat brain as determined by quantitative autoradiography. *J. Neurosci.*, 5: 2909–2919.

Monyer, H., Sprengel, R., Schoepfer, R., Herb, A., Higuchi,

M., Lomeli, H., Burnashev, N., Sakmann, B. and Seeburg, P.H. (1992) Heteromeric NMDA receptors: molecular and functional distinction of subtypes. *Science*, 256: 1217–1220.

Mori, M., Kose, A., Tsujino, T. and Tanaka, C. (1990) Immunocytochemical localization of protein kinase C subspecies in the rat spinal cord: light and electron microscopic study. *J. Comp. Neurol.*, 299: 167–177.

Moriyoshi, K., Masu, M., Ishii, T., Shigemoto, R., Mizuno, N. and Nakanishi, S. (1991) Molecular cloning and characterization of the rat NMDA receptor. *Nature*, 354: 31–37.

Munro, F., Young, M., Fleetwood-Walker, S., Parker, R. and Mitchell, R. (1993) Receptor and cellular mechanisms involved in mustard oil-induced activation of dorsal horn neurons. In: G.F. Gebhart, D.L. Hammond, and T.S. Jensen (Eds.), *Proceedings of the 7th World Congress of Pain, Progress in Pain Research and Management, Vol. 2*, IASP Press, Seattle, WA, pp. 337–346.

Nakanishi, N., Axel, R. and Shneider, N.A. (1992) Alternative splicing generates functionally distinct N-methyl-D-aspartate receptors. *Proc. Natl. Acad. Sci. USA*, 89: 8552–8556.

Neugebauer, V., Lucke, T. and Schaible, H.G. (1994) Requirement of metabotropic glutamate receptors for the generation of inflammation-evoked hyperexcitability in rat spinal cord neurons. *Eur. J. Neurosci.*, 6: 1179–1186.

Nishizuka, Y. (1986) Studies and perspectives of protein kinase C. *Science*, 233: 305–312.

Randic, M., Jiang, M.C. and Cerne, R. (1993) Long-term potentiation and long-term depression of primary afferent neurotransmission in the rat spinal cord. *J. Neurosci.*, 13: 5228–5241.

Rane, S.G., Walsh, M.P., McDonald, J.R. and Dunlap, K. (1989) Specific inhibitors of protein kinase C block transmitter-induced modulation of sensory neuron calcium current. *Neuron*, 3: 239–245.

Raymond, L.A., Blackstone, C.D. and Huganir, R.L. (1993) Phosphorylation of amino acid neurotransmitter receptors in synaptic plasticity. *Trends Neurosci.*, 16: 147–153.

Ren, K., Hylden, L.K., Williams, G.M., Ruda, M.A. and Dubner, R. (1992) The effects of a non competitive NMDA receptor antagonist, MK-801, on behavioral hyperalgesia and dorsal horn neuronal activity in rats with unilateral inflammation. *Pain*, 50: 331–344.

Seeburg, P.H. (1993) The molecular biology of mammalian glutamate receptor channels. *Trends Neurosci.*, 16: 359–365.

Sommer, B., Keinänen, K., Verdoorn, T.A., Wisden, W., Burnashev, N., Herb, A., Köhler, M., Takagi, T., Sakmann, B. and Seeburg, P.H. (1990) Flip and flop: a cell-specific functional switch in glutamate-operated channels of the CNS. *Science*, 249: 1580–1585.

Stevens, C.F. (1992) No pain, no gain. *Curr. Biol.*, 2: 497–499.

Swartz, K.J., Merritt, A., Bean, B.P. and Lovinger, D.M. (1993) Protein kinase C modulates glutamate receptor inhibition of Ca^{2+} channels and synaptic transmission. *Nature*, 361: 165–168.

Tingley, W.G., Roche, K.W., Thompson, A.K. and Huganir, R.L. (1993) Regulation of NMDA receptor phosphorylation by alternative splicing of the C-terminal domain. *Nature*, 364: 70–73.

Tölle, T.R., Castro-Lopes, J.M. and Zieglgänsberger, W. (1989) Transient increase in excitability ("wind-up") of somatosensory neurons might be linked to NMDA receptor-mediated processes. In: N. Elsner and W. Singer (Eds.), *Dynamics and Plasticity in Neuronal Systems*, Thieme, Stuttgart, p. 249.

Tölle, T.R., Berthele, A., Zieglgänsberger, W., Seeburg, P.H. and Wisden, W. (1993) The differential expression of 16 NMDA and non-NMDA receptor subunits in the rat spinal cord and in periaqueductal gray. *J. Neurosci.*, 13: 5009–5028.

Tölle, T.R., Ableitner, A., Castro-Lopes, J.M. and Zieglgänsberger, W. (1994) C-Fos protein, prodynorphin mRNA and protein kinase C are altered with distinct spatial and temporal patterns in the spinal cord of monoathritic rats. In: G.F. Gebhart, D.L. Hammond, and T.S. Jensen (Eds.), *Proceedings of the 7th World Congress of Pain, Progress in Pain Research and Management, Vol. 2*, IASP Press, Seattle, pp. 409–422.

Tölle, T.R., Berthele, A., Castro-Lopes, J.M., Wisden, W. and Zieglgänsberger, W. (1995a) Temporal and spatial analyses of molecular changes in the spinal cord of monoarthritic rats: NMDA and non-NMDA receptor subunit genes and protein kinase C. *J. Neurol.*, 242 (Suppl. 2): 135.

Tölle, T.R., Berthele, A., Laurie, D.J., Seeburg, P.H. and Zieglgänsberger, W. (1995b) Cellular and subcellular distribution of NMDAR1 splice variant mRNA in the rat lumbar spinal cord. *Eur. J. Neurosci.*, 7:1235–1244.

Tölle, T.R., Berthele, A., Zieglgänsberger, W., Seeburg, P.H. and Wisden, W. (1995c) Flip and Flop variants of AMPA receptors in the rat lumbar spinal cord. *Eur. J. Neurosci.*, 7:1414–1419.

Urban, L., Thompson, S.W. and Dray, A. (1994) Modulation of spinal excitability: co-operation between neurokinin and excitatory amino acid neurotransmitters. *Trends Neurosci.*, 17: 432–438.

Wenthold, R.J., Yokotani, N., Doi, K. and Wada, K. (1992) Immunochemical characterization of the non-NMDA glutamate receptor using subunit-specific antibodies – evidence for a hetero-oligomeric structure in rat brain. *J. Biol. Chem.*, 267: 501–507.

Westbrook, G.L. (1994) Glutamate receptor update. *Curr. Opin. Neurobiol.*, 4: 337–346.

Wisden, W. and Morris, B.J. (1994) *In situ* hybridization with synthetic oligonucleotide probes. In: W. Wisden and B.J. Morris (Eds.), *In Situ Hybridization: Protocols for the Brain*, Academic Press, London, pp. 9–34.

Wisden, W. and Seeburg, P.H. (1993) Mammalian ionotropic glutamate receptors. *Curr. Opin. Neurobiol.*, 3: 291–298.

Wo, Z.G. and Oswald, R.E. (1994) Transmembrane topology

of two kainate receptor subunits revealed by N-glycosylation. *Proc. Natl. Acad. Sci. USA*, 91: 7154–7158.

Wo, Z.G. and Oswald, R.E. (1995) Unraveling the modular design of glutamate-gated ion channels. *Trends Neurosci.*, 18: 161–168.

Worley, P.F., Baraban, J.M. and Snyder, S.H. (1986) Heterogeneous localization of protein kinase C in rat brain: autoradiographic analysis of phorbol ester receptor binding. *J. Neurosci.*, 6: 199–207.

Yashpal, K. and Coderre, T.J. (1993) Contribution of nitric oxide, arachidonic acid and protein kinase C to persistent pain following tissue injury in rats. *Abstracts 7th World Congress on Pain*. IASP Publications, Seattle, WA, p. 224.

Yashpal, K., Pitcher, G.M., Parent, A., Quiron, R. and Coderre, T.J. (1995) Noxious thermal and chemical stimulation induce increases in ^{3}H-phorbol 12,13-dibutyrate binding in spinal cord dorsal horn as well as persistent pain and hyperalgesia, which is reduced by inhibition of protein kinase C. *J. Neurosci.*, 15: 3263–3272.

Zieglgänsberger, W. and Tölle, T.R. (1993) The pharmacology of pain signalling. *Curr. Opin. Neurobiol.*, 3: 611–618.

Zimmermann, M. (1983) Ethical guidelines for investigations of experimental pain in conscious animals. *Pain*, 16: 109–110.

CHAPTER 16

Neurobiology of spinal nociception: new concepts

J. Sandkühler

II. Physiologisches Institut, Universität Heidelberg, Heidelberg, Germany

Introduction

The classical concept of acute nociception describes a serial neuronal system which transmits information from nociceptors to second order neurons in the spinal or trigeminal dorsal horn and to higher order neurons in the thalamus and cortex. Information transfer is mediated by classical neurotransmitters such as excitatory amino acids leading to changes of membrane potential in a time window of milliseconds to seconds. As a consequence, discharge rates of nociceptive neurons increase as long as nociceptors are excited (Foerster, 1913; Zimmermann, 1977; Besson and Chaouch, 1987). Since many nociceptors may converge onto second order neurons and because of the putative stochastic nature of ion channel kinetics and neurotransmitter release it is generally assumed that discharges of nociceptive neurons constitute a simple renewal process with very high number of degrees of freedom (Steedman and Zachary, 1990). Thus, in most of the previous electrophysiological studies nociception was evaluated on the basis of mean discharge rates of single neurons and background activity was considered to be stochastic noise, which decreases the signal-to-noise ratio. In support of this classical concept it was shown that discharge rates are often correlated in a linear fashion to the intensity of noxious stimulation and antinociception is consequently defined as a reduction in discharge rates of nociceptive neurons (Zimmermann, 1977).

This classical concept well describes many features of acute nociception but fails to explain phenomena seen during prolonged, chronic pain states.

For example, chronic pain may outlast the period of nociceptor discharge and may persist when discharge rates in higher order neurons have declined (Lenz et al., 1989). The classical concept also provides no explanation for plastic changes in nociception which may last for hours, days or longer.

Recent work has largely extended our concepts about the nociceptive system. Not only have powerful endogenous antinociceptive mechanisms been identified which may depress nociceptive discharges by segmental (Melzack and Wall, 1965), propriospinal (Sandkühler et al., 1993) and supraspinal, descending (Fields and Basbaum, 1978) systems, also new hypotheses have been proposed about the mechanisms underlying chronicity of pain.

Here, our concepts on spinal processing of nociceptive information under pathophysiological conditions are described; some of the aspects are depicted in Fig. 1. We are interested in answering the following questions:

(1) What are indicators for spinal nociception under pathophysiological conditions?
(2) What are the neurophysiological mechanisms underlying plastic changes of nociception and which neuromediators are involved in the spinal cord?
(3) What are possible molecular mechanisms leading to long-term changes of spinal nociception?

Non-linear dynamics of discharges of nociceptive neurons

From clinical research we have learned that

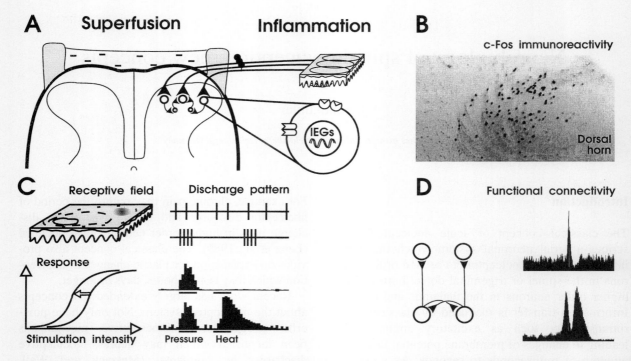

Fig. 1. Summary of some aspects of the experimental design. We employ the controlled superfusion of the rat lumbar spinal cord dorsum to study the effects of extrasynaptic neuromediators on the function of the intact spinal neuronal network. We determine the discharge properties of single dorsal horn neurons (C), the temporal correlation of discharges of simultaneously recorded neurons (cross-correlograms in D), synaptic strength of excitatory neurotransmission in primary afferent C-fibers and the expression of immediate early genes such as c-*fos* (B).

pathological processes are often characterized by a drastic reduction in the complexity of the underlying dynamical process (Pool, 1989; Elbert et al., 1994). For example, during epileptic seizure the complexity of the EEG signal is strongly reduced and beat-to-beat variability of the heart may be diminished in patients which later suffer from a heart attack or a sudden infant death (Schechtman et al., 1992). Finally, the variability of whole blood cell counts may be lower in patients with leukaemia as compared to normal subjects. We have adopted these ideas and tested the hypothesis that non-linear dynamics of the discharges of nociceptive neurons differ under physiological and pathophysiological conditions.

In pentobarbital anaesthetized rats we have recorded extracellularly in the lumbar spinal dorsal horn from multireceptive neurons. General experimental conditions and procedures have been described in detail elsewhere (Sandkühler and Eblen-Zajjur, 1994) (see also Fig. 1). In brief, all neurons had low threshold mechano-receptive fields at the glabrous skin of the ipsilateral hindpaw and responded to noxious radiant skin heating and to electrical stimulation of A- and C-fibers in afferent nerves. Thus, neurons were typical multireceptive or wide-dynamic range neurons. About two-thirds of the neurons displayed background activity in the absence of intentional stimulation, about one-third of the neurons had a monosynaptic projection to the contralateral thalamus. Inflammation of the glabrous skin at or close to the cutaneous receptive fields was produced by long lasting radiant heat stimulation (three heat pulses of 48°C for 100 s each).

To determine non-linear dynamics of background activity, stationary discharges consisting of 2000–5000 action potentials were analysed. As a graphical tool to evaluate dynamics of discharges we plotted phase-space portraits and for a quantitative description we calculated the D_2 correlation dimension of the discharges according to the algorithm of Grassberger and Procaccia (1983). We constructed the phase-space portrait by determining the differences (D_n) of neighbouring interspike intervals and plotting D_{n+1} on the ordinate versus D_n on the abscissa (Debus and Sandkühler, 1996; see Fig. 2) Fig. 2 illustrates that small changes in discharge pattern may result in pronounced changes in phase-space portrait. A random or highly complex discharge pattern is characterized by a cloud of points in the phase-space portrait without any detectable inner structure (Fig. 3A). The background activity of 66 of 90 nociceptive spinal dorsal horn neurons analysed so far had clearly deterministic patterns in their phase-space portraits (see Fig. 3B–D and Fig. 4A for examples). We have also quantitatively determined the complexity of the discharge patterns by calculating the D_2 correlation dimension. The D_2 correlation dimension provides a direct measure of the complexity of a dynamical process and the minimal number of independent variables which underly the process (Eckman and Ruelle, 1985). It may therefore be used as sensitive indicator for changes in a dynamical process. The D_2 correlation dimensions of background activity of those 66 neurons with a deterministic pattern in the phase-space portrait were typically not higher than five under control conditions (see Fig. 3C,D and Fig. 4A for examples). This suggests, that background activity of these neurons is not a simple renewal process or stochastic noise but results from deterministic processes with low number of degrees of freedom, contradicting previous assumptions (Steedman and Zachary, 1990).

Acute noxious skin heating destroyed the deterministic patterns in the phase-space portraits of 90% of the neurons, as illustrated for one typical example in Fig. 4A. Skin heating drastically enhanced the D_2 correlation dimension often to values higher than ten.[1] Thus, the number of degrees of freedom of the discharges increased, probably because a large number of heat-sensitive nociceptors has been activated which independently from each other generate action potentials and directly or indirectly evoke excitatory postsynaptic potentials in the spinal neuron under study.

The effects of skin inflammation on non-linear dynamics of discharges of multireceptive spinal dorsal horn neurons were qualitatively different from the effects of acute skin heating. Fig. 4B illustrates that the complexity of background discharge of the neuron was strongly reduced about one hour after induction of skin inflammation. This change in the non-linear dynamics was observed in 9 of 11 (81.8%) multireceptive neurons and typically persisted throughout the recording period of up to 6 h and outlasted changes (mainly increases but also decreases) in discharge rates. The increase in the number of degrees of freedom of the neuronal discharges during acute noxious skin heating and the decrease during inflammation of the skin suggests that underlying neurophysiological mechanisms are fundamentally different. Apparently, the decrease in the D_2 correlation dimension may be used as an indicator for spinal nociception under pathological (inflammatory) conditions. This is in line with previous reports showing that disordered dynamical processes may lose their normal complexity (Pool, 1989).

Extrasynaptic mediators of long-term changes of spinal nociception

It is likely that composition and distribution of chemical signals which are released in the spinal dorsal horn during acute, non-damaging noxious stimuli differ from the mixture of neuroactive substances released in the spinal cord during inflammation of skin or other peripheral tissues. Following inflammation or strong long lasting noxious stimulation neuropeptides including substance P, neurokinin A and somatostatin are detected extrasynaptically throughout the superficial and in some cases also throughout the deep dorsal horn ipsi- and contralateral to the stimulation site (Kuraishi et al., 1985; Duggan and Hendry, 1986; Morton et al., 1989; Duggan et al., 1990). An extrasynaptic

---

[1] Beyond this limit the Grassberger–Procaccia algorithm becomes inaccurate and the D_2 correlation dimension is classified as "HIGH".

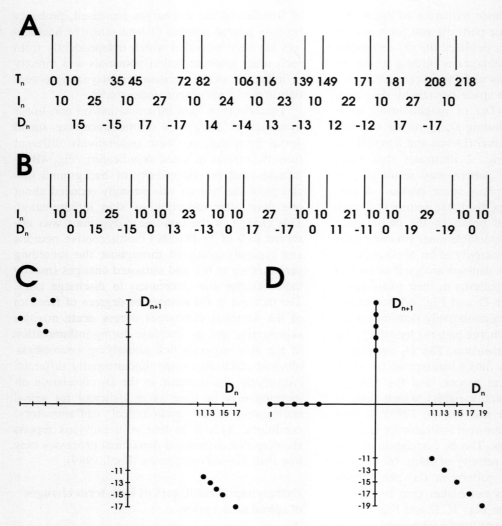

Fig. 2. Phase-space portraits of neuronal discharges are used as a graphical tool to evaluate non-linear dynamics. Normpulses of constructed discharges are displayed in (A) and (B). T_n indicates time points of the discharges. I_n are the resulting inter-spike intervals. The differences between neighbouring inter-spike intervals $(I_{n+1} - I_n)$ are nominated D_n (this abbreviation has nothing to do with the D_2 correlation dimension). In these constructed examples short intervals of identical length (10 ms) are followed by longer and more variable intervals. The phase-space portraits are constructed by plotting D_{n+1} on the ordinate versus D_n on the abscissa. The trajectories, i.e. the connecting lines between sequential points are not shown for simplicity. The example illustrates that a minor change in discharge pattern may lead to a major change in the phase-space portrait of the discharge. Thus, this graphical method may be used as a sensitive indicator for changes in discharge patterns. The phase-space portrait is not affected by scaling operations and is therefore not changed by variations in mean discharge rates.

spread of substance P in spinal dorsal horn is in line with a mismatch between the location of release sites and binding sites for substance P in the spinal cord (Liu et al., 1994). In contrast, an extra-synaptic spread of neuropeptides has not been described following brief noxious stimuli. The extra-synaptic spread of chemical signals has been termed "volume transmission" (Agnati et al., 1986). This mode of chemical signaling clearly does not mediate the punctate fast synaptic transmission which may convey information about a brief sensory stimulus in a somatotopically ordered fashion. The functional roles of extrasynaptic neuropeptides in the spinal cord are currently under investigation.

We have developed and quantitatively described a method for the controlled superfusion of the spinal cord dorsum of cats (Sandkühler and Zimmermann, 1988) and rats (Sandkühler et al., 1991) and we have determined the distribution and the tissue concentration of [^{125}I]neurokinin A following superfusion of the rat lumbar spinal cord (Beck et al., 1995). A well is formed directly on the intact arachnoidea with a specially synthesized silicone rubber (see Fig. 1). The pool may have any size and any shape and provides complete sealing (Sandkühler et al., 1991). Fifteen, 30 and 60 min after beginning of superfusion with 50 μM neurokinin A the concentration of the peptide in the superficial dorsal horn at a depth of 250 μm was approximately 2000 pmol/g, 1600 pmol/g and 525 pmol/g respectively. At a depth of 750 μm the concentrations were 170, 330 and 170 pmol/g. In the ventral horn the concentrations were always lower than 20 pmol/g at any of these time points (Beck et al., 1995). These dorso-ventral concentration gradients are similar to those achieved after spinal release of neuropeptides. This superfusion model may therefore be used to study the effects of chemical mediators which are present extrasynaptically in the spinal dorsal horn, e.g. under some pathophysiological conditions. Of course, the technique may also be used for pharmacological studies, e.g. to determine the effects of spinal analgesic substances on single neurons in the dorsal horn (Sandkühler et al., 1990) (Fig. 1).

Substance P is one of the neuropeptides which are released in the spinal dorsal horn during skin inflammation. Bath application of substance P not only depolarizes many spinal dorsal horn neurons and increases their excitability in vitro (Randić and Miletic, 1977, 1978), substance P may also selectively enhance the efflux of glutamate in spinal cord slices (Kangrga et al., 1990; Kangrga and Randić, 1990). Thus, extrasynaptic substance P might act as a neuromodulator which modifies the effects of classical neurotransmitters and thereby the synaptic strength and functional connectivity within the spinal neuronal network. If this also happens following spinal release of substance P, then spinal nociception could not only be modified by changes in the excitability of neurons, but also by changes of their functional connectivity. This could, for example, lead to changes in synchronization of discharges. Synchronization of converging neurons constitutes a powerful mechanisms by which information transfer may be strengthened. To test whether synchronization of discharges of nociceptive neurons is affected during skin inflammation, we have recorded simultaneously from two or more multireceptive spinal dorsal horn neurons, some of which projected to the same site in the contralateral thalamus. Thus, some of these neurons may have converged onto the same target neurons.

Multi-neuron recordings were made through tungsten microelectrodes or glass microelectrodes with a carbon fiber (impedance at 1000 Hz, 2–4 MΩ). Action potentials of individual neurons were identified by principal component analysis of

---

Fig. 3. Some aspects of the analysis of background activity of four different multireceptive neurons (A–D) recorded in laminae IV–VI of the lumbar enlargement. A hardcopy from the computer screen of a program written in Turbo Pascal for disc operating system (DOS) computers is shown. (A) Top right: cumulative number of ISIs (total number 12 000) is plotted versus recording time (0–611.25 s). Middle right: interspike interval histogram including the shortest ISI (1.5 ms) and the longest ISI (155 ms) of the spike train. Bottom right: the spike train is divided into two halves. Mean ISIs and the coefficient of dispersion (CD = variance/mean) of ISIs, which is an indicator for burst-like discharges, is calculated separately for each half (Mean1, Mean2, CD1 and CD2) and for the complete spike train (Mean and CD). Only spike trains with at least 5000 interspike intervals (ISIs) and with stationary discharge rates were analysed further. Left: phase-space portrait of spike train was constructed as described in legend to Fig. 2. $\tau = 1$ indicates that D_{n+1} is plotted against D_n. Axes are from −21 to +21 ms not showing all data points (zoomfactor is five). The phase-space portrait of this spike train shows no inner structure and the D_2 correlation dimension was higher than ten (D_2 = HIGH). (B) Results from a neuron with a harmonic oscillation (fundamental frequency 25 Hz) and D_2 = HIGH. (C) A deterministic pattern is seen in the phase-space portrait of this neuron, $D_2 = 0.17 \pm 0.01$. (D) Results from another neurons with $D_2 = 2.65 \pm 0.04$.

212

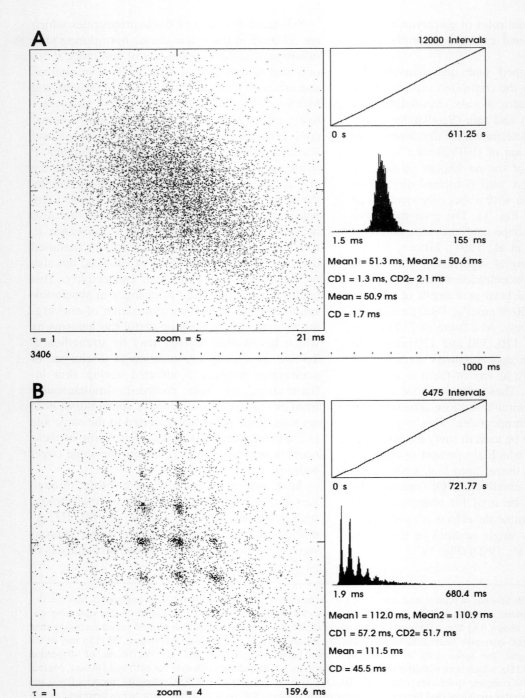

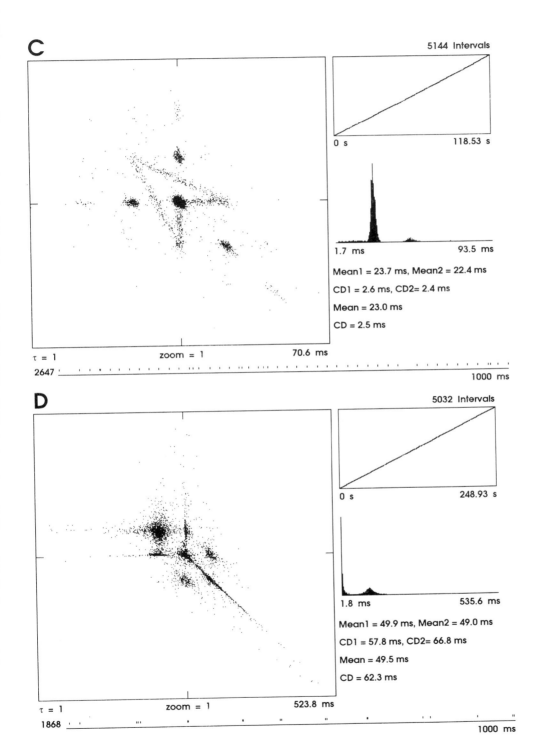

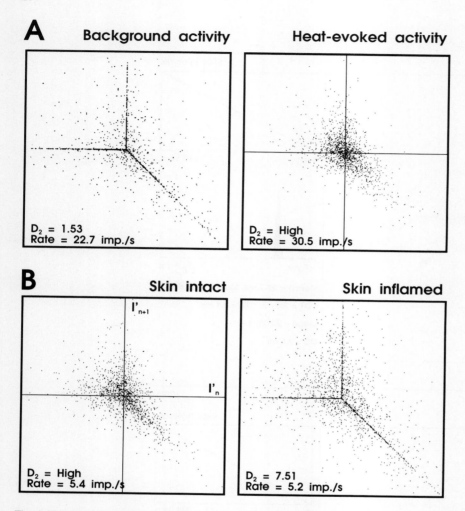

Fig. 4. Non-linear dynamics of discharges of multireceptive spinal dorsal horn neurons are differentially affected by acute noxious skin heating and inflammation of skin. Phase-space portraits of two different neurons are shown in (A) and (B). Phase-space portraits of background activity recorded in the absence of any skin stimulation are shown on the left. Noxious skin heating (46°C for 100 s) destroyed the deterministic pattern in the phase space portrait (A) and strongly enhanced the complexity of the discharge pattern as revealed by the increase in the D_2 correlation dimension from 1.53 to a value larger than ten (High). Inflammation of the skin had qualitatively different effects as shown for one representative example in (B). Phase-space portraits of background activity before and one hour after noxious radiant heat-induced inflammation of the skin within the neurons receptive field are shown (from Sandkühler et al., 1994).

their shape. As a sensitive indicator for synchronized discharges we used cross-correlograms which here indicate the total number of impulses in a neuron (neuron one) generated 250–0 ms before (negative latency) or 0–250 ms after an action potential was discharged by a simultaneously recorded reference neuron (neuron two) (see Figs. 5 and 6). Flat cross-correlograms, as illustrated in Fig. 5C, therefore indicate that no temporal correlation exists between the discharges of the two neurons on this time scale. This was seen in the background activity of only 12 of 65 pairs of simultaneously recorded multireceptive neurons (Fig. 5 shows examples in B and in C). The remaining neurons displayed background discharges which were correlated in time. Most cross-correlograms (41/65 pairs) had a central peak which indicates that the two neurons discharged action potentials synchronously within a narrow time window (width of the peak) and with a high probability (height of the peak/height of the background level), see Figs. 5A and Fig. 6A for examples. Acute noxious skin heating failed to induce qualitative changes in the patterns of the cross-correlograms of 30 of 37 pairs of neurons but affected the pattern in the remaining seven, mainly by changes from a central peak (indicating synchronized discharges) to flat (indicating non synchronous discharges) (see Fig. 5A). Discharges were never synchronized by acute noxious skin heating (Sandkühler et al., 1994).

The effects of inflammation of the skin were qualitatively different from the effects of acute skin heating. In eight of 26 pairs of neurons the pattern of cross-correlograms changed qualitatively but desynchronizations were never observed. In contrast, neuronal pairs with previously flat cross-correlograms did develop synchronous discharges during the first hour after onset of inflammation (Fig. 5B) or cross-correlograms with a central peak changed to a pattern with bilateral peaks suggestive of opening reverberant circuits (Fig. 6B). Synchronizations typically lasted for 1–2 h in one case until the end of the recording period of 4 h (Sandkühler et al., 1994). In computer simulations in which cellular and network properties of the superficial dorsal horn were implemented we could illustrate that strengthening of synaptic pathways which leads to an opening of reverberant circuits results in bilateral peaks in the cross-correlograms of the corresponding neuronal discharges (Fig. 6A,B).

To test the hypothesis that extrasynaptic substance P mediates some of the plastic changes in the spinal dorsal horn we have superfused the cord dorsum at the recording segment for 30 min with substance P at 10 or 100 μM. Substance P superfusions significantly enhanced the mean level of background activity of 18 multireceptive laminae I and II neurons tested (from 2 to 8 imp/s), increased the mean size of the cutaneous receptive fields of 12 neurons tested from 360 to 465 mm^2 and enhanced mean responses of 15 neurons to mechani-

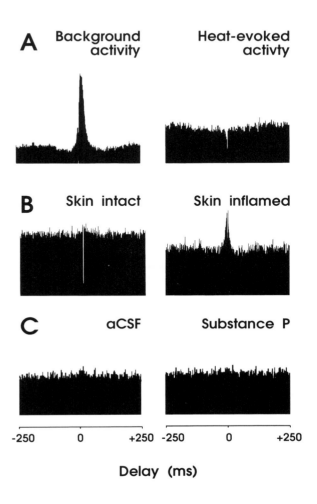

Fig. 5. Synchronization of discharges of simultaneously recorded multireceptive neurons is differentially affected by acute noxious skin heating, skin inflammation and extrasynaptic substance P. Cross-correlograms (bin width 1 ms) of background activity of three different pairs of neurons which were recorded under control conditions (skin intact, superfusion with artificial cerebrospinal fluid, ACSF) are shown on the left. Cross-correlograms of the same pairs of neurons recorded during acute noxious skin heating (46°C for 100 s, A), 1 h after induction of skin inflammation (B) or during superfusion with substance P at 100 μM (C) are depicted on the right. Synchronized discharges of the neuronal pair in (A) were desynchronized during encoding of an acute noxious stimulus, while neurons with previously non-correlated discharges may synchronize during skin inflammation (B). Substance P alone is not sufficient to trigger changes in the temporal correlation of discharges (C) (from Sandkühler et al., 1994).

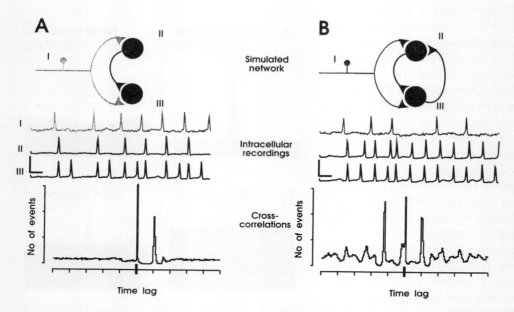

Fig. 6. Results from a computer simulation of a simple neuronal network (top) using the GENESIS script language and a SUN workstation. Cell parameters were chosen to mimic the properties of a fine primary afferent (neuron I) and those of lamina II neurons (neurons II and III). Intracellular recordings of the three neurons are shown in middle part and the resulting cross-correlograms of neuron III with neuron II as the reference are shown at the bottom. In (A) and (B) the primary afferent provides common input to both lamina II neurons with the same latency. This leads to the central peak in the cross-correlograms centered at zero time lag. In addition neuron II excites neuron III which produces the peak to the right. In (A) the excitatory connection from neuron III to neuron II is anatomically present but not effective. In (B) this connection has become suprathreshold for triggering action potentials in neuron II. The temporal correlation of discharges is not readily apparent in the traces from intracellular recordings, it can, however, easily be detected in the cross-correlogram (bottom) which now shows bilateral peaks.

cal skin probing with a calibrated von Frey hair (6.8 g) from 90 to 147 imp/15 s (Liu and Sandkühler, 1995c). In contrast, substance P failed to change qualitatively the patterns of the cross-correlograms of 23 of 25 pairs of multireceptive neurons in the superficial dorsal horn (Fig. 5C). In two pairs synchronous discharges became non-synchronous after substance P (Sandkühler et al., 1994). Thus, extrasynaptic substance P may produce changes in the excitability of multireceptive spinal dorsal horn neurons, similar to those seen after inflammation of peripheral tissues. In contrast, substance P alone did not change the temporal correlation of discharges of simultaneously recorded pairs of neurons, apparently other neuromediators must be involved (Sandkühler et al., 1994). Possibly calcitonin-gene-related-peptide which modifies the release of a number of amino acids in the spinal cord (Kangrga et al., 1990) may play a role. Exogenous applications of neuropeptides (and of all other substances) either by superfusion of spinal cord in vivo or in vitro, by microinjections or by the iontophoretic application near the recording site are always extrasynaptic. Thus, from these studies the synaptic effects of neuropeptides can not be deduced. However, the fact that strong or long lasting conditioning afferent stimulations but not mild or brief stimulations evoked long-term changes of spinal nociception suggests that under pathophysiological conditions an extrasynaptic spread of neuropeptides may be involved.

Long-term potentiation of C-fiber-evoked potentials in the superficial spinal dorsal horn

The neurobiological mechanisms which lead to an increase in the level of background activity, ex-

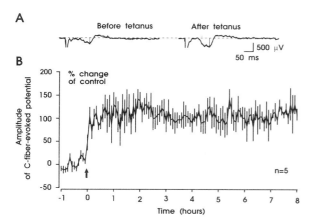

Fig. 7. In (A) original recordings of spinal field potentials from one animal are shown. Potentials were evoked by sciatic nerve stimulation at 15 V before (left hand site) and one hour after tetanic stimulation of sciatic nerve. Recordings started 40 ms prior to electrical nerve stimulation. In (B) mean amplitudes of C-fiber-evoked field potentials are plotted versus time of the experiment. Potentials were evoked at 5 min intervals by electrical test stimuli of the sciatic nerve (20 V, 0.5 ms pulses) before (controls, −1 to 0 h) and after (0 to 8 h min) conditioning tetanic stimulation of sciatic nerve (indicated by the arrow) (from Liu and Sandkühler, 1995b).

pansion of receptive fields, enhanced responses to sensory stimulation and to changes in discharge patterns and in temporal correlation of discharges are not known. Long-term modifications in the synaptic strength within the neuronal network of the spinal dorsal horn may underly at least some of these lasting changes. In the hippocampus long-term potentiation of synaptic transmission is considered to be a fundamental mechanism of learning and memory. We have now shown for the first time that long-term potentiation of C-fiber-evoked potentials exists in the superficial spinal dorsal horn (Liu and Sandkühler, 1995b). Field potentials which were recorded in laminae I and II of the lumbar spinal dorsal horn of rats under urethane anesthesia were evoked by electrical stimulation of the ipsilateral sciatic nerve. The potentials were characterized by high thresholds (7 V), long latencies (90–130 ms, see Fig. 7A, corresponding to conduction velocities of less than 1.2 m/s), they were not affected by spinalization at C5–C6 or muscle relaxation with pancuronium. From pulse duration curves the chronaxy was found to be 1.2 ms, thus, these potentials are considered to be evoked by primary afferent C-fibers. Tetanic, high frequency, high intensity electrical stimulation (100 Hz, 30 V, 0.5 ms pulses given in four trains of 1 s duration at 10 s intervals) of sciatic nerve always induced long-term potentiation of late potentials. The increase ranged between +71% and +174% of controls and always lasted until the end of the recording period (4–9 h, see Fig. 7B). Superfusion of the spinal cord dorsum at the recording segment with a NMDA-receptor antagonist (D-CPP at 500 nM) abolished induction of long-term potentiation (Fig. 8). Superfusion of the cord dorsum with NMDA at 1, 10 or 50 μM failed to induce a long-term potentiation in six out of eight rats tested (Liu and Sandkühler, 1995b). Thus, like in hippocampus, activation of NMDA-receptors is essential but not sufficient for induction of long-term potentiation in the superficial dorsal horn of the intact spinal cord. Further, blockade of spinal neurokinin 1 or neurokinin 2 receptors also abolished induction of long-term potentiation (Liu and Sandkühler, 1995a) in the spinal dorsal horn indicating that substance P acting on neurokinin 1 receptors and neurokinin A acting on neurokinin 2 receptors may also be required. These tachykinins are known to produce long-lasting depolarizations

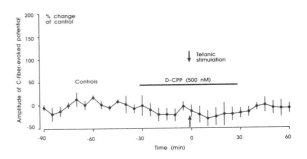

Fig. 8 illustrates that superfusion of the spinal cord with the NMDA receptor antagonist D-CPP at 500 nM abolishes induction of long-term potentiation. Mean amplitudes of C-fiber-evoked field potentials before superfusion (−90 to −30 min) served as controls. The interval of superfusion is represented by the horizontal bar. The time point of conditioning tetanic stimulation of sciatic nerve is indicated by the arrow (from Liu and Sandkühler, 1995b).

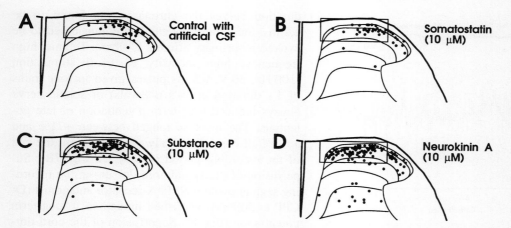

Fig. 9. Superfusion of the rat spinal cord in vivo with tachykinins substance P or neurokinin A but not with somatostatin induce c-*fos* in neurons of the dorsal horn. Representative transverse sections through the fourth lumbar segment are shown. Each dot represents one strongly labelled cell with c-Fos immunoreactivity in a 40 μm thick section. In the medial part of the superficial dorsal horn (directly underneath the superfused cord surface) the number of c-Fos positive cells was affected by the superfusion. Thus, immunoreactivity was analysed quantitatively in an area indicated by the square. c-Fos immunoreactivity following superfusion with ACSF is probably due to the preparation as virtually no constitutive c-Fos expression exists in spinal cord of untreated animals.

of nociceptive spinal dorsal horn neurons. This may be necessary to expel Mg^{2+} from NMDA receptor coupled ion channels to allow influx of Ca^{2+} ions into the cell upon binding of L-glutamate. High frequency, high intensity stimulation was found to be more effective than low frequency (3 Hz), high intensity stimulation to induce long-term potentiation of C-fiber-evoked potentials in the spinal dorsal horn. This is in line with the hypothesis that large dense-cored vesicles, which are believed to contain neuropeptides, are present in terminals of Aδ- and C-fibers and can be released most readily by high frequency, burst-like discharges.

Descending control of long-term potentiation in spinal cord

We have recently shown that intact tonic descending pathways may inhibit induction of long-term potentiation of spinal C-fiber-evoked potentials by nerve injury, by inflammation of skin or subcutaneous tissue (Liu and Sandkühler, 1996) or by superfusion of the spinal cord with substance P (Liu and Sandkühler, 1995a). Strong inflammation of the glabrous skin at a hindpaw was induced by radiant skin heating (60°C, given 4–6 times for 30 s at 60 s intervals). This inflammation never induced long-term potentiation of C-fiber-evoked field potentials in rats with the spinal cord intact. However, in rats with a spinal transection at a high cervical level, inflammation produced a mean potentiation to $183 \pm 21.4\%$ of control ($n = 5$) which lasted for at least 4 h and which was not affected by cutting the sural nerve distal to the site of electrical stimulation. Subcutaneous formalin injections into a hindpaw also induced long-term potentiation (to $205 \pm 15.3\%$ of control) in all five spinalized rats tested. Noxious mechanical skin stimulation at a hindpaw with an artery clamp (ten times for 5 s at 10 s intervals) also induced long-term potentiation (to $210 \pm 20\%$ of controls) which lasted for 1–4 h in four spinal rats tested. Nerve injury by squeezing the sural nerve with forceps (four times) distal to the stimulation electrode induced long-term potentiation in all five spinalized rats tested (to $250 \pm 35\%$ of control) for 2–4 h. Superfusion of the spinal cord at the recording segment with substance P at 10 μM for 1 h also induced long-term potentiation which always lasted until the end of the recording period (up to 4 h). In rats with the spinal cord intact identical conditioning stimuli never induced long-term potentiation.

Thus, supraspinal, tonic descending systems do not only modify discharge rates of nociceptive spinal dorsal horn neurons but may also prevent the long-lasting potentiation of synaptic transmission between primary afferent C-fibers and neurons in the superficial spinal dorsal horn which can be induced by natural skin stimuli, nerve injury or extrasynaptic substance P.

Induction and blockade of expression of immediate-early genes in spinal neurons

What are the molecular mechanisms which lead to long-term changes of spinal nociception? A hypothesis which is discussed presently suggests that the expression of immediate-early genes could change the phenotype of spinal nociceptive neurons and thereby the function of the nociceptive system (Hunt et al., 1987; Morgan and Curran, 1991). It was shown that plasticity inducing stimuli such as strong noxious stimulation or inflammation of peripheral tissues not only evoked the release of neuropeptides such as substance P, neurokinin A and somatostatin but also triggered the expression of immediate-early genes in neurons of the spinal dorsal horn. Up-to-today it has, however, not been shown that extrasynaptic neuropeptides may induce immediate-early gene expression in spinal neurons, or that the expression of any of these genes plays a role for long-term changes of nociception.

To address these questions we have superfused the cord dorsum either with artificial cerebrospinal fluid (ACSF) or with the neuropeptides substance P, neurokinin A or somatostatin at 1, 10 or 100 μM for 30 min. Ninety minutes after the end of the superfusions the animals were perfused and the tissues were processed for immunohistochemistry to detect the c-Fos and Jun B proteins in transverse sections through the superfused segment. Substance P and Neurokinin A at 10 or 100 μM induced c-*fos* and *jun*-B in cells of the superficial dorsal horn, while somatostatin did not trigger immediate-early gene expression (Sandkühler and Beck, unpublished observations) (see Fig. 9). Following superfusion with artificial cerebrospinal fluid the mean number of c-Fos positive cells in the medial aspect of laminae I/II in 40 μm thick transverse sections through the lumbar spinal cord was 19. Following superfusion with substance P at 10 μM a mean of 45 cells was c-Fos positive and following superfusion with neurokinin A a mean of 36 cells. Somatostatin at 10 or 100 μM failed to induce c-Fos. This differential effect of neuropeptides on the expression of immediate-early genes is in line with qualitatively different effects of substance P and somatostatin on the excitability of spinal neurons, as substance P may enhance (Murase and Randić, 1984) and somatostatin may depress the excitability (Randić and Miletic, 1978; Sandkühler et al., 1990). These results further support the notion that inhibitory agents and hyperpolarizations do not induce c-*fos*.

The observation that the same stimuli (inflammation of the skin or superfusion of the spinal cord with substance P) may induce long-lasting increases in excitability of spinal dorsal horn neurons and the expression of immediate-early genes is consistent with the hypothesis that some of these genes could play a role for plastic changes of nociception. Of course, these correlative findings do not proof in any way that a causal relationship does exist. Thus, the key questions remain whether the expression of individual immediate-early genes is required and/or sufficient for the induction and/or maintenance of long-term changes of spinal nociception. To determine whether the massive expression of immediate-early genes in neurons of the superficial spinal dorsal horn is followed by long-term changes of nociception in behaving animals we have compared nociceptive thresholds before and after conditioning intrathecal stimulations. Intrathecal injections of NMDA (25 nmol) or electrical stimulation of dorsal roots (15 V, 0.5 ms pulses given at 3 Hz for 15 min) but not intrathecal injection of ACSF or low intensity electrical stimulation of dorsal roots (3 V) led to a massive expression of c-Fos protein in neurons of the superficial and, to a lesser extent, also in neurons of the deep dorsal horn of the sacral and lumbar spinal cord. Thermal and mechanical nociceptive thresholds were, however, not different between sham treated and stimulated animals before and for up to

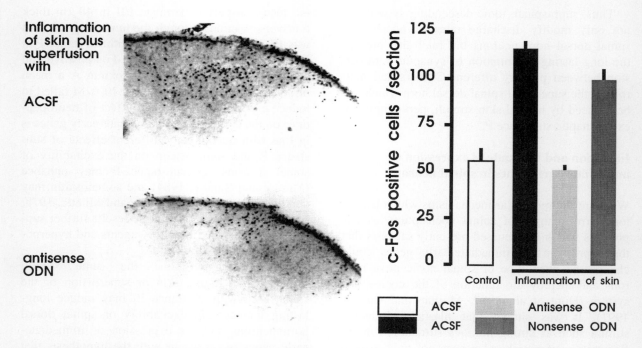

Fig. 10. Inflammation of the skin induced by immersion of both hindpaws into hot water (52°C for 20 s, 10 times in intervals of 2 min) induced c-*fos* in neurons of the superficial dorsal horn of the lumbar enlargement. The section at the top illustrates c-Fos immunoreactivity in one representative, 40 μm thick transverse section. Superfusion of the cord for 6 h with antisense oligodeoxynucleotides complementary to parts of the c-*fos* mRNA reduced the number of c-Fos positive cells (section at the bottom). Superfusion with a random sequence of oligodeoxynucleotides (nonsense ODNs) did, however, not affect c-*fos* expression (from Gillardon et al., 1994).

14 days after stimulations as determined by hotplate and tail-flick latencies and by von Frey thresholds at hindpaws (Sandkühler et al., 1996). Thus, the massive expression of c-Fos protein in spinal dorsal horn neurons is not necessarily followed by lasting changes in spinal nociception.

To evaluate whether the expression of immediate-early genes is necessary for the maintenance of long-lasting changes in nociception, it will be required to selectively block the expression of individual immediate-early genes following plasticity-inducing stimuli. We have therefore used our superfusion technique to apply phosporothioate modified antisense deoxynucleotides (Krieg, 1994) directly to the spinal cord 6 h prior to prolonged noxious skin heating (Gillardon et al., 1994). Superfusion with a random sequence of oligodeoxynucleotides did not affect the expression of c-Fos or jun B, while superfusion with 75 μM oligodeoxynucleotides complementary to c-*fos* mRNA significantly reduced the number of c-Fos immunoreactive neurons in the superfused spinal cord segment from a mean of 119 cells per section to 54 cells per section ($n = 4$) (see Fig. 10). The expression of Fos B and c-Jun was not affected (Gillardon et al., 1994). Thus, superfusion of the spinal cord in vivo with phosphorothioate modified antisense oligodeoxynucleotides may be a useful tool to study the role of individual immediate-early genes for functional changes in the spinal dorsal horn. This will open the opportunity to go beyond correlative studies and to search for causal relationships between gene expression and spinal cord function.

Conclusions and perspectives

The majority of nociceptive neurons which have

been recorded from in the spinal dorsal horn respond to both noxious and innocuous stimuli (wide-dynamic range, class 2 or multireceptive neurons). However, the role of multireceptive neurons for nociception is still not unequivocally clear. Single cell recordings and conventional counting statistics such as mean discharge rates and interspike interval distributions cannot answer the long-standing question whether discharge patterns and/or the temporal relationship of discharges in an assembly of neurons contribute to the encoding of nociceptive information by multireceptive neurons.

New analytical tools derived from concepts of chaos research are now beginning to be used to evaluate dynamical processes in medicine (Pool, 1989; Denton et al., 1990) and physiology (Nicolis, 1983; Elbert et al., 1994) and they have already provided new insights into mechanisms of health and disease. Here, we have applied these tools to analyse the complexity and the number of degrees of freedom of discharges of nociceptive neurons. Our results suggest that encoding of a brief noxious stimulus may be characterized by an increase in the number of degrees of freedom and that this is qualitatively different from the reduced complexity of discharge patterns under pathophysiological conditions (i.e. during inflammation of the skin). This suggests that the underlying mechanisms are fundamentally different. It should be rewarding to perform further studies which determine non-linear dynamics of responses of multireceptive neurons encoding stimuli of different modality or which process nociceptive information under different pathophysiological conditions. Single cell recordings should be complemented by simultaneous recordings from multiple individual nociceptive neurons (Sandkühler and Eblen-Zajjur, 1994; Sandkühler et al., 1995) to determine the temporal correlation of their discharges. The present results have provided evidence that synchronization of discharges and opening of reverberant pathways in the spinal cord may play a role during inflammation of peripheral tissues. Encoding of a brief "physiological" noxious stimulus was, however, characterized by desynchronizations. One might speculate that synchronization of discharges

it is a more economical (less cell energy consuming) way to achieve a long-lasting increase in information transfer (e.g. during skin inflammation) than prolonged increases in discharge rates. Longterm potentiation of synaptic transmission in primary afferent C-fibers exists in the superficial spinal dorsal horn and may underly long-lasting changes in spinal nociception. These plastic changes in spinal dorsal horn do not solely depend upon the level and the pattern of the afferent barrage, they are also under the control of tonically active pathways descending from supraspinal sites.

Quick increases in discharge rates which do not require any changes in synaptic strength may, in contrast, be more appropriate to encode a brief noxious stimulus.

Before drawing any firm conclusions it is, of course, mandatory to show that changes in nonlinear dynamics, in temporal correlation of discharges and in synaptic strength can be decoded by the nervous system. Because of the complexity and the non-linearities involved it has been impossible in most cases to predict theoretically the dynamics of single cells and neuronal networks. It has now become feasible to test the possible role of nonlinear dynamics and temporal correlation of discharges by modelling single neurons and networks (Grillner et al., 1991; Selverston, 1993). Detailed maps about the anatomical structure of nerve cells and their connectivity as well as quantitative information about passive membrane properties and membrane conductances can now be implemented into computer models which allow realistic simulations of some of the integrative properties of single cells and networks (Wallen et al., 1992; De Schutter and Bower, 1994). Computer models have already become an important tool to formulate and test hypotheses about information processing in the nervous system. Results from computer models can help to guide experimenters to design their experiments and may lead to the formulation of new hypotheses which then can be tested experimentally.

We have suggested that fundamentally different mechanisms in the spinal cord may be active during noxious skin heating and inflammation of the skin since non-linear dynamics and temporal correlation of discharges are affected qualitatively different under these conditions. Most likely the composition of chemical signals released into the spinal cord and the mode of signal transfer are different. Fast synaptic transmission is generally assumed to account for the quick and short lasting increase in discharge rates during acute noxious stimuli. In contrast, an extrasynaptic spread of neuromediators which has been termed "volume transmission" may be responsible for long-lasting modifications of the functions of the neuronal network, e.g. by enhancing the synaptic strength in selected pathways. Recent work has revealed insights into some of the cellular and molecular mechanisms involved in long-term changes of nociception. In electrophysiological studies (Cerne et al., 1993; Randić et al., 1993) and in behavioral experiments (Kolhekar et al., 1993; Mao et al., 1993; Meller and Gebhart, 1993; Meller et al., 1994) the causal involvement of the putative retrograde messenger NO, of membrane receptors such as the NMDA-receptor, second messengers such as free cytosolic calcium ions and protein kinases such protein kinase C have been established. An appealing hypothesis suggests that, in addition, the expression of immediate-early genes such as c-*fos* may also play a role for long-term changes of nociception (Hunt et al., 1987; Menetrey et al., 1989; Herdegen et al., 1991; Morgan and Curran, 1991; Lucas et al., 1993). However, despite of the large number of studies which were conducted after the initial report by Hunt and his colleagues in 1987, a causal relationship between altered gene expression and changes in nociception has, up to today, not been shown. Further electrophysiological and behavioral studies will be necessary to clarify this question.

References

Agnati, L.F., Fuxe, K., Zoli, M., Ozini, I., Toffano, G. and Ferraguti, F. (1986) A correlation analysis of the regional distribution of central enkephalin and β-endorphin immunoreactive terminals and of opiate receptors in adult and old male rats. Evidence for the existence of two main types of communication in the central nervous system: the volume transmission and the wiring transmission. *Acta Physiol. Scand.*, 128: 201–207.

Beck, H., Schröck, H. and Sandkühler, J. (1995) Controlled superfusion of the rat spinal cord for studying non-synaptic transmission: an autoradiographic analysis. *J Neurosci Methods*, 58: 193–202.

Besson, J.M. and Chaouch, A. (1987) Peripheral and spinal mechanisms of nociception. *Physiol. Rev.*, 67: 67–186.

Cerne, R., Rusin, K.I. and Randić, M. (1993) Enhancement of the *N*-methyl-D-aspartate response in spinal dorsal horn neurons by cAMP-dependent protein kinase. *Neurosci. Lett.*, 161: 124–128.

Debus, S. and Sandkühler, J. (1996) Low dimensional attractors in discharges of sensory neurons in the rat spinal dorsal horn are maintained by supraspinal descending systems. *Neuroscience*, 70: 191–200.

De Schutter, E. and Bower, J.M. (1994) An active membrane model of the cerebellar Purkinje cell. I. Simulation of current clamps in slice. *J. Neurophysiol.*, 71: 375–400.

Denton, T.A., Diamond, G.A., Helfant, R.H., Khan, S. and Karagueuzian, H. (1990) Fascinating rhythm: a primer on chaos theory and its application to cardiology. *Am. Heart J.*, 120: 1419–1437.

Duggan, A.W. and Hendry, I.A. (1986) Laminar localization of the sites of release of immunoreactive substance P in the dorsal horn with antibody-coated microelectrodes. *Neurosci. Lett.*, 68: 134–140.

Duggan, A.W., Hope, P.J., Jarrott, B., Schaible, H.-G. and Fleetwood-Walker, S.M. (1990) Release, spread and persistence of immunoreactive neurokinin A in the dorsal horn of the cat following noxious cutaneous stimulation. Studies with antibody microprobes. *Neuroscience*, 35: 195–202.

Eckman, J.P. and Ruelle, D. (1985) Ergodic theory of chaos and strange attractors. *Rev. Mod. Phys.*, 57: 617–656.

Elbert, T., Ray, W.J., Kowalik, Z.J., Skinner, J.E., Graf, K.E. and Birbaumer, N. (1994) Chaos and physiology: Deterministic chaos in excitable cell assemblies. *Physiol. Rev.*, 74: 1–47.

Fields, H.L. and Basbaum, A.I. (1978) Brainstem control of spinal pain-transmission neurons. *Annu. Rev. Physiol.*, 40: 217–248.

Foerster, O. (1913) Vorderseitenstrangdurchschneidung im Rückenmark zur Beseitigung von Schmerzen. *Berlin. Klin. Wochenschr.*, 50: 1499.

Gillardon, F., Beck, H., Uhlmann, E., Herdegen, T., Sandkühler, J., Peyman, A. and Zimmermann, M. (1994) Inhibition of c-Fos protein expression in rat spinal cord by antisense oligodeoxynucleotide superfusion. *Eur. J. Neurosci.*, 6: 880–884.

Grassberger, P. and Procaccia, I. (1983) Characterization of strange attractors. *Am. Phys. Soc.*, 50: 346–349.

Grillner, S., Wallén, P., Brodin, L. and Lansner, A. (1991) Neuronal network generating locomotor behavior in lamprey: circuitry, transmitters, membrane, properties and simulation. *Annu. Rev. Neurosci.*, 14: 169–199.

Herdegen, T., Tölle, T.R., Bravo, R., Zieglgänsberger, W. and Zimmermann, M. (1991) Sequential expression of JUN B, JUN D and FOS B proteins in rat spinal neurons: cascade of transcriptional operations during nociception. *Neurosci. Lett.*, 129: 221–224.

Hunt, S.P., Pini, A. and Evan, G. (1987) Induction of *c-fos*-like protein in spinal cord neurons following sensory stimulation. *Nature*, 328: 632–634.

Kangrga, I., Larew, J.S.A. and Randić, M. (1990) The effects of substance P and calcitonin gene-related peptide on the efflux of endogenous glutamate and aspartate from the rat spinal dorsal horn in vitro. *Neurosci. Lett.*, 108: 155–160.

Kangrga, I. and Randić, M. (1990) Tachykinins and calcitonin gene-related peptide enhance release of endogenous glutamate and aspartate from the rat spinal dorsal horn slice. *J. Neurosci.*, 10: 2026–2038.

Kolhekar, R., Meller, S.T. and Gebhart, G.F. (1993) Characterization of the role of spinal N-methyl-D-aspartate receptors in thermal nociception in the rat. *Neuroscience*, 57: 385–395.

Krieg, A.M. (1994) Uptake and efficacy of phosphodiester and modified antisense oligonucleotides in primary cell cultures. *Clin. Chem.*, 39: 710–712.

Kuraishi, Y., Hirota, N., Sato, Y., Hino, Y., Satoh, M. and Takagi, H. (1985) Evidence that substance P and somatostatin transmit separate information related to pain in the spinal dorsal horn. *Brain Res.*, 325: 294–298.

Lenz, F.A., Kwan, H.C., Dostrovsky, J.O. and Tasker, R.R. (1989) Characteristics of the bursting pattern of action potentials that occurs in the thalamus of patients with central pain. *Brain Res.*, 496: 357–360.

Liu, X.-G. and Sandkühler, J. (1995a) Long-term potentiation of C-fiber-evoked potentials in the rat spinal dorsal horn. *Pflügers Arch. Eur. J. Physiol.*, 429, Suppl. 6: 40.

Liu, X.-G. and Sandkühler, J. (1995b) Long-term potentiation of C-fiber-evoked potentials in the rat spinal dorsal horn is prevented by spinal N-methyl-D-receptor blockage. *Neurosci. Lett.*, 191: 43–46.

Liu, X.-G. and Sandkühler, J. (1995c) The effects of extrasynaptic substance P on nociceptive neurons in laminae I and II in rat lumbar spinal dorsal horn. *Neuroscience*, 68: 1207–1218.

Liu, X.-G. and Sandkühler, J. (1996) Long-term potentiation of spinal C-fiber-evoked potentials induced by skin inflammation or nerve injury. *IASP Abstr. 8th World Congress on Pain*, 40.

Liu, H., Brown, J.L., Jasmin, L., Maggio, J.E., Vigna, S.R., Mantyh, P.W. and Basbaum, A.I. (1994) Synaptic relationship between substance P and the substance P receptor: light and electron microscopic characterization of the mismatch between neuropeptides and their receptors. *Proc. Natl. Acad. Sci. USA*, 91: 1009–1013.

Lucas, J.J., Mellström, B., Colado, M.I. and Naranjo, J.R. (1993) Molecular mechanisms of pain: serotonin$_{1A}$ receptor agonists trigger transactivation by c-fos of the prodynorphin gene in spinal cord neurons. *Neuron*, 10: 599–611.

Mao, J., Mayer, D.J., Hayes, R.L. and Price, D.D. (1993) Spatial pattern of increased spinal cord membrane-bound protein kinase C and their relation to increases in ^{14}C-2-deoxyglucose metabolic activity in rats with painful peripheral mononeuropathy. *J. Neurophysiol.*, 70: 470–481.

Meller, S.T. and Gebhart, G.F. (1993) Nitric oxide (NO) and nociceptive processing in the spinal cord. *Pain*, 52: 127–136.

Meller, S.T., Cummings, C.P., Traub, R.J. and Gebhart, G.F. (1994) The role of nitric oxide in the development and maintenance of the hyperalgesia produced by intraplantar injection of carrageenan in the rat. *Neuroscience*, 60: 367–374.

Melzack, R. and Wall, P.D. (1965) Pain mechanisms: a new theory. *Science*, 150: 971–979.

Menetrey, D., Gannon, A., Levine, J.D. and Basbaum, A.I. (1989) Expression of c-fos protein in interneurons and projection neurons of the rat spinal cord in response to noxious somatic, articular and visceral stimulation. *J. Comp. Neurol.*, 285: 177–195.

Morgan, J.I. and Curran, T. (1991) Stimulus-transcription coupling in the nervous system: involvement of the inducible proto-oncogenes *fos* and *jun*. *Annu. Rev. Neurosci.*, 14: 421–451.

Morton, C.R., Hutchison, W.D., Hendry, I.A. and Duggan, A.W. (1989) Somatostatin: evidence for a role in thermal nociception. *Brain Res.*, 488: 89–96.

Murase, K. and Randić, M. (1984) Actions of substance P on rat spinal dorsal horn neurons. *J. Physiol. (London)*, 346: 203–217.

Nicolis, J.S. (1983) The role of chaos in reliable information processing. In E. Basar, H. Flohr, H. Haken and A.J. Mandell, (Eds.), *Synergetics of the Brain*. Springer Verlag, Berlin, pp. 102–121.

Pool, R. (1989) Is it healthy to be chaotic? *Science*, 243: 604–607.

Randić, M. and Miletić, V. (1977) Effects of substance P in cat dorsal horn neurones activated by noxious stimuli. *Brain Res.*, 128: 164–169.

Randić, M. and Miletić, V. (1978) Depressant actions of methionine-enkephalin and somatostatin in cat dorsal horn neurones activated by noxious stimuli. *Brain Res.*, 152: 196–202.

Randić, M., Jiang, M.C. and Cerne, R. (1993) Long-term potentiation and long-term depression of primary afferent neurotransmission in the rat spinal cord. *J. Neurosci.*, 13: 5228–5241.

Sandkühler, J. and Eblen-Zajjur, A. (1994) Identification and characterization of rhythmic nociceptive and non-nociceptive spinal dorsal horn neurons in the rat. *Neuroscience*, 61: 991–1006.

Sandkühler, J. and Zimmermann, M. (1988) Neuronal effects of controlled superfusion of the spinal cord with monoaminergic receptor antagonists in the cat. In: H.L. Fields and J.-M. Besson (Eds.), *Pain Modulation, Progress in Brain Research*, Vol. 77, Elsevier, Amsterdam, pp. 321–327.

Sandkühler, J., Fu, Q.-G. and Helmchen, C. (1990) Spinal somatostatin superfusion *in vivo* affects activity of cat nociceptive dorsal horn neurons: comparison with spinal morphine. *Neuroscience*, 34: 565–576.

Sandkühler, J., Fu, Q.-G. and Stelzer, B. (1991) Propriospinal neurons are involved in the descending inhibition of lumbar spinal dorsal horn neurons from the mid-brain. In: M.R. Bond, J.E. Charlton and C.J. Woolf (Eds.), *Proceedings of the VIth World Congress on Pain*. Elsevier, Amsterdam: pp. 313–318.

Sandkühler, J., Stelzer, B. and Fu, Q.-G. (1993) Characteristics of propriospinal modulation of nociceptive lumbar spinal dorsal horn neurons in the cat. *Neuroscience*, 54: 957–967.

Sandkühler, J., Eblen-Zajjur, A.A. and Liu, X.-G. (1994) Differential effects of skin inflammation, extrasynaptic substance P and noxious skin heating on rhythmicity, synchrony and nonlinear dynamics in rat spinal dorsal horn. In: G.F. Gebhart, D.L. Hammond and T.S. Jensen (Eds.), *Progress in Pain Research and Management, Vol. 2, Proceedings of the 7th World Congress on Pain*. IASP Publications, Seattle, WA, pp. 347–356.

Sandkühler, J., Eblen-Zajjur, A., Fu, Q.-G. and Forster, C. (1995) Differential effects of spinalization on discharge patterns and discharge rates of simultaneously recorded nociceptive and nonnociceptive spinal dorsal horn neurons. *Pain*, 60: 55–65.

Sandkühler, J., Treier, A.-C., Liu, X.-G. and Ohnimus, M. (1996) The massive expression of c-Fos protein in spinal dorsal horn neurons is not followed by long-term changes in spinal nociception. *Neuroscience*, 73: 657–666.

Schechtman, V.L., Raetz, S.L., Harper, R.K., Garfinkel, A., Wilson, A.J., Southall, D.P. and Harper, R.M. (1992) Dynamic analysis of cardiac R-R intervals in normal infants and in infants who subsequently succumbed to the sudden infant death syndrome. *Pediatr. Res.*, 31: 606–612.

Selverston, A.I. (1993) Modelling of neural circuits: What have we learned? *Annu. Rev. Neurosci.*, 16: 531–546.

Steedman, W.M. and Zachary, S. (1990) Characteristics of background and evoked discharges of multireceptive neurons in lumbar spinal cord of cat. *J. Neurophysiol.*, 63: 1–15.

Wallén, P., Ekeberg, Ö., Lansner, A., Brodin, L., Travén, H. and Grillner, S. (1992) A computer-based model for realistic simulations of neural networks. II. The segmental network generating locomotor rhythmicity in the lamprey. *J. Neurophysiol.*, 68: 1939–1950.

Zimmermann, M. (1977) Encoding in dorsal horn interneurons receiving noxious and non noxious afferents. *J. Physiol. (Paris)*, 73: 221–232.

CHAPTER 17

Balances between excitatory and inhibitory events in the spinal cord and chronic pain

Anthony H. Dickenson

Department of Pharmacology, University College, Gower Street, London, WC1E 6BT, UK

Actions of the excitatory amino acid glutamate

Glutamate has been recognized as a key transmitter in neurones for many years and is one of the main excitatory transmitters in the central nervous system, not only conveying synaptic excitation from neurone to neurone but being of great importance in synaptic plasticity (Collingridge and Singer, 1990). Given this widespread action it is therefore not surprising that the recent development of selective drugs has revealed an important role of glutamate in both the spinal and thalamic transmission of nociception and pain. In particular, activation of one of the receptors for glutamate, the *N*-methyl-D-aspartate (NMDA) receptor, in the spinal cord seems to be a critical step in the production of a number of pain states such as those leading to post-operative pain but also in chronic pains such as those caused by inflammation and neuropathy (Dickenson, 1990; Dubner and Ruda 1992; Dickenson, 1994a; Price et al., 1994a).

Many nociceptive sensory fibres themselves contain glutamate which can coexist with peptides such as substance P and CGRP (Battaglia and Rustioni, 1988). There is much evidence for a release of glutamate after noxious stimuli together with peptides. In addition to these afferent sources of glutamate it is likely to be the main transmitter in the output neurones which conveys messages to the thalamus, reticular formation and beyond. Thus in a number of clinically relevant pains there is likely to be a post-synaptic activation of a number of peptide receptors together with the receptors for the excitatory amino acids on spinal neurones. The interplay between these receptors will determine the excitatory component of spinal pain transmission (Dray et al., 1994).

NMDA receptors and the transmission of pain

C-fibre stimulation at low frequency produces an acute constant response of spinal cord neurones to each stimulus. The NMDA receptor does not appear to participate in this baseline response to a noxious stimulus which appears to be transmitted via another receptor for glutamate, the α-amino-3-hydroxy-5-methyl-isoxazole propionate (AMPA) receptor. Thus in the presence of 6-nitro-7-sulphamoylbenzo[*f*]quinoxaline-2,3-dione (NBQX), a selective AMPA receptor antagonist, there is an abolition of the baseline constant response of spinal nociceptive neurones whereas this activity is unaltered by NMDA antagonists (see Fig. 1). However, if the stimulus is given at higher frequency the baseline response is suddenly transformed to a much higher level of activity which is NMDA receptor dependent. Marked increases in the neuronal response, up to 20-fold, occur despite the constant afferent stimulus. Wind-up is the term for these dramatic increases in both the duration and the magnitude of the cell responses that occur although the input into the spinal cord remains constant (Dickenson, 1990) and can be demonstrated in human psychophysical studies (Price et al., 1994). Fig. 1 depicts the independent AMPA/NMDA components of wind-up. There is

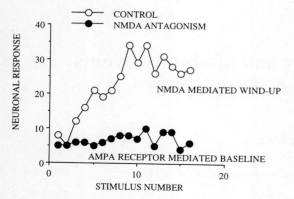

Fig. 1. Wind-up of a spinal nociceptive neurone. The response of a dorsal horn neurone to a repeated constant 0.5 Hz stimulus (electrical transcutaneous C-fibre strength) is enhanced by wind-up. Note how in the presence of an NMDA receptor/channel antagonist (50 µg of intrathecal MK801) only a constant response to the constant stimulus occurs. This baseline response is AMPA receptor mediated (see Dickenson, 1994).

good evidence that cooperation between the peptides and glutamate released in response to a painful stimulus allows the NMDA receptor to be activated to produce this wind-up. The NMDA receptor channel is plugged by normal physiological levels of magnesium and it is likely that the actions of peptides on their receptors removes this channel block and allows the receptor–channel complex to operate (see Dray et al., 1994).

Wind-up in animals and man are sensitive to NMDA receptor antagonists which have been applied intrathecally, demonstrating a spinal site of action, or systemically, with important clinical implications for the control of chronic pains (McQuay and Dickenson, 1990; Dickenson, 1990; Dubner and Ruda, 1992; Dickenson, 1994a; Price et al., 1994a; Woolf, 1994). Many persistent pains are characterized by hyperalgesia, and NMDA mediated wind-up is thought to be critical for this. The idea is that a low level of afferent barrage can, via wind-up, be transformed to a high level of spinal output and thereby underlies hyperalgesia (Dickenson, 1994a). Thus wind-up could be looked upon as a key to central hypersensitivity. Activation of C fibres results in a marked and prolonged increase in the flexion withdrawal reflex in rats which was the first evidence for the existence of central hypersensitivity in pain states and is now known to be NMDA receptor mediated (Woolf, 1983; Woolf and Thompson, 1991).

The NMDA receptor then could be proposed to play a crucial role in more prolonged pain states where functional alterations in central transmission processes such as hyperalgesia and allodynia are seen, and this has turned out to be the case.

NMDA mediated events have physiological consequences in that NMDA antagonists have clear effectiveness in inflammatory pain, neuropathic pain, and experimental models of allodynia (Yaksh, 1989; Haley et al., 1990; Seltzer et al., 1991a,b; Schaible et al., 1991; Stanfa et al., 1992; Mao et al., 1993). In all these situations the neuronal and behavioural hyperalgesic/allodynic responses are abolished by NMDA antagonists. However, since the baseline responses are not NMDA mediated, these antagonists are not analgesics in the conventional sense but antihyperalgesics. In inflammation, NMDA receptor activation induces the enhanced pain states (Haley et al., 1990; Schaible et al., 1991) to both brief and more prolonged tissue damage and then maintain this elevated pain condition. Injury discharges caused by nerve damage transmitted into the spinal cord appear to evoke wind-up and NMDA activation and then contribute to the subsequent pain states (Seltzer et al., 1991a,b; Yamamoto and Yaksh, 1992; Mao et al., 1994). Continued NMDA receptor activation then also maintains the pain. In these models spsinal application of NMDA antagonists have beneficial effects both when given before the injury or inflammation where they can prevent the establishment of the enhanced pain but are also very effective weeks after induction of the injury against the established hyperalgesia and spontaneous pain (Dickenson, 1994a). Finally, in animal models of allodynia, NMDA antagonists are effective against the tactile evoked nociception whereas morphine is not (Yaksh, 1989).

Thus much evidence supports a role of NMDA receptors in the induction and maintenance of pain following tissue damage, nerve dysfunction and surgery and so NMDA antagonism may well be of clinical relevance to the treatment of pain. Side effects of NMDA antagonists could include amnesia, sedation and psychotomimesis. Spinal routes of administration could be used to avoid side effects mediated by brain sites. In fact, there is now evidence from a number of clinical studies to support the concepts from the animal models, in that ketamine has been shown to be effective in prolonged clinical pains.

The roles of inhibitory systems in chronic pain

A problem addressed in this chapter is: why, since the NMDA receptor is clearly implicated in both inflammatory pains and neuropathic pains, is the ability to treat these pain states so different in humans? In the former case, post-operative pain is very clearly amenable to effective treatment by opioids yet with neuropathic pain it is clear that morphine effectiveness is reduced, although there is some range of opinion as to whether it is unresponsive to morphine or whether morphine dose escalation will be required to produce some relief.

In addition to circumstances leading to the activation of excitatory systems, the lifting of or interference with inhibitory controls is also important. In chronic pains, particularly neuropathic states, a reduction in the activity of inhibitory controls is very likely to be an important factor (Zimmermann, 1991; Dickenson, 1994b). Allodynia can be produced by pathological or pharmacological interference with spinal GABA or glycine inhibitions (Yaksh, 1989). This occurs in normal animals, so that blocking these inhibitory controls can mimic the effects of peripheral pathology. The resultant touch evoked responses are NMDA dependent showing that lifting of inhibitions controlling this receptor in otherwise normal animals causes aberrant sensory processing in that touch now elicits pain-like behaviour (Yaksh, 1989). These inhibitory systems may control not only nociception but prevent A-fibre access to pain transmitting systems (Fig. 2). There is evidence for a reduction in the levels of spinal GABA after nerve damage and since GABA is found in interneurones it is of interest to note the evidence for interneuronal malfunction after nerve section. Here, signs of interneuronal damage can be seen

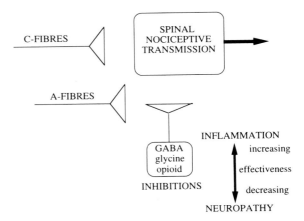

Fig. 2. The control of transmission in the spinal cord by GABA and glycine; increases and decreases in these transmitter levels occur differentially in inflammation and neuropathy, respectively. See text for further details.

after nerve section which may be one of the substrates for loss of inhibitions in neuropathic states (see Castro-Lopes et al., 1993).

Neuropathic damage can also reduce opioid receptor numbers, particularly the pre-synaptic receptors made in C-fibres (see Dickenson, 1994b). In addition to this loss of inhibitory opioid receptors, the peptide cholecystokinin reduces the ability of morphine to produce analgesia at the spinal level. Increases in the level of cholecystokinin, induced by nerve damage, have been shown to be responsible for at least part of the reduction in morphine analgesia seen after nerve section (Stanfa et al., 1994a).

Consequently, in neuropathic states, a number of key spinal inhibitory controls are reduced. Thus even if the level of excitability remained the same as in normal animals, the level of transmission would be higher due to the lesser inhibitions and the ability to control the pain would also be reduced.

By complete contrast, in inflammation, a number of inhibitory transmitter systems show increased activity. Subsequent to the induction of greater levels of excitation there is evidence, in models of acute inflammation lasting for hours, for adaptive increases in spinal inhibitions and interestingly, some time later, in additional reduced

NMDA receptor and peptide mediated excitation (Dray et al., 1994). These upregulations in inhibitions could represent physiological attempts to counter the increased peripheral drive arriving in the spinal cord. Within 1–3 h of inflammation, neurones show changes in excitability which are directly related to the original level of excitability of these cells. Neurones with low excitability, as expected, become more excitable but those with a high degree of NMDA mediated wind-up become less active after inflammation indicating that inhibitory controls are induced once a certain level of excitation is produced (Stanfa et al., 1992). The substrate for these inhibitions is yet unknown but there are a number of candidates. Thus, GABA levels are increased (Castro-Lopes et al., 1994), descending alpha-2 activity is also increased (Stanfa et al., 1994), the levels of adenosine, acting at the inhibitory A1 receptor are enhanced (Reeve and Dickenson, 1995) and the ability of morphine to inhibit nociception is also increased (Stanfa et al., 1994b); all these changes occur within 1–3 h of the induction of inflammation.

The increased effects of morphine are due to two factors: (a) the inflammation rapidly inducing a novel peripheral opioid site of action and (b) an enhancement of the spinal effects of morphine (see Dickenson, 1994b). In the second case, there is good evidence that this is not due to any change in spinal opioid receptors but results from a reduction in the spinal levels of the peptide cholecystokinin (CCK). CCK, via actions on the CCK-B receptor can interfere with morphine, probably by opposing the intracellular events following opioid receptor activation. The reduced levels of CCK seen after inflammation is akin to removal of a physiological brake on opioid analgesia and leads to the observed enhanced effects of morphine (Stanfa et al., 1993). Recall that the opposite happens in neuropathic states (see Stanfa et al., 1994b). Thus morphine analgesia is enhanced after inflammation yet reduced in neuropathic states.

Difficulties in controlling pains where NMDA receptors are active may, on the one hand, arise from the resultant high levels of excitation in the nociceptive circuitry (Chapman and Dickenson, 1992; Yamamoto and Yaksh, 1992) but could equally arise from failed inhibitions. When these NMDA mediated central events leading to hypersensitivity are active, such as in the shorter term models of tissue damage (wind-up, formalin and the hypersensitive reflex) there is a reduced sensitivity to opioids (Chapman and Dickenson, 1992). In the case of these models dose-escalation can overcome the reduced opioid sensitivity but clinically, side-effects may complicate this tactic. By contrast, NMDA antagonists such as ketamine and dextromethorphan have the potential not to abolish pain, but to prevent or block central hypersensitive states. They have no effect on the inputs onto spinal neurones but abolish wind-up so resetting the potentiated response to a steady response (Dickenson, 1990).

The spinal neuronal responses to the peripheral injection of formalin provide a measure of neuronal activity produced by peripheral inflammation. What is seen is an acute phase of firing followed by a vigorous and very rapid spinal NMDA receptor mediated response as inflammation occurs. Formalin induced activity differs from other inflammatory models such as carrageenan inflammation, by having a reduced sensitivity to opioids. These responses to formalin, lasting for about 1 h, may differ from the slower developing inflammatory states in that the peripheral activity arrives at the spinal cord and activates the NMDA receptor before the more slowly developing compensatory increases in inhibitory controls and CCK-mediated enhanced opioid effects are effective. The NMDA receptor mediation of activity in the shorter term models unbalanced by the induction of the slower developing inhibitory changes may underlie the poorer opioid inhibitions seen in this condition (Chapman et al., 1994). However, as the inflammation continues, the formalin response, both neuronal and behavioural, declines and finishes although the peripheral inflammation continues. Again, the presumed increased spinal inhibitions that attenuate this response are yet unknown. These events are depicted in Fig. 4. Now at about 3 h after the start of the inflammation, with carrageenan as the promoter, compensatory increases in spinal opioid sensitivity (via altered CCK) and other inhibitions means that the longer term in-

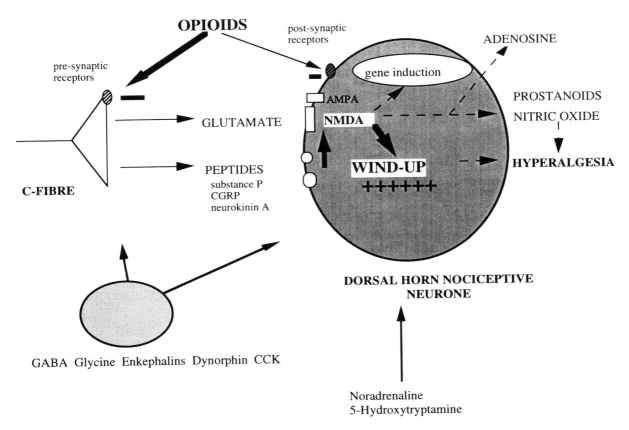

Fig. 3. An overview of the spinal pharmacology of pain.

flammatory pains respond well to opioids, partly to the enhanced opioid effectiveness per se but also because the increased non-opioid inhibitory systems (GABA, adenosine and α-2 adrenoceptor mediated) will reduce the NMDA driven level of excitability. Thus as time goes on, the excitations

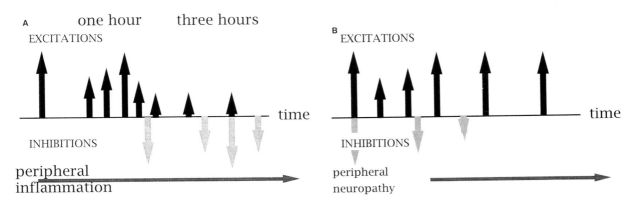

Fig. 4. A schematic diagram of the balance between excitations and inhibitions after inflammation (A) and neuropathic damage (B) and how these change with time. See text for further implications.

produced by peripheral inflammation become balanced by increased inhibitory controls. As discussed above, failure of these and other inhibitory controls may be factors leading to the poor control and chronicity of neuropathic pain (Dickenson 1994a,b; Price et al., 1994a). Fig. 3 shows some of these controls acting on spinal transmission.

Pre-emptive analgesia

A logical approach is to pre-empt the induction of increases in central excitability (Woolf, 1994). The idea here is to prevent the changes in excitability from becoming established, for example, by giving analgesics prior to, or early in a surgical procedure. The clinical use of this tactic has been recently reviewed in the context of post-operative pain and whether non-steroidal anti-inflammatory drugs, local anaesthetics or opioids are given, advantages of this approach have been minimal. In general, analgesic requirements or pain ratings are very similar whether analgesics are given prior to or after the surgery (McQuay, 1994). A problem that can confound this approach is the clinical time scale over which the effects of these pre- (and post-treatments) are gauged. Post-operative pain can be considered as generally involving inflammation and tissue damage. As has been discussed, a few hours will encompass the short term and then the longer term acute inflammatory states so that within the spinal cord, both the excitations and opioid and non-opioid inhibitions are altering due to plasticity. Plastic changes within spinal transmitter systems leads to compensatory increased central inhibitions over a few hours. Subsequently, pre-emptive approaches to analgesia may prevent both this beneficial plasticity as well as the target of central excitatory hypersensitivity mechanisms. If this is the case, as the pre-emptive agent wears off, the pain may return in the absence of compensatory inhibitions so that the pain intensity would be heightened for a while before the inhibitions, now delayed, start to compensate (see Fig. 4). By contrast, in the neuropathic states, since we believe inhibitions start to fail with time, aggressive pre-emptive approaches may have long term beneficial effects.

Thus, much emphasis has rightly been given to excitatory amino acid mechanisms in the spinal cord but it is clear that these events are themselves controlled by inhibitions. Consequently, the balance between excitations and inhibitions will determine the final level of transmission. It would appear that inflammatory pains can be self-limiting since they promote physiological inhibitory controls. By contrast, in pathological and experimental models of nerve damage, inhibitory controls appear to be lost and this may facilitate the chronicity of pain (Zimmermann, 1991).

References

Battaglia, G. and Rustioni, A. (1988) Coexistence of glutamate and substance P in dorsal root ganglion cells of the rat and monkey. *J. Comp. Neurol.*, 277: 302–312.

Castro-Lopes, J.M., Tavares, I. and Coimbra, A. (1993) GABA decreases in the spinal cord after peripheral neurectomy. *Brain Res.*, 620: 287–291.

Castro-Lopes, J.M., Tavares, I., Tolle, T.R. and Coimbra, A. (1994) Carrageenan-induced inflammation of the hind foot provokes a rise of GABA-immunoreactive cells in the rat spinal cord that is prevented by peripheral neuroectomy or neonatal capsaicin treatment. *Pain*, 56: 193–201.

Chapman, V. and Dickenson, A.H. (1992) The combination of NMDA antagonism and morphine produces profound antinociception in the rat dorsal horn. *Brain Res.*, 573: 321–323.

Chapman, V., Haley, J.E. and Dickenson, A.H. (1994) Electrophysiological analysis of preemptive effects of spinal opioids on NMDA mediated events. *Anesthesiology*, 81: 1429–1435.

Collingridge, G. and Singer, W. (1990) Excitatory amino acid receptors and synaptic plasticity. *Trends Pharmacol. Sci.*, 11: 290–296.

Dickenson, A.H. (1990) A cure for wind-up: NMDA receptor antagonists as potential analgesics. *Trends Pharmacol. Sci.*, 11 :307–309.

Dickenson, A.H. (1994a) NMDA receptor antagonists as analgesics. In: H.L. Fields and J.C. Liebeskind (Eds.), *Progress in Pain Research and Management.* Vol. 1. IASP Press, Seattle, WA, pp. 173–187.

Dickenson, A.H. (1994b) Where and how opioids act. In: G.F. Gebhart, D.L. Hammond and T. Jensen (Eds.), *Progress in Pain Research and Management,* Vol. 2. IASP Press, Seattle, WA, pp. 525–552.

Dray, A., Urban, L. and Dickenson, A.H. (1994) Pharmacology of chronic pain. *Trends Pharmacol. Sci.*, 15: 190–197.

Dubner, R. and Ruda, M.A. (1992) Activity dependent neuronal plasticity following tissue injury and inflammation. *Trends Neurol. Sci.*, 115: 96–103.

Haley, J.E., Sullivan, A.F. and Dickenson, A. H. (1990) Evidence for spinal N-methyl-D-aspartate receptor involvement in prolonged chemical nociception in the rat. *Brain Res.*, 518: 218–226.

Mao, J., Price, D.D., Hayes, R.L., Lu, J., Mayer, D.J., and Frank, H. (1993) Intrathecal treatment with dextrophan or ketamine potently reduces pain-related behaviours in a rat model of peripheral mononeuropathy. *Brain Res.*, 605: 164–168.

McQuay, H. (1994b) Do preemptive treatments provide better pain control? In: G.F. Gebhart, D.L. Hammond and T. Jensen (Eds.), *Progress in Pain Research and Management*, Vol. 2. IASP Press, Seattle, WA, pp. 709–723.

McQuay, H. and Dickenson, A.H. (1990) Implications of central nervous system plasticity for pain management. *Anaesthesiology*, 45: 101–102.

Price, D.D., Mao, J. and Mayer, D.J. (1994a) Central neural mechanisms of normal and abnormal pain states. In: H.L. Fields and J.C. Liebeskind (Eds.), *Progress in Pain Research and Management*, Vol. 1. IASP Press, Seattle, WA, pp. 61–84.

Price, D.D., Mao, J., Frenk, H. and Mayer, D.J. (1994b) The N-methyl-D-aspartate antagonist dextromethorphan selectively reduces temporal summation of second pain. *Pain*, 59: 165–174.

Reeve, A.J. and Dickenson A.H. (1995) The roles of spinal adenosine receptors in the control of acute and more persistent nociceptive responses of dorsal horn neurones in the anaesthetized rat. *Br. J. Pharmacol.*, 116: 2221–2228.

Schaible, H.-G., Grubb, B.D., Neugebauer, V. and Oppmann M. (1991) The effects of NMDA antagonists on neuronal activity in cat spinal cord evoked by acute inflammation in the knee joint. *Eur. J. Neurosci.*, 3: 981–991.

Seltzer, Z., Beilin, B.Z., Ginzburg, R., Paran, Y. and Shimko, T. (1991a) The role of injury discharge in the induction of neuropathic pain behavior in rats. *Pain*, 46: 327–336.

Seltzer, Z., Cohn, S., Ginzburg, R. and Beilin, B.Z. (1991b) Modulation of neuropathic pain in rats by spinal disinhibition and NMDA receptor blockade of injury discharge. *Pain*, 45: 69–76.

Stanfa, L.C., Sullivan, A.F. and Dickenson, A.H. (1992) Alterations in neuronal excitability and the potency of spinal mu, delta and kappa opioids after carrageenan induced inflammation. *Pain*, 50: 345–354.

Stanfa, L.C. and Dickenson A.H. (1993) Cholecystokinin contributes to the enhanced potency of spinal morphine following carrageenan inflammation. *Br. J. Pharmacol.*, 108: 967–973.

Stanfa, L.C. and Dickenson, A.H. (1994) Enhanced alpha-2 adrenergic controls and spinal morphine potency in inflammation. *NeuroReport*, 5: 469–472.

Stanfa, L.C., Dickenson, A.H., Xu, X.-J. and Wiesenfeld-Hallin, Z. (1994) Cholecystokinin and morphine analgesia: variations on a theme. *Trends Pharmacol. Sci.*, 15: 65–66.

Woolf, C.J. (1983) Evidence for a central component of post-injury pain hypersensitivity. *Nature*, 306: 686–688.

Woolf, C.J. (1994) A new strategy for the treatment of inflammatory pain, *Drugs*, Suppl. 5: 1–9.

Woolf, C.J. and Thompson, S.W.N. (1991) The induction and maintenance of central sensitization is dependent on N-methyl-D-aspartic acid receptor activation: implications for the treatment of post-injury hypersensitivity states. *Pain*, 44: 293–299.

Yaksh, T.L. (1989) Behavioural and autonomic correlates of the tactile evoked allodynia produced by spinal glycine inhibition: effects of modulatory receptor systems and excitatory amino acid antagonists. *Pain*, 37: 111–123.

Yamamoto, T. and Yaksh, T.L. (1992) Studies on the spinal interaction of morphine and the NMDA antagonist MK-801 on the hyperesthesia observed in a rat model of sciatic mononeuropathy. *Neurosci. Lett.*, 135: 67–70.

Zimmermann, M. (1991) Central nervous mechanisms modulating pain related information: Do they become deficient after lesions of the peripheral or central nervous system? In: K.L. Casey (Ed.), *Pain and Central Nervous System Disease: The Central Pain Syndromes*. Raven Press, New York, pp. 183–199.

CHAPTER 18

Plasticity of the nervous system at the systemic, cellular and molecular levels: a mechanism of chronic pain and hyperalgesia

M. Zimmermann and T. Herdegen

II. Physiologisches Institut, Universtät Heidelberg, Im Neuenheimer Feld 326, D-69120 Heidelberg, Germany

Introduction

The ideas of "neuronal plasticity" and "supersensitivity" as mechanisms that enhance nervous system excitability in a deleterious manner to induce chronic pain were presented presumably for the first time in 1978 at the 2nd World Congress on Pain in Montreal (Zimmermann, 1979, p. 25 f.). Of course, the concept of nervous system plasticity had been existing long before, being associated, however, with problems of development, memory formation and repair following damage. In regard to the sensory and motor functions of the adult nervous system, the term "plasticity" was associated with the postlesional restitution of function by fibre sprouting, activation of silent synapses, opening of alternative (redundant) pathways and remapping of somatotopically ordered projections and connections (see review by Wall, 1976), but not with the active induction of chronic pain.

The dominant theme at those times (i.e. the mid-1970s) was "pain modulation" with the focus of research and clinical interest on endogenous inhibitory systems that could be activated to induce analgesia, and may function continually to control the sensitivity and responsiveness of the pain related nervous system. Thus, a shift occurred in the prevailing neuroscientific views related to pain, moving away from the assumption of a stable nociceptive sensory system. The phenomena subsumed under the term "pain modulation" also can be regarded as cases of (short lasting and reversible) nervous system plasticity, and this interpretation was offered for the first time in a letter to *Lancet* (Walker and Katz, 1979).

In contrast, pain research of the last 10 years has revealed a great variety of slow and, in part, irreversible processes in the nervous system induced by peripheral trauma and noxious stimuli, with time courses ranging from hours to months. These processes imply long lasting modifications in nervous system function that may contribute to persistent hyperalgesia and progressive chronicity of clinical pain. The mechanisms involved imply system physiology, biochemistry and cellular and molecular biology, including the level of induced gene transcription. In the following we present our view of a highly flexible and plastic nervous system with adaptive but often deleterious responses to almost any kind of peripheral trauma resulting in long term sensitization and proneness against chronic pain.

The new concepts of pain emerging from a neuroscience of plasticity corroborate with increasing clinical evidence that nervous system reactions to pain may aggravate the pain and eventually result in a process termed pain chronification (Flor et al., 1994). Clinical studies have identified a number of somatic and psychosocial risk factors that underline the involvement of the nervous system in the chronicity of pain, at various levels of function.

In the first part of the chapter we select examples of neuroplasticity in peripheral and spinal systems that result in sensitization, i.e. the enhanced excitation of peripheral or central neurons in response to a noxious test stimulus. In the second part we review the potential of nerve trauma or noxious stimulation to induce transcriptional processes in the nuclei of nerve cells, which result in far reaching plasticity of nervous system function in relation to nociception and pain.

Plasticity in primary afferent nociceptive neurons

An early study of plasticity of nociceptive neurons was performed in our laboratory with Ingve Zotterman as a visiting scientist in 1975 (Dickhaus et al., 1976). Here, the plantar nerves of the cat were severed by repeated pinching so that the nerve fibres distal to the lesion underwent Wallerian degeneration, followed by regeneration of the proximal nerve stumps. After about 3 weeks functional tests revealed that regenerating nociceptive nerve fibres had reinvaded the original territory in the hind paw (Fig. 1A). Regenerated nociceptors could be functionally identified that showed graded responses to noxious skin heating of varying temperatures. The stimulus response functions were linear, very much like in nociceptors found in normal skin of the cat's hind paw. However, when the thresholds were determined of a population of regenerated heat sensitive nociceptors, they were lower by 3.7°C compared to a random sample of nociceptors recorded in normal nerve (Fig. 1B). This finding was later confirmed for nociceptors that had redeveloped in the rabbit ear skin after nerve transection (Shea and Perl, 1985).

Thus, nociceptors that redeveloped following nerve trauma show sensitization, and this state of altered responsiveness of nociceptors probably persists for weeks or months after the nerve lesion. This change is a case of sensory plasticity which is part of a complex cell body response probably aimed at repair and restitution of function of the severed neuron. Teleologically the decreased threshold to skin heating may be seen as an adaptive response, where a partial loss of fibres is compensated by an increase in sensitivity of those fibres which regain function. This compensatory adjustment may accelerate the recovery of the no-

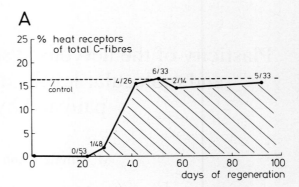

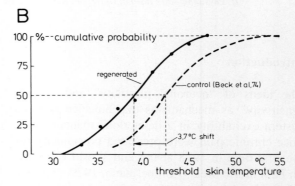

Fig. 1. Sensitization of regenerated cutaneous nociceptors. (A) The plantar nerves were repeatedly crushed in anaesthetized cats. Up to 90 days later animals were re-anaesthetized and single C-fibres were dissected proximal to the site of crush. After 21 days an increasing proportion of C-fibres were recorded which responded to radiant skin heating and had the characteristics of nociceptors. (B) The threshold temperature was determined from the sample of heat sensitive nociceptors shown in (A). The ordinate plots the cumulative probability of threshold skin temperature for the excitation of these regenerated nociceptors (continuous line). The interrupted line shows the corresponding cumulative probability distribution of control nociceptors recorded in normal skin of the cat's hind foot, as established in the same laboratory (Beck et al., 1974). The data on regenerated nerves are from Dickhaus et

ciceptive system towards normal. Although this is a biological advantage for the animal with nerve damage to regain the protective mechanism of nociception, the same process may result in hyperalgesia and ongoing pain, a most disturbing situation for the patient who is capable of suffering from this pain.

Ample evidence shows that nociceptors become sensitized by inflammation, and that normally silent (inexcitable) nociceptors are switched on by inflammation. These processes of functional plasticity will not be detailed here, they have recently been reviewed (Schmidt et al., 1994). Sensitization of nociceptors by inflammation has been reported in muscle, joint, skin and viscera. However, it is not clear whether the sensitization outlasts the direct effect of inflammatory substances, i.e. whether there is a long-lasting or persistent change in nociceptor function.

Development of abnormal chemosensitivity in regenerating nerve fibres

The cell body response after nerve transection results also in altered chemical responsiveness of regenerating sensory fibres. We have studied the responsiveness to bradykinin of regenerating nerve sprouts, and found a sensitizing interaction between adrenaline and bradykinin at the sprouting sensory C-fibres (Zimmermann et al., 1987).

In cats the sural nerve was transected and ligated. A neuroma had developed 2 weeks later when we performed single unit recording from sural A- and C-fibres. The neuroma was desheathed and placed in a perspex chamber for superfusion with solutions containing various neuromediator substances. Superfusion of the neuroma with bradykinin resulted in long lasting impulse discharges in 26% of C-fibres, but not in A-fibres. The discharge frequency increased with the bradykinin concentration in the superfusion fluid, the threshold being about 1 μmol/l. Superfusion of an intact (desheathed) nerve with bradykinin never elicited any responses. Thus, chemosensitivity to bradykinin develops fairly early in regenerating C-fibres. Many nociceptors in normal skin also respond to bradykinin (Beck and Handwerker, 1974).

When adrenaline was given to the superfusion fluid before bradykinin, the subsequent response to bradykinin was much enhanced, to up to ten times the response in the absence of adrenaline (Fig. 2). Thus, in the regenerating C-fibre an enormous sensitization of a chemosensory response occurs when adrenaline is present. These findings may be explained by the development of bradykinin receptors at the nerve terminals with either an adrener-

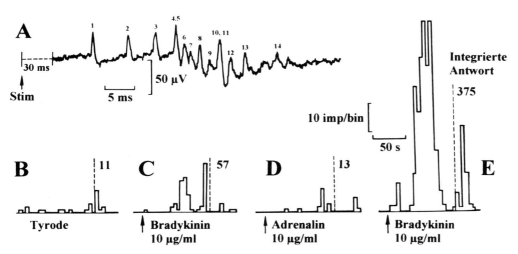

Fig. 2. Discharges of neuroma C-fibres evoked and sensitized by chemical substances. (A) Recording from a filament of a cat's sural nerve, after electrical stimulation 40 mm distally. C-fibre action potentials are labelled 1 to 14. (B–E) Histograms of C-fibre discharges recorded from a filament during superfusion of the neuroma with Tyrode solution (B), with a solution containing bradykinin (C) or adrenaline (D) and then again bradykinin (E). Superfusions lasted 5 min, starts were indicated by arrows. The numbers at the vertical broken lines indicate the integrated responses from the start of superfusion. The time separation between (C) and (D) was 10 min. Bradykinin superfusion in (E) was started immediately at the end of the 5 min superfusion with adrenaline in (D). From Zimmermann et al. (1987).

gic modulators site or an interaction with a closely associated adrenoceptor. We were not able to clarify this further. Sensitization of the bradykinin response by adrenaline is not seen in normal C-fibre nociceptors of the skin or muscle and adrenaline does not have any excitatory effect on nociceptors, neither alone nor in combination with other chemicals.

There are other pathophysiological conditions where adrenaline produces excitatory effects onto nociceptive afferents, including several modes of nerve lesions. The primary pathophysiology has been associated with excitatory interactions of the sympathetic nervous system onto afferent nociceptive neurons at the site of nerve trauma and at the dorsal root ganglion (Jänig and Schmidt, 1992; Bossut and Perl, 1995; Xie et al., 1995). It is likely that both central and peripheral mechanisms cooperate to induce and maintain the self-stabilizing vicious circle of sympathetic reflex dystrophy that long ago has been claimed to exist (Zimmermann, 1979).

As the neuroma contains both afferent C-fibres and efferent postganglionic sympathetic C-fibres it seems that an enhanced sensitivity of the regenerating sprouts towards the detection of nociceptive substances can actively be turned on in situations of increased levels of sympathetic activity. The functional meaning of this sympathetically controlled sensitization may be to compensate for losses of nociceptive sensory function after a nerve lesion. Thus, in situations of emergency and stress the peripheral nociceptive system might show higher responsiveness. This could mean an evolutionary advantage in terms of biological function, although at the same time the ensuing hyperalgesia and persistent pain will have a negative value in a species that can suffer from these sensory changes.

Plasticity of neuropeptide expression in primary afferent neurons upon noxious events

Dorsal root ganglion neurons with non-myelinated fibres contain various "sensory" neuropeptides such as substance P, calcitonin gene-related peptide (CGRP), vasoactive intestinal peptide (VIP), somatostatin (Weihe, 1990). These sensory neuropeptides are particularly abundant in neurons with non-myelinated C-fibres and subserve several functions in relation to pain (Fig. 3):

- They are neuromediator substances at synapses to spinal dorsal horn neurons, where substance P, CGRG and VIP are excitatory and galanin is inhibitory. For somatostatin both excitatory and inhibitory functions have been reported.
- Sensory neuropeptides are released from the peripheral nerve endings in the skin, muscle, joint or viscera. Here, they mediate neurogenic inflammation (Chahl et al., 1984; Maggi and Meli, 1988) which subsumes vasodilatation, enhanced microvascular permeability and oedema formation.
- Sensory neuropeptides result in activation and proliferation of mast cells, fibroblasts, macrophages and other inflammation related cells, which are part of the non-specific immune response and of the neurogenic inflammation.

There is much evidence that gene expression of neuropeptides in dorsal root ganglion (DRG) neurons can be profoundly modified by several pathophysiologic conditions. There are two basic types of changes:

- Substance P and CGRP expressions are greatly enhanced by an experimental inflammation (Levine et al., 1984, 1993; Schmidt et al., 1994) and by UV-irradiation of the skin (Gillardon et al., 1992; Benrath et al., 1995; Eschenfelder et al., 1995).
- After nerve lesions, on the other hand, the gene expression of substance P and CGRP are significantly reduced whereas the expressions of galanin, VIP and nitric oxide synthase (NOS) are greatly upregulated in dorsal root ganglion neurons (see below).

We have recently performed a thorough study of the nervous system involvement in the sunburn reaction and related pain and hyperalgesia following ultraviolet irradiation (Gillardon et al., 1992; Benrath et al., 1995; Eschenfelder et al., 1995). Irradiation with UV-B of small patches of human skin resulted in elevated skin blood flow for more than 5 days with two peaks at 12 and 36 h. The

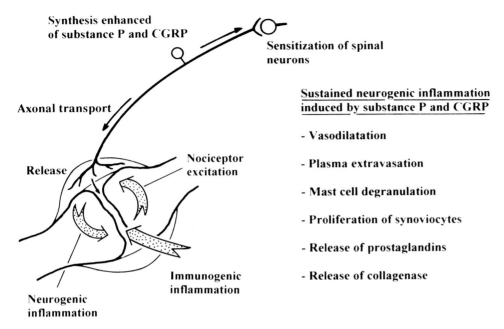

Fig. 3. Plasticity of neuropeptides in primary afferent neurons. Part of somatosensory nociceptive neurons contain substance P and/or calcitonin gene-related peptide (CGRP) which are released at the peripheral and central (spinal) axonal endings. The peripheral release induces neurogenic inflammation of the innervated organ. The components of this inflammation are listed in the figure. At the central synapse substance P and CGRP act as excitatory transmitters or neuromodulators. When a prolonged inflammation is induced in a joint, substance P and CGRP become up-regulated in the dorsal root ganglion cell, resulting in enhanced availability and release at both peripheral and central axonal sites. The ensuing prolonged and intensified neurogenic inflammation contributes to the maintenance of total joint inflammation. A vicious circle may be initiated of inflammation–nociceptor excitation–neuropeptide up-regulation–prolonged neurogenic inflammation–nociceptor excitation and sensitization and so on, which is a mechanism of prolonged inflammation and pain due to a plasticity response of the peripheral nervous system.

peak at 36 h is associated with increased contents and/or releases of substance P and CGRP in cutaneous sensory nerve fibres, as could be concluded from animal experiments and pretreatment of irradiated human skin with capsaicin.

Neuropeptide levels are enhanced both at the peripheral and central endings of the sensory neurons involved (Figs. 4 and 5). We found an increase to 150% of normal (Fig. 4) of CGRP immunoreactivity in the spinal dorsal horn of rats after UV-B irradiation of the hindpaws. It is important to note that a large part of CGRP containing fibres in the spinal cord section investigated may be related to non-irradiated skin areas, otherwise the relative increase in CGRP level might have been much larger. The enhanced contents in CGRP (and presumably also substance P) in presynaptic terminals in the spinal dorsal horn may result in enhanced excitatory transmission to second order dorsal neurons, thus contributing to spinal sensitization and hyperalgesia, in addition to the sensitization of nociceptors in the inflamed skin.

The long lasting increases of substance P and CGRP levels result in a prolonged neurogenic inflammation that could contribute to chronicity. Thus, it seems that a vicious circle is induced by a persistent neurogenic inflammation (Colpaert et al., 1983) that facilitate the maintenance and perpetuation of a condition of chronic inflammation (Fig. 3). There is increasing evidence that a neurogenic component is involved in chronic inflammatory joint diseases such as chronic polyarthritis: for example, Menkes et al. (1993) have found that substance P levels were enhanced in synovial fluid samples from patients with osteoarthritis. Synovio-

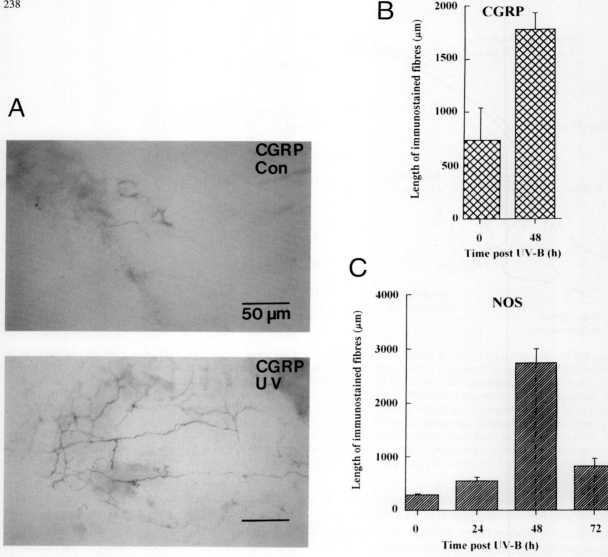

Fig. 4. Increase in CGRP and NOS immunoreactivity in cutaneous nerve fibres of rat. (A) CGRP-immunoreactivity hindpaw nerve fibres in rat skin before and 48 h after irradiation with UV-B, skin "whole-mount" preparation. (B) Total length (μm) of fibres immunostained for CGRP, in a skin area of $100 \times 150\ \mu$m. (C) Total length (μm) of fibres immunostained for NOS, in a skin area of $100 \times 150\ \mu$m. From Benrath et al. (1996).

cytes from inflamed joints in human patients are induced to proliferate by substance P (Lotz et al., 1987).In fibromyalgia patients substance P levels are elevated in spinal fluid, indicating an enhanced spinal release as an excitatory transmitter (Vaeroy et al., 1988). Overexpression of substance P and CGRP in perivascular nociceptive trigeminal fibres on meningeal and other intracranial vasculature and facilitated release is considered an important step in the sequence of events of the migraine attack (Goadsby et al., 1990; Buzzi et al., 1991; Goadsby and Edvinsson, 1993; Moskovitz and

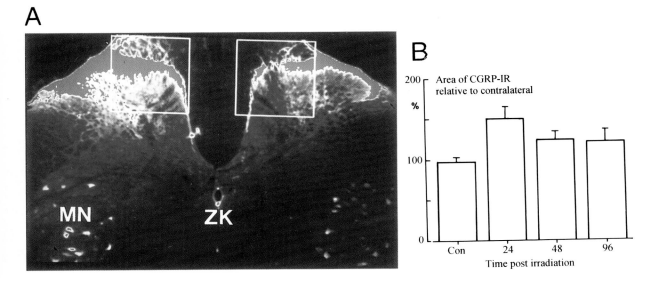

Fig. 5. Increase in CGRP content in intraspinal primary afferent terminals. (A) CGRP immunoreactivity 24 h after UV-B irradiation of the left hind paw (left site) compared with non-irradiated contralateral (right) site. The areas of CGRP immunoreactivity were evaluated by digital image analysis. (B) Total area of immunoreactivity in the spinal dorsal horn of a sample of rats after UV-B irradiation of the hind paw, relative to the non-irradiated site of the same animals. From Gillardon et al. (1992).

MacFarlane, 1993). Many drugs with a potential to terminate a migraine attack have been shown to block the release of substance P and CGRP from trigeminal afferent fibres.

Plasticity and sensitization of the spinal nociceptive system due to deficits in inhibition after lesions of spinal cord, dorsal root or peripheral nerve

The contributions to this volume by Meller et al., Dickenson, Mense and Hoheisel, Sandkühler, Schaible and Schmidt, Willis focus mainly on sensitization of excitatory transmission that is related to NMDA-receptors or the cooperation of NMDA-receptors with AMPA-receptors, metabotropic glutamate receptors or NK_1 receptors. However, it is clear that sensitization of central neurons may also be due to decreased inhibitory controls, particularly following neuropathic lesions (Zimmermann, 1991).

The neurosurgeon Otfrid Foerster, in his pioneer book *Die Leitungsbahnen des Schmerzgefühls* (1927), made the suggestion that a descending (corticofugal) inhibitory system had become defective in some pain patients, and that the ensuing disinhibition of spinal mechanisms caused the pain. This contention is supported by recent experimental studies on spinal control systems, and these are reviewed below.

To assess the acute effect of a spinal lesion on descending inhibition we used reversible blocks of descending axons by locally cooling the cord at the thoracic level (Handwerker et al., 1975; Dickhaus et al., 1985). The responsiveness of dorsal horn neurons to afferent input was greatly enhanced during these blocks, indicating that a tonic descending inhibition was at work even under the condition of general anesthesia. Thus the stimulus-response-functions (SRFs) to graded skin heating characteristically showed parallel shifts towards higher neuronal responses and lower thresholds of skin temperature (see Fig. 6).

It is well established that lesions of the nervous system can result in autotomy which is considered a pain related behaviour (Wall et al., 1979; Coderre et al., 1986a). Few investigations have addressed the question what the effects are of neuro-

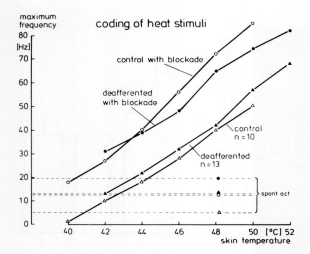

Fig. 6. Effects of previous dorsal root section and reversible spinal blockade on the responsiveness of lumbar dorsal horn neurons in anaesthetized cats. Mean intensity functions of the responses of dorsal horn neurons (ordinate scale) to noxious skin heating at varying temperature (abscissa scale) were measured in partially deafferented (filled symbols) and control (open symbols) animals. The upper two intensity functions were obtained from the same 2 samples of neurons after blockade of the supraspinal influences by a reversible cold block at L1 segmental level. The dotted lines indicate the levels of mean spontaneous activity of the 2 samples of neurons (i.e. from deafferented and control animals, respectively), in the absence (lower values) or during (upper values) spinal cold block at L1. From Brinkhus and Zimmermann (1983).

pathic lesions on control of afferent transmission in the central sensory pathways, and on the excitability of central sensory nerve cells.

We have studied the effects of lumbar dorsal root sections in the cat on the excitability, afferent responsiveness and tonically active component of descending inhibition of dorsal horn neurons (Brinkhus and Zimmermann, 1983). Three to six weeks after transection of several adjoining roots we recorded from dorsal horn neurons located in the transitional zone of deafferentation. The mean rate of ongoing discharges was higher by a factor of 2.5 in the neurons of the deafferented cords compared to controls (Fig. 6), which reflects a hyperexcitability of these partially deafferented neurons. No significant differences in the (mean) heat evoked stimulus-response-functions were found between the lesioned and control animals, indicating that the neurons, although partially deafferented, had sufficient afferent input from heat-sensitive cutaneous nociceptors to be driven to normal discharge frequencies. This apparently normal range of discharges upon a decreased input can also be considered a sign of increased excitability of the partially deafferented neurons.

This finding of deafferentation induced hyperexcitability is in line with reports, e.g., by Albe-Fessard and colleagues (Lombard et al., 1979; Albe-Fessard and Lombard, 1983) who observed bursting discharges in dorsal horn and thalamic neurons after dorsal root transections. Similar phenomena of hyperexcitability of thalamic neurons have been reported in human patients after lesions of the spinal cord (Lenz et al., 1988).

The hyperexcitability of dorsal horn neurons in our experiments upon dorsal root sections (Brinkhus and Zimmermann, 1983) might in part be the result of decreased efficiency of a tonic descending inhibition. Our results are in line with this contention as was seen when a reversible cold block was performed at the thoracic level which transiently abolishes all descending inhibitory control of spinal neurons (Fig. 6): In both normal and deafferented spinal cords the ongoing discharges of dorsal horn neurons increased after the spinal cold block, indicating release of the neurons from a tonic descending inhibition. The stimulus-response-functions to noxious skin heating showed parallel shifts towards higher discharge frequencies, also indicating release from tonic inhibition. However, the mean increments of both the ongoing and heat evoked neuronal activities during spinal cold block were smaller in the deafferented than in control animals, i.e. this tonically active component of descending inhibition had became weaker after the dorsal root sections. This difference in the efficiency of tonic inhibitory control was more pronounced at the higher intensities of noxious skin stimulation (Fig. 6).

After nerve lesions rats show excessive self care behaviour of the denervated extremity that may result in self-mutilation (autotomy) and this may be associated with central hyperexcitability (Bennett, 1994). However, experimental evidence suggests that in these conditions descending inhibitory systems are still functioning tonically: in these neuropathic rats, lesions of the locus coeruleus (Coderre et al., 1986b), the parabrachial nucleus or the ipsilateral or contralateral spinal dorsolateral funiculus (Wall et al., 1988) resulted in an accelerated onset and enhanced degree of autotomy. These observations are best interpreted by the contention that descending inhibitory control of spinal neuronal processing was still functional in rats showing autotomy as a result of nerve section, although the studies did not address the question whether the descending control had the same efficacy as in the normal animals.

When we manipulated serotonin synthesis (Carstens et al., 1981) by pretreating the animals with parachlorophenylalanine (PCPA), a toxin known to block 5-HT synthesis and thus resulting in depletion of this neurotransmitter, the descending inhibition from the PAG to the dorsal horn was significantly reduced, indicating that 5-HT is a mediator of the descending control from PAG. This adds to a wealth of findings which revealed that part of the descending bulbo-spinal inhibitory system is serotoninergic (Rivot et al., 1987; Saito et al., 1990).

The decreased potency of descending inhibition in animals treated with PCPA suggests that some hyperalgesia and chronic pain might be due to pathophysiological dysfunction of serotoninergic inhibitory systems. This concept has been used therapeutically in an attempt to treat chronic pain by dietary supplementation of pain patients with L-tryptophan, a biochemical precursor of 5-HT (Seltzer et al., 1981).

It has been repeatedly shown that tricyclic antidepressant drugs are particularly effective in treating neuropathic pain in patients (Kishore et al., 1990). The analgesic effect of tricyclic antidepressants has been attributed to the ability of these drugs to block the reuptake of 5-HT (and/or noradrenaline which is also a transmitter involved in descending inhibition) thus resulting in a higher level of available inhibitory transmitter. In animals autotomy due to a nerve lesion was greatly reduced during daily treatment with amitriptyline, a tricyclic antidepressant drug (Abad et al., 1989; Seltzer et al., 1989). These findings indicate that part of the neuropathic pain may be due to a deficient monoaminergic inhibitory system.

This contention was directly studied by measurements of spinal monoamine levels in rats who developed autotomy behaviour after dorsal rhizotomy at cervical segments (Colado et al., 1988). These animals showed considerably decreased contents of dopamine and noradrenaline in the cervical spinal cord, whereas 5-HT showed a maximum decrease at spinal segments caudal to the rhizotomy. These data suggest a concomitant decrease in potency of inhibitory systems related to monoaminergic transmitters. In an attempt to supplement shortage of spinal catecholamines small pieces of adrenal medulla were grafted into the subarachnoid space of rats in which transections of the sciatic and saphenous nerves were performed subsequently (Ginzburg and Seltzer, 1990). The results clearly showed that autotomy as a sign of chronic neuropathic pain was much suppressed and delayed in onset in the grafted animals.

GABA is another inhibitory transmitter that has often been associated with the control of pain sensitivity. After nerve transection spinal GABA goes down (Castro-Lopes et al., 1993) which suggests that release from GABA inhibition may be part of the behavioral hyperalgesia seen after nerve lesions.

Opioid functions after nerve lesions

A clinically important question relates to the function of the endogenous opioid system after a nervous system lesion. Terenius (1979) has reported that endorphin-like immunoreactivity in cerebrospinal fluid of pain patients was lower with neuropathic than with other pain mechanisms. In rats, the incidence of autotomy after sciatic nerve section was correlated with a decrease of β-endorphin content in brain and spinal cord (Panerai et al., 1987). Opiate receptor binding was reduced in the spinal cord of rats 2 weeks and 4 months after a dorsal rhizotomy (Zajac et al., 1989). After trauma of the spinal cord the expression of prodynorphin and its messenger RNA were increased

(Przewlocki et al., 1988). These biochemical results indicate that the endogenous opioids show long-lasting responses to lesions of the nervous system.

Lombard and Besson (1989) have assessed the functional meaning of these findings in a study on the inhibitory effects of morphine on dorsal horn neurons in rats. They found that in arthritic rats the spontaneous activity of dorsal horn cells was suppressed to 46% of control by systemic morphine (2 mg/kg i.v.). In rats deafferented by a multiple dorsal rhizotomy, the ongoing discharges of a sample of dorsal horn neurons was not changed by the same dose of morphine. With an additional injection of 4 mg/kg of morphine the ongoing activity in the deafferented rats was reduced to 69% of control. These findings indicate that the opioid inhibitory system was less sensitive in the deafferented compared to the arthritic animals, and much higher doses were required in the deafferented animals to induce an appreciable inhibitory effect on dorsal horn neurons.

A dramatic decrease in efficacy of opioid analgesia following a nerve lesion was reported by Mao and his colleagues (Mao et al., 1995; Mayer et al., 1995). They induced a neuropathy in rats by transecting one sciatic nerve. Before and after the lesion they assessed the responsiveness of the spinal opioid system by measuring the paw withdrawal latency to a noxious heat stimulus. The antinociceptive effect of intrathecal administration of morphine was assessed, in percent of the maximum analgesic effect (Fig. 7). A dose effect curve yielded an effective dose range between 3 and 20 μg of morphine to obtain antinociception. When a nerve lesion was produced 8 days before the morphine test, the dose effect curve of antinociception induced by morphine was shifted to higher doses by a factor of 6 (Fig. 7), indicating that the spinal opioid system was greatly attenuated. Most interesting, this shift of the effective doses range could be prevented by pretreatment of the animals with MK-801, an NMDA-receptor antagonist. The authors interpret these findings by uncoupling of the μ-opiate-receptor from a G protein and/or changing opioid-receptor-gated ion channel activity, and that this processes requires

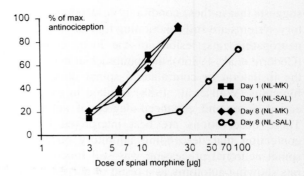

Fig. 7. Antinociception by spinal morphine in rats affected by a nerve lesion. In rats a paw withdrawal reaction was measured to noxious heating. Morphine was administered via an intrathecal catheter implanted beforehand. Antinociception by spinal morphine is expressed in percent of maximum antinociceptive effect (ordinate). At 8 days after transection of one sciatic nerve the dose-response curve for the antinociceptive effect was shifted to higher morphine doses (○). This shift could be prevented by daily systemic administrations of the NMDA antagonist MK-801 (10 nmol) subsequent to the nerve lesion (◆). From Mao et al. (1995).

phosphorylation at various intracellular sites including an NMDA-receptor (Mayer et al., 1995). The authors claim that the same type of neuroplasticity is a mechanisms of opioid tolerance observed under similar conditions.

These findings corroborate the clinical observations that neuropathic pain does not respond well to opiates (Arnér and Meyersson, 1988). The finding by Mao et al. (1995) suggest that neuropathic lesions result in insensitivity rather than ineffectiveness of the opioid system, which can be overcome by increasing the dose to provide pain relief in neuropathic pain patients.

Expression and function of AP-1 transcription factors in neurons following noxious stimulation

From the available body of evidence, it is conceivable that virtually every long-term change in nervous system function implies modified gene expression in nerve cells. At the DNA level, processes of transcription control can be assessed by studying the expression of immediate-early genes (IEGs) the master switches of the transcription machinery

(Morgan and Curran, 1989). Hunt et al. (1987) were the first to show by immunocytochemistry that expression of c-Fos protein occurs in the nuclei of dorsal horn neurons after noxious stimulation. This finding has repeatedly been confirmed. More recently, it was shown that nociceptive input to spinal neurons also results in the expression of other IEGs, such as krox-24, c-jun, junB, junD, and fosB. The pronounced expression of a multitude of IEGs after noxious events suggests that induced gene transcription may be a mechanism involved in the long term plasticity reported in the preceding sections.

The protein products of IEGs induced by noxious cutaneous stimulation have also been found in the brain. The distribution pattern of these immunoreactivities (IR) in the CNS was only partially homologous to what is generally considered the central pain system (cf. Fig. 2 in Zimmermann and Herdegen, 1994). However, areas with definite neuronal excitation by acute somatosensory noxious stimulation, such as the cerebellum and the hippocampus, do not express IEGs. After noxious stimulation, IEGs are also visible in nuclei of the limbic system and hypothalamic areas, whereas they are absent in the ventrobasal complex of

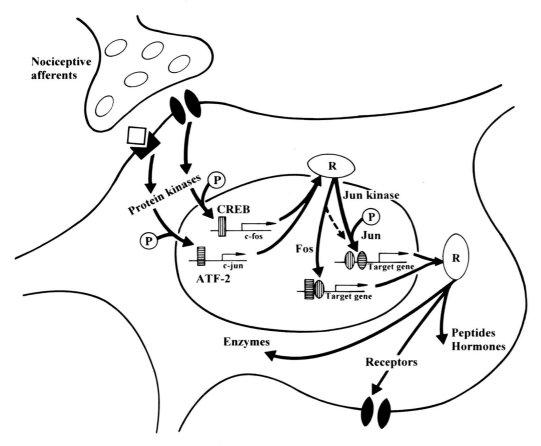

Fig. 8. Induced transcription of immediate-early genes c-fos and c-jun following nociceptive stimulation of spinal neurons. An intracellular signal cascade is activated by postsynaptic receptors (e.g. NMDA and NK-1) that induces phosphorylation of transcription proteins CREB and/or ATF-2. These bind to promoter elements of the c-fos and c-jun genes. c-Fos and c-Jun proteins dimerize and bind to promoter or enhancer elements of other genes ("target genes") which may include genes for various peptides, hormones, receptors or enzymes. The transactivation of target genes by the inducible transcription factor proteins c-Fos and c-Jun involves phosphorylation, e.g. via Jun kinase.

thalamus and in the dorsal column nuclei. It is conspicuous that basal IEG expression can be seen in many areas that show elevated levels of IEG expression following noxious peripheral stimulation. Thus, IEG encoded proteins are not just markers for neuronal activity, they specifically indicate that processes have been activated at the nuclear level in selected neuronal populations (Fig. 8) which, according to our view, are meaningful in relation to long term plasticity in the pain system.

Expression of c-Jun in axotomized neurons

Transection of peripheral somatosensory, somatomotor and cranial nerve fibres as well as preganglionic sympathetic nerve fibres results in the selective expression of c-Jun and, to a minor extent, of JunD proteins in the nuclei of the axotomized neurons in dorsal root ganglion, spinal cord and brainstem. In contrast, JunB, c-Fos, FosB, Krox-20 and Krox-24 proteins are not expressed (Herdegen et al., 1991b; Leah et al., 1991; Jenkins et al., 1993; de Leon et al., 1995).

Following sciatic nerve cut, c-Jun immunoreactivity (IR) increases over basal levels in both primary afferent neurons and motoneurons of small and large diameter type between 10 h and 15 h post-axotomy, and reaches its maximal expression (both number of labelled neurons and intensity of immunoreactivity) between 48 h and 72 h. JunD parallels the temporo-spatial pattern of c-Jun by a delay in onset of few hours, but its increase over basal levels is not as strong as that of c-Jun. This moderate increase of JunD is probably due to the substantial basal (probably constitutive) expression of JunD. The persistence of Jun proteins depends on the success of regeneration: after re-establishment of the neuron-target axis Jun proteins return to basal levels between 30 and 50 days. In contrast, elevated levels of c-Jun are visible in small diameter dorsal root ganglion neurons up to 15 months if successful regeneration is prevented by ligation of the proximal nerve stump. In large diameter neurons, c-Jun returns to basal levels within 2 months post-axotomy (Herdegen et al., 1992; Jenkins et al., 1993).

A second transection of sciatic nerve, 100 days after the first transection and ligation, evoked a dramatic up-regulation of c-Jun in axotomized neurons which exceeded the intensity of c-Jun-IR visible after the first transection (Herdegen et al., 1995). This re-induction of c-Jun parallels the findings about enhanced regenerative efforts of repeatedly damaged neurons and demonstrates that the slow decrease of c-Jun following the first lesion is neither due to neuronal cell death nor to restriction of the neuronal protein synthesis.

The increased level of c-Jun protein is the consequence of persistent transcription of the c-jun gene as shown by in situ hybridisation, Northern blotting and polymerase chain reaction (reviewed by Herdegen and Zimmermann, 1994). Retrograde neuronal tracing by fast blue applied through the transected sciatic nerve stump demonstrated that c-Jun expression is confined to the retrogradely labelled, i.e. axotomized, dorsal root ganglion neurons and motoneurons (Leah et al., 1991). It is conceivable that this long lasting activation of transcription processes in lesioned neurons is related also to plasticity responses resulting in neuropathic pain.

Expression of c-Jun is part of the cell body response to axonal injury

It is important to realize that the c-Jun expression is closely related to the cell body response (also called axon response; Liebermann, 1971) and not to the mere transection of nerve fibre. In the peripheral nervous system, dorsal rhizotomy of sciatic nerve fibres is neither effective for c-Jun expression nor for stimulation of the cell body-response (Jenkins et al., 1993), and proximal transection of sciatic nerve evokes both a stronger c-Jun synthesis and cell-body response compared to distal transection. Similarly, axotomy of retinal ganglion cells, hippocampal, cortical or rubropsinal neurons at sites distant from their perikaryon, does neither evoke c-Jun synthesis nor the cell body response nor induction of regeneration-associated genes (reviewed by Herdegen and Zimmermann, 1994).

An intriguing feature of the cell-body response

is its pivotal positive correlation to both the increased regenerative potency and the enhanced risk for neuronal cell death of the damaged neurons. Thus, juvenile neurons of the peripheral nervous system show both an enhanced capacity for sprouting and axonal elongation but also a higher vulnerability for dysfunctional regeneration, and a high rate of cell death when compared to adult neurons. Interestingly, expression of c-Jun protein and its mRNA is also enhanced in juvenile rats compared to adult rats. This behaviour could also be observed in the central neurons, e.g. in axotomized retinal ganglion cells that regrow only through a peripheral nerve graft if the optic nerve is cut closely to the retina, but this type of injury also kills the majority of retinal ganglion cells (Schaden et al., 1994). Again, c-Jun is dramatically expressed in proximally but not distally axotomized neurons (Schaden et al., 1994).

c-Jun expression reflects the regenerative propensity of injured neurons

As discussed above, c-Jun is closely related with the cell body response that can be considered an indicator of regenerative propensity of the injured neurons. There is also positive correlation between c-Jun expression and the axonal elongation of axotomized neurons as it could be demonstrated following transection of optic nerve fibres in the rat and goldfish, respectively (Herdegen et al., 1993a). In the goldfish, Jun-IR persisted up to 30 days in the injured retinal ganglion cells and declined when the optic nerve fibres had successfully re-innervated the tectum. In the rat, the selective increase of c-Jun started within 24 h and reached its maximal levels between 2 and 5 days. After 8 days, Jun expression had rapidly declined, and this decrease preceded the abortive sprouting and the death of axotomized retinal ganglion cells. In the rat, grafting of peripheral nerve fibres between the optic nerve stump and the tectum allows a successful axonal elongation of the retinal neurons that re-innervates their target. Importantly, retinal ganglion cells that have grown into the graft express the c-Jun protein over weeks until their fibres have reached the tectum, and the regenerating retinal ganglion cells exhibited a clear colocalization between c-Jun and GAP-43, a major protein for axonal elongation and sprouting (Schaden et al., 1994).

Target genes of the c-Jun transcription factor

Related to their functions as transcription factors and to the appearance of cell-body response, Jun proteins might control the change of effector gene encoded proteins associated with the efforts for axonal elongation and synaptic re-modelling such as GAP-43 growth associated protein, tubulins and cytoskeleton proteins (Mikucki and Oblinger, 1991). This raises the hypothesis that the expression of c-Jun is an intrinsic molecular-genetic prerequisite for the induction of regeneration-associated genes.

c-Jun has the capacity to bind to CRE and AP-1 response elements and to associate with a variety of transcription factor proteins (reviewed by Herdegen, 1996). This attributes to c-Jun a wide operative range to modulate gene transcription. One possibility for the elucidation of transcription factor-dependent gene induction is the study of colocalization, since the coincident presence of transcription factor and putative target gene is a basic prerequisite for the assumption of such a transcription control.

The c-Jun protein precedes the so far known changes in gene expression following axotomy and this suggests that c-Jun presents the first genetic alteration. Following sciatic nerve cut, c-Jun precedes and co-varies with the expression of nitric oxide synthase (NOS) and galanin in dorsal root ganglion neurons and with CGRP in motoneurons (Fiallos-Estrada et al., 1993; Herdegen et al., 1993c). The promoters of the *galanin, CGRP* and *GAP-43* genes have effective CRE or AP-1 response elements that represent high affinity binding sites for c-Jun.

Galanin

Axotomy induces the expression of the neuropeptide galanin in some neuronal population of the peripheral and central nervous system. Thus,

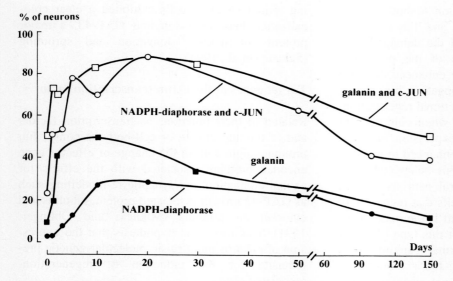

Fig. 9. Time course of relative numbers of dorsal root ganglion neurons (L4 and L5) in the rat expressing NADPH diaphorase or galanin, and colocalization with c-Jun, after transection and ligation of the sciatic nerve. From Fiallos-Estrada et al. (1993), Herdegen et al. (1993b).

galanin is upregulated in sciatic dorsal root ganglion neurons (Fiallos-Estrada et al., 1993) (Fig. 9), mammillary body and medial septum. Interestingly, these selective populations are congruent with those neurons that show an increased or persistent expression of nitric oxide synthase following axotomy (Brecht et al., 1996). It has been shown that galanin inhibits the excitation of deafferented spinal neurons (Xu et al., 1990; see Wiesenfeld-Hallin, this volume) thereby counteracting the hyperalgesia following peripheral nerve fibre lesion. In addition, galanin might also have neurotrophic effects that could be beneficial for axotomized neurons.

Calcitonin gene related peptide (CGRP)

A co-localisation of c-Jun above random probability was also observed with CGRP in axotomized sciatic motoneurons (Fiallos-Estrada et al., 1993). The synthesis of CGRP triggers the reformation of acetylcholine receptor in muscle fibres (New and Mudge, 1986) resulting in the functional re-establishment of neuromuscular transmission.

The deafferentated neurons of spinal cord: a further site of neuroplastic changes underlying pain induced by nerve fibre transection

As summarized above, galanin and NOS are upregulated and co-expressed with c-Jun in axotomized dorsal root ganglion neurons. By transport into the presynaptic terminals, galanin and NOS can act at the synapse to second order spinal neurons resulting in lasting changes of the transynaptic impulse transfer. There is evidence that multiple transynaptic processes of plasticity can be induced by a peripheral nerve lesion, including the activation of IEGs in spinal and higher order neurons.

The neuronal enzyme NOS catalyzes the transformation of arginin into citrulline thereby producing the gas radical NO (Moncada et al., 1991). The enzymatic activity of NOS can be visualized by the NADPH-diaphorase reaction (NADPHd). It has been previously demonstrated that NADPHd-labelled neurons are selectively resistant against ischemic, toxic and degenerative insults (reviewed by Gass and Herdegen, 1996). Therefore, we have been interested in the questions (a) whether the expression of NOS is altered by axotomy and (b) whether changes in NOS/NADPHd are also visible in the deafferentated dorsal horn.

(a) Following sciatic nerve cut, NOS-IR and NADPHd increased in the axotomized DRG neurons between 5 and 10 days, and returned to basal levels between 30 and 50 days (Fiallos-Estrada et al., 1993; Verge et al., 1992) (Fig. 9). In contrast, axotomized sciatic motoneurons show neither change of their weak basal NOS-IR nor do they develop NADPH-diaphorase activity (Fiallos-Estrada et al., 1993). DRG neurons labelled for NOS/NADPHd showed an almost complete co-expression of c-Jun (Fig. 9). A similar close and lasting association of c-Jun and NOS was also seen in the CNS neurons such as neurons of mammillary body (Herdegen et al., 1993b) and the medial septum (Brecht et al., 1995).

(b) Axotomy increases the NADPHd reactivity of the deafferented dorsal horn following sciatic nerve cut (Fiallos-Estrada et al., 1993) (Fig. 10). Assuming that levels of NADPHd closely reflect increase in NOS and, more importantly, in released NO, axotomy would enhance the excitatory input and hyperalgesia via enhanced production of NO. Recently it was shown that NO is also involved in the expression of c-Fos in spinal cord neurons following peripheral noxious stimulation (Lee et al., 1992). Moreover, we could demonstrate that noxious cutaneous stimulation induced NOS expression and a substantial colocalization of NOS and AP-1 proteins (Herdegen et al., 1994).

Deafferentation evokes a new pattern of c-Fos expression in otherwise non-responsive spinal cord neurons

Stimulation of sciatic Aβ-fibres is not effective for induction of c-Fos protein in the neurons of second order, i.e. spinal cord neurons of lamina III and ncl. gracilis. In contrast, stimulation at C-fibre intensity evokes a dramatic expression of c-Fos in the superficial and deep layers of dorsal horn except layer III (Herdegen et al., 1991a; Hunt et al., 1987; Molander et al., 1992). It is an intriguing question whether lamina III neurons cannot express c-Fos or whether the stimulation of intact Aβ-fibres cannot provide the adequate transynaptic stimulus.

This question could be answered following electrical stimulation of the proximal sciatic nerve stump 2 weeks following a sciatic nerve fibre transection. Under these experimental conditions, c-Fos was strongly expressed in lamina III neurons and ncl. gracilis (Molander et al., 1992). This indicates that the quality of synaptic transmission can be substantially altered and this could result in induction of otherwise non-responsive genes. Expression of c-Fos in lamina III neurons was also observed when rats were allowed to move for hours in a running wheel or move freely (Jasmin et al., 1994).

This alteration is not only restricted to c-Fos expression. Since induction of the c-fos gene is the final step of multi-enzymatic cascades, the transection-induced c-Fos expression has to be preceded by activation of otherwise quiescent kinases. Moreover, the presence of the c-Fos transcription factor will profoundly change the process of gene transcription. Recently, we could provide evidence that deafferentated neurons enhance their synthesis of the neuropeptides dynorphin and enkephalin (Herdegen and Zimmermann, 1995).

Transynaptic expression of IEGs following noxious stimulation

Complex patterns and prolonged time courses of activation of several IEGs in spinal and other central neurons were observed histochemically after acute and chronic noxious stimulations. Following repeated application of noxious heat or injection of formalin into one hindpaw, we observed individual spatiotemporal patterns of IEG expression in anaesthetized (noxious heat) and awake (formalin) rats (Herdegen et al., 1991c, 1994): within 2 h, a maximal expression of c-Jun, JunB, c-Fos, and Krox-24 proteins was visible in 50–150 nuclei per 50 μm section of the ipsilateral dorsal horn (Fig. 11). These immunoreactivities declined to base levels after 10 h. JunD and FosB reached their

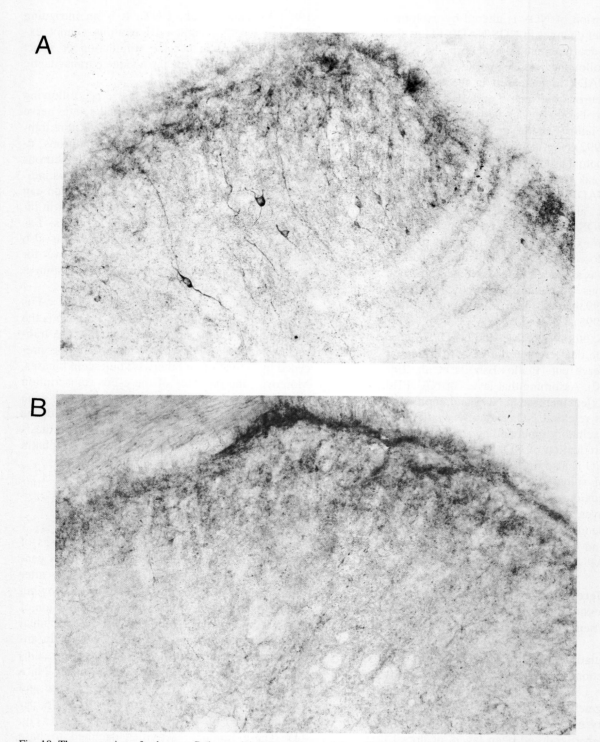

Fig. 10. The expression of substance P (brown label) decreases and that of NADPH diaphorase (blue) increases in the dorsal horn after sciatic nerve transection (A) compared to control (B). From Fiallos-Estrada et al. (1993).

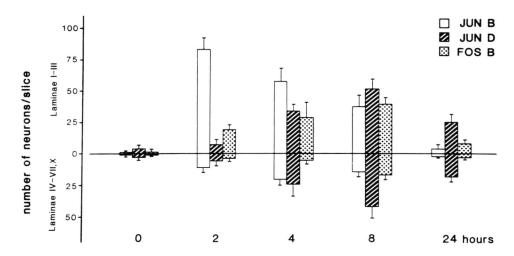

Fig. 11. Temporal and spatial patterns of IEGs in the rat spinal cord following noxious skin heating. Under anesthesia (halothane) one hindpaw was repeatedly immersed in hot water (52°C, 10 times for 20 s each, at 90 s intervals). Animals were killed 2–24 h after the end of stimulation, and the lumbar spinal cord was processed for immunohistochemistry of JunB, JunD and FosB. The mean numbers of labelled neurons (±SD) are given per 25 μm slice. The means were calculated from 15 slices (5 slices each of 3 rats). The upper part gives the number of labelled neurons in ipsilateral laminae I–III, the lower part in laminae IV–VII and X. From Herdegen et al. (1991c).

maximal expression between 5 and 10 h and were still present after 24 h.

The numbers of neuronal nuclei labelled for c-Fos, JunD, and Krox-24 were fairly equally distributed between superficial and deep dorsal horn, whereas c-Jun and FosB were predominantly expressed in the superficial layers. Thus, varying compositions are possible of AP-1 complexes from the available transcription factor proteins, depending on the specific temporo-spatial expression of the individual IEG proteins: within the first 2 h, dimers containing c-Jun, JunB, and c-Fos can be formed in the superficial layers and JunB:c-Fos dimers in the deep layers, whereas between 10 and 24 h, AP-1 complexes may consist of FosB:JunD dimers. It is noteworthy that c-Jun and FosB which have the highest DNA binding affinity and transcriptional activity (Ryseck and Bravo, 1991), are expressed in a fairly low number of neurons compared to c-Fos or JunB. This suggests a meaningful neural control of transcription factors. Similarly, lasting differential expression of IEG-encoded proteins with putatively specific formation of AP-1 complexes could also be observed following chronic somatic inflammation (Lanteri-Minet et al., 1993).

Repeated electrical stimulation of the sciatic nerve was performed for 10 min (Herdegen et al., 1991a). Stimulation of large A-fibres alone did not result in the expression of IEG-encoded proteins beyond base expression in spinal dorsal horn neurons in the area of termination of sciatic nerve fibres. However, when stimulation (at 5 Hz) included C-fibres, many neurons of the ipsilateral dorsal horn showed nuclear labelling for these six proteins, with an onset of detection at about 0.5 h after the start of stimulation and a maximum of labelled neurons between 1 and 4 h. The expression declined to base levels between 8 and 16 h, except for JunD and FosB, which were still expressed after 32 h. After 4 h, the six proteins were also seen in some neurons of the contralateral dorsal horn. The IEG expression patterns following electrical stimulation at C-fibre strength of sciatic nerve were fairly congruent with those following stimulation of hindlimb nociceptors by noxious heating or formalin injection.

In the brain, the IEG-encoded proteins studied

showed stimulation dependent increases in the numbers of labelled neurons almost bilaterally in numerous brain nuclei, such as the lateral reticular formation of medulla, locus coeruleus, dorsal raphe and pontine nuclei, periaqueductal grey, parabrachial nucleus, supramammillary nucleus, midline nuclei of thalamus and hypothalamus, amygdala, and lateral habenular nucleus. Again, c-Fos, JunD, and Krox-24 showed the greatest numbers of labelled neurons, whereas c-Jun, JunB, and FosB were absent in some areas with induced c-Fos and JunD, labelling. Thus, similar to the spinal cord, different AP-1 compositions are formed in individual brain areas resulting in differential transcriptional operations. IEGs were absent or not altered in nucleus gracilis and nucleus cuneatus, ventrobasal complex of thalamus, and the cerebellum.

The differential spatiotemporal patterns of stimulation induced IEG expression suggest that the IEGs do not just reflect neuronal activity. Rather, IEG expression may label specific neuronal populations which alter their neuronal program and their de novo protein synthesis in response to a noxious condition. Since changes in protein synthesis are a major attribute of plasticity, IEGs can be used to visualize those neurons reacting with plasticity to somatosensory stimulation.

The distribution and amount of IEG expression increase with stimulus duration. Three seconds of peripheral stimulation causes c-Fos labelling of a few cells that is maximum at 1 h and is gone at 4 h. Twenty minutes stimulation gives 9 times as many labelled cells 2 h later (i.e. at the time of maximum expression), and after 4 h of continuous stimulation there are twice as many cells again with increasing expression in the deeper laminae of spinal cord (Bullitt et al., 1992). This increasing expression of c-Fos in the deeper laminae with increasing stimulus duration also occurs with the continuous stimulation produced by inflammation in one ankle of the rat.

The number of c-Fos expressing neurons in the spinal cord and medulla also varies with stimulation intensity. The expression is induced by activating non-myelinated nociceptive C-fibres, but only weakly, if at all, by activating the myelinated Aβ-fibres that respond to non-noxious stimulation (Herdegen et al., 1991a; Hunt et al., 1987; Molander et al., 1992).

The distribution of c-Fos expressing cells in the spinal cord is in part related to the modality of the noxious stimulus. c-Fos is expressed in the laminae I–II after noxious thermal stimulation of a small patch of hindpaw skin (Lima et al., 1993). In deep laminae neurons c-Fos appears only after noxious mechanical stimulation, and with different time courses, e.g. in laminae III–IV 30 min and laminae VII 2 h after the stimulation (Lima et al., 1993). Whether such modality-specific expression occurs generally remains to be investigated.

Habituation of IEG expression following chronic stimulation: loss of plasticity at the level of gene transcription?

In rats with an arthritis in one ankle there is continuous C-fibre activity in the affected sensory nerves leading to spontaneous activity and hyperexcitability of dorsal horn neurons (reviewed by Coderre et al., 1993). Following induction of the arthritis there is an initial, strong expression of Fos, Jun and Krox transcription factors throughout the ipsilateral dorsal horn (Lanteri-Minet et al., 1993; Leah et al., 1996). Between 2 and 7 days after the onset of arthritis, however, the IEG expression disappears from all laminae, and does not re-appear during the subsequent 5 weeks, when pain behaviour and afferent C-fibre activity where at maximum levels. This decrease of c-Fos expression in mono-arthritic rats is surprising because here many spinal neurons continue to show spontaneous bursting and remain hyperexcitable (reviewed in Lanteri-Minet et al., 1993). Moreover, there are increased numbers of NK-1, NK-3 and glutamate receptors on the spinal neurons as well as an increased glutamate, substance P and calcitonin gene-related peptide (CGRP) content and release from the affected C-fibre afferents, all of which would excite dorsal horn neurons and potentiate their responses to further stimulation. Such changes would be expected to at least maintain the level of c-Fos expression after the initial

48 h. It is not known whether this decrease is caused by physiological inputs such as descending inhibition from supraspinal nuclei, or by events at the second messenger or gene level. In contrast, with polyarthritis, where several joints are inflamed, the number of c-Fos expressing spinal neurons increased over 3 weeks and thereafter declined over the next weeks (Abbadie and Besson, 1992; Lanteri-Minet et al., 1993; Leah et al., 1996). Both cases impressively demonstrate the dissociation of electrophysiological activity and IEG expression in the stimulated neurons.

Two days after continuous stimulation from an arthritis the number of labelled cells in the deep laminae is greater than that produced by a brief noxious stimulus (Abbadie and Besson, 1992; Leah et al., 1996). With continuous stimulation the deep laminae cells may accrue greater amounts of second messengers (Igwe and Ning, 1994) that enhance transcription of c-Fos. The increased c-Fos expression could also result from some signals that are transported axonally to the spinal cord along sensory afferents from the arthritis and/or from neurons in the superficial laminae, and which arrive several hours after induction of monoarthritis. Such a signal could be substance P that is known to induce c-Fos in spinal cord (Leah et al., 1989a,b) since the level and axonal transport of substance P is increased in sensory and spinal neurons in response to persisting inflammation (Kantner et al., 1986), and presynaptic terminals with substance P immunoreactivity are in contact with c-Fos labelled spinal neurons (Pretel and Piekut, 1991a,b).

Spinal neurons in rats with a (mono- or poly-) arthritic ankle show an enhanced expression of c-Fos in the ipsilateral superficial and deep laminae as well as in the contralateral deep laminae, when additional noxious stimulation is given to that ankle (Abbadie and Besson, 1992; Leah et al., 1996). When a noxious stimulus is given to the normal ankle there is also an increased expression in the ipsilateral superficial and deep laminae, and in the contralateral deep laminae (Leah et al., 1996). This increase probably results from a sensitization of the neurons affected by the arthritis (Leah et al., 1992).

Thus, chronic events can result in loss of responsiveness and plasticity on a genetic level. The decreasing inducibility could result in fixation of gene expression, possibly contributing to disease progression which might solidify chronic diseases and counteract therapeutic access. These processes of desensitization at the genetic and transcriptional levels might be related to the loss of novelty of those stimuli which provoked IEG expression following their first application.

It is generally accepted that early postnatal socialization, including noxious experiences, generate lasting engrams for the development of the individual "personality". These engrams must be genetically fixed. Therefore, it is interesting that noxious, but not non-noxious, stimulation evokes a massive IEG expression in juvenile rats which exceeds the expression level produced by the same stimulus in adult rats. Noxious stimulation can be also considered as a novel experience for laboratory animals, and this component of novelty might also contribute to IEG expression.

Potentiation of c-Fos expression by coincident painful events

Stimulation of C-fibres in a hindlimb sensory nerve causes c-Fos expression in ipsilateral but not in contralateral spinal dorsal horn neurons. However, if C-fibres in the contralateral nerve are stimulated 1–12 h later more than twice as many cells express c-Fos than would have without the first nerve stimulation, i.e. there is a potentiation of c-Fos expression (Leah et al., 1992; Williams et al., 1991). This potentiation has two components, an initial phase with a maximum at 1 h after the first stimulus, and a second phase lasting from 3 to 12 h after the first stimulus. This potentiation parallels the increased sensitivity of spinal neurons to noxious stimulation (hyperalgesia) that occurs following noxious stimulation (reviewed by Coderre et al., 1993). The potentiation is enhanced when NMDA or kainate agonists are applied to the contralateral neurons during the initial sciatic nerve stimulation. Conversely, the potentiation is reduced by NMDA or AMPA antagonists applied to the contralateral neurons at the time of the initial sciatic nerve stimulation. This potentiation of IEG expression is likely to have some mechanisms in common with long-term potentiation of synaptic transmission in the hippocampus.

Plasticity and stability at the level of gene transcription

The nervous system is characterized by a great complexity of cellular interactions, morphological diversification and functional plasticity. This complexity has its equivalent at the level of gene expression: it is estimated that up to 30–50% of all genes of the individual mammalian genome are expressed by cells of the embryonic and/or adult nervous system. This variety in gene expression requires a corresponding variety in proteins for coordinated gene transcription. Between 1% and 10% (i.e. 1000–10 000) of all the genes expressed in the nervous system encode for transcriptionally operating proteins. This large pool of transcription factors demonstrates the importance of transcription control and the subsequent de novo protein synthesis in the brain. For example, stimulation of dentate gyrus by generalized seizures provokes within 60 min, the transcription of 500–1000 genes encoding for different proteins such as enzymes, secretory proteins, tissue-specific proteins, and numerous transcription factors (Nedivi et al., 1993).

This nuclear response can be considered as a form of plasticity at the genetic level and is defined via reactive gene expression in response to extra- and intracellular stimulation including circulating factors, toxins, axonally transported compounds or tissue injury. In contrast, "everyday" stimuli might not be effective, or might be only weakly effective as a trigger for de novo protein synthesis above basal levels. It has to be carefully distinguished whether the stimulus per se (i.e. the pre- or post-synaptic events) or the intracellular second messenger cascades cannot evoke (detectable) changes in protein synthesis.

Plasticity has to be realized under the primacy of stability: for example, the somatotopic representation of the body surface and visceral organs should be relatively fixed for life, and thus, the capacity for synaptic re-arrangements and reactive gene expression in the cells making up this representation must be fixed so that they do not result in alterations of the body map. On the other hand, formation of memory is tightly coupled with, and probably depends on, de novo protein synthesis in specific compartments of the limbic system. Thus, plasticity has also to be closely controlled at the level of gene induction.

Intraneuronal mechanism that contribute to, activate or restrict the expression of induced transcription factors

The expression of a particular gene can be induced by different intraneuronal mechanisms as demonstrated in the following example. Adult, but not neonatal dorsal root ganglion neurons that are depolarized with a subsequent calcium influx and an increase in cAMP concentration express VIP (Dobson et al., 1994). This finding demonstrate that the signal pathways leading to the de novo synthesis of a protein (preceded by activation of transcription factors) do not statically depend on transynaptic excitation or depolarization, but can fluctuate over time. Conversely, there must exist powerful intraneuronal mechanisms that prevent the expression of induced transcription factors (ITF) such as the proteins encoded by many IEGs. The inhibition of a particular ITF might vary across neuronal subtypes and/or the normal or pathophysiological contexts. It is thus not easy to determine whether the absence of expression of an ITF (e.g. c-Fos in axotomized neurons) is due to an active repression or to the absence of those intraneuronal cascades which otherwise mediate ITF induction.

Induction of ITF-encoding genes also depends on upstream intracellular events such as the activation of kinases, or elevation of calcium or cAMP levels (reviewed by Gosh and Greenberg, 1995). Thus, neurons which have a particular capacity for calcium sequestration, e.g. by calcium binding proteins, would be expected to have a higher threshold for calcium-dependent gene expression, as compared to neurons with lower concentrations of calcium binding proteins. For example, alterations in the levels of calmodulin in the cell nucleus parallel the induced levels of *c-fos* following application of convulsants (Vendrell et al., 1992), and calmodulin-antagonists can prevent the induction of c-fos mRNA in primary cortical neurons (Barron et al., 1995).

Calcium binding proteins could scavenge any depolarization-triggered rise in calcium and prevent activation of transcription factors. Consequently, neuronal stimulation would only result in ITF expression after saturation of the scavenger system, i.e. after strong but not weak neuronal stimulation. This mechanism would protect neurons from large alterations in gene expression that might shift the balance between plasticity and stability at the level of gene expression.

Initial transient changes in transcription factor expression might result in late lasting functional alterations

The transient expression of ITF evokes the question for the duration of their effects. It is known from oncology and cell biology that Fos and Jun proteins are potent modifiers of cellular functions. The transcription of ITF is tightly regulated since their dysregulation can result in cell death, malignant transformation, growth arrest or deregulated differentiation. In the case of *c-fos* transcription, instability elements, negative cis-autoregulation of the c-Fos protein and instability of the c-Fos protein result in a temporally limited function of c-Fos. However, the initial transient blast of AP-1 protein expression might be sufficient to initiate lasting ongoing alterations on the level of gene expression with subsequent permanent functional changes. For example, c-Fos can induce the expression of Fra-2 and, in turn, Fra-2 protein is present for weeks on an suprabasal level following a short neuronal stimulation as it has been shown following olfactory stimulation.

Block of c-Fos expression by antisense-oligodeoxynucleotides

The identification of the target genes of the ITFs is essential in understanding the cascade of gene expressions in neurons following noxious stimulation. As shown in the preceding paragraphs we have used histochemical double labelling to screen the colocalization and covariance of ITF encoded proteins and some neuropeptides or NOS in the same neurons, as a prerequisite for a causal relationship to exist in transcriptional processes related to these proteins.

For a more direct assessment of links in transcription control we have blocked c-fos translation by the antisense technique. A 20-mer end-capped phosphorothioate antisense oligodeoxynucleotide (ODN) complementary to the translation initiating sequence of c-fos mRNA was applied by superfusion to the exposed spinal cord dorsum of rats (Gillardon et al., 1994). After 6 h of superfusion with ODN to reach an equilibrium concentration in the spinal dorsal horn, both hindpaws were repeatedly heated to a noxious temperature (52°C) in order to induce expression. After 1.5 h c-Fos and c-Jun labelled neurons were histochemically assessed on spinal sections. In dorsal horns superfused with c-fos antisense ODN the numbers of neurons showing intensive nuclear c-Fos expression due to the noxious stimulation were reduced to less than 50% of values found in sham treated animals or after superfusion with a random sequence ODN. The expression of c-Jun was unchanged by superfusion with c-fos antisense ODN. This experiment shows that gene expression of c-fos can be effectively controlled by the antisense technique.

In a second experiment (Gillardon et al., 1996) we have studied the effect of c-fos antisense ODN on the expression of dynorphin A following noxious stimulation of the hindpaws (Fig. 12). The ODN was administered systemically with an implanted miniosmotic pump. After 3 days of infusion of antisense ODN against c-fos translation or random ODN noxious stimulation of one hindpaw was performed. Afterwards, the number of dorsal horn neurons showing c-Fos immunoreactivity was reduced in a dose dependent manner by up to 50% of values obtained after infusion of vehicle or random sequence ODN (Fig. 12). In parallel, the density of dynorphin A label induced in the dorsal horn by the noxious stimulation was reduced by

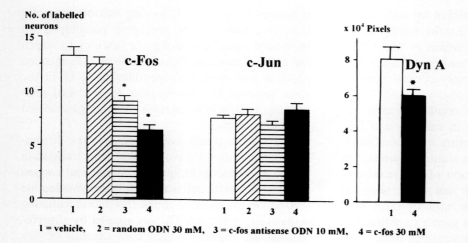

Fig. 12. Block of gene expression by administration of antisense oligodeoxynucleotides (ODN). In rats osmotic minipumps were implanted subcutaneously releasing antisense ODN specifically designed against c-fos mRNA. Carrageenan injection was applied to one hind paw as a noxious stimulus. After 3 days immunoreactivities of c-Fos, c-Jun and Dyn A were assessed by immunohistochemistry. The c-fos antisense ODN resulted in reduced numbers of neurons expressing c-Fos protein, and this reduction was more pronounced at a higher dose of antisense ODN. A random sequence ODN was without effect. The expression of c-Jun was not affected by c-fos antisense ODN. Dyn A immunohistochemistry assessed by digital image analysis was also reduced by 25% after infusion of c-fos antisense ODN. (Gillardon et al., 1996).

25%. This covariation indicates that c-Fos is involved in the control of transcription of the dynorphin A gene. The decrease in Dynorphin A label was less than that of c-Fos. This suggests that c-Fos may be one of a set of redundant transcription controlling proteins at the AP-1 site of the dynorphin A gene.

Conclusions

We have presented two classes of phenomena that are related to neuroplasticity after noxious stimuli or nerve lesions: direct observations of changed neuronal function and the induction of gene transcription. We hypothesize that the responses at the genetic and functional levels are interrelated and both reflect the plasticity of the nervous system that can result in enhanced sensitivity of the pain system.

The observation that nervous activity can alter gene transcription at first seemed paradoxical, as adult nerve cells do not undergo mitosis and normally do not show signs of growth. Why, then, should the nuclear transcription machinery and other processes related to nervous system development be activated by physiological nerve cell stimulation?

Inducible transcription factors (ITFs) might be an essential part of the mechanisms that enable nerve cells to modify their working range at the molecular level in response to the requirements of changing conditions. This would best correspond to the dominant principle of the nervous system, i.e. to enable the organism to continuously adapt to the external world. Many nervous system adaptations are related to learning, and learning is mostly a long-term modification that might last for the entire life span. Hence, learning and memory research, since Hyden, has had a great interest in assessing what happens at the level of gene transcription, although the changes that may occur in single neurons during a learning experiment might be below the detection level of currently available method.

We do not yet know what determines the outcome of ITF induction by transient trauma or noxious stimulation. The same c-Fos/c-Jun transcription complex can activate a variety of genes, and the number of ITFs, and other transcription controlling molecules, currently known is too small to provide specificity of control in relation to the subsequent processes. However, we do not know what happens beyond the few transcription factors we have been able to observe so far, and it might well be that what we can see today is a mere fraction of what will be discovered in our nervous system in the future.

Future research in this field will have to examine in detail the molecular genetic processes in the nervous system. However, many questions aimed at manipulating the system and its therapeutic implications are being posed, and we are confident that there will be intensive research into therapeutic possibilities at the level of transcription control, as the research methods used in the field suggest highly selective approaches. Thus, if we consider that some kinds of chronic pain are diseases of acquired transcription failures, the therapeutic and preventive strategies to be followed might not be too different from those now being developed for other diseases, which are due to inherited or acquired gene defects or transcription deficiencies.

References

Abad, F., Feria, M. and Boada, J. (1989) Chronic amitriptyline decreases autotomy following dorsal rhizotomy in rats. *Neurosci. Lett.,* 99: 187–190.

Abbadie, C. and Besson, M. (1992) C-fos expression in rat lumbar spinal cord during the development of adjuvant-induced arthritis *Neuroscience,* 48: 985–993.

Albe-Fessard, D. and Lombard, M.C. (1983) Use of an animal model to evaluate the origin of and protection against deafferentation pain. In: J.J. Bonica, U. Lindblom and A. Iggo (Eds.), *Advances in Pain Research and Therapy,* Vol. 5. Raven Press, New York, pp. 691–700.

Arnér, S. and Meyerson, B.A. (1988) Lack of analgesic effect of opioids on neuropathic and idiopathic forms of pain. *Pain,* 33, 11–23.

Barron, S., Tussel, J.M., Sola, C. and Serratosa, J. (1995) Convulsant agents activate c-fos induction in both a calmodulin-dependent and calmodulin-independent manner. *J. Neurochem.,* 65: 1731–1739.

Beck, P.W. and Handwerker, H.O. (1974) Bradykinin and serotonin effects on various types of cutaneous nerve fibres. *Pflügers Arch.,* 347: 209–222.

Beck, P.W., Handwerker, H.O. and Zimmermann, M. (1974) Nervous outflow from the cat's foot during noxious radiant heat stimulation. *Brain Res.,* 67: 373–386.

Bennett, G.J. (1994) Animal models of neuropathic pain. In: G.F.Gebhart, D.L. Hammond, D.S. Jensen (Eds.). *Proceedings of the 7th World Congress on Pain Research and management,* Vol.2. IASP Press, Seattle.pp. 495-510.

Benrath, J., Eschenfelder, C., Zimmermann, M. and Gillardon, F. (1995) Calcitonin gene-related peptide, substance P and nitric oxide are involved in cutaneous inflammation following ultraviolet irradiation. *Eur. J. Pharmacol. Environ. Toxicol. Pharmacol.,* 293: 87–96.

Bossut, D.F. and Perl, E.R. (1995) Effects of nerve injury on sympathetic excitation of A-delta mechanical nociceptors. *J. Neurophysiol.,* 73: 1721–1723.

Brecht, S., Martin-Villalba, A., Zuschratter, W., Bravo, R. amd Herdegen, T. (1995) Transection of rat fimbria-fornix induces lasting expression of c-Jun protein in axotomized septal neurons immunonegative for choline acetyltransferase and nitric oxide synthase. *Exp. Neurol.,* 134: 112–125.

Brecht, S., Buschmann, T., Grimm, S. and Herdegen, T. (1996) Persisting expression of galanin in axotomized mammillary and septal neurons of the adult rat labelled for c-Jun and NADPH-diaporase. *Mol. Brain Res.,* in press.

Brinkhus, H.B. and Zimmermann, M. (1983) Characteristics of spinal dorsal horn neurons after partial chronic deafferentation by dorsal root transection. *Pain,* 15: 221–236.

Bullitt, E., Lee, C.L., Light, A.R. and Willcockson, H. (1992) The effect of stimulus duration on noxious-stimulus induced c-fos expression in the rodent spinal cord. *Brain Res.,* 580: 172–179.

Buzzi, G.M., Moskowitz, M.A. and Agnoli, A. (1991) The trigeminovascular system and neurogenic inflammation: a possible pathogenetic mechanism for migraine headaches. *New Trends Clin. Neuropharmacol.,* 5: 5–18.

Carstens, E., Fraunhoffer, M. and Zimmermann, M. (1981) Serotonergic mediation of descending inhibition from midbrain periaqueductal gray, but not reticular formation, of spinal nociceptive transmission in the cat. *Pain,* 10: 149–167.

Castro-Lopes, J.M., Tavares, I. and Coimbra, A. (1993) GABA decreases in the spinal cord after peripheral neurectomy. *Brain Res.,* 620: 287–291.

Chahl, L.A., Szolcsanyi, J. and Lembeck, F. (Eds.) (1984) *Antidromic Vasodilatation and Neurogenic Inflammation.* Akademiai Kiado, Budapest.

Coderre, T.J., Grimes, R.W. and Melzack, R. (1986a) Deafferentation and chronic pain in animals: an evaluation of evidence suggesting autotomy is related to pain. *Pain,* 26: 61–84.

Coderre, T.J., Grimes, R.W. and Melzack, R. (1986b) Autotomy after nerve sections in the rat is influenced by tonic descending inhibition from locus coeruleus. *Neurosci. Lett.,* 67: 82–86.

Coderre, T.J., Katz, J., Vaccarino, A.L. and Melzack, R. (1993) Contribution of central plasticity to pathological pain: review of clinical and experimental evidence. *Pain,* 52: 259–285.

Colado, M.I., Arnedo, A., Peralta, E. and Del-Rio, J. (1988)

Unilateral dorsal rhizotomy decreases monoamine levels in the rat spinal cord. *Neurosci. Lett.*, 87: 302–306.

Colpaert, F.C., Donnerer, J. and Lembeck, F. (1983) Effects of capsaicin on inflammation and on the substance P content of nervous tissues in rat with adjuvant arthritis. *Life Sci.*, 32: 182–183.

De Leon, M., Nahin, R.L., Molina, C.A., De Leon, D.D. and Ruda, M.A. (1995) Comparison of c-jun, junB, and junD mRNA expression and protein in the rat dorsal root ganglia following sciatic nerve transection. *J. Neurosci. Res.*, 42: 391–401.

Dickhaus, H., Zimmermann, M. and Zotterman, Y. (1976) The development in regenerating cutaneous nerves of C-fibre receptors responding to noxious heating of the skin. In: Y. Zotterman (Ed.), *Sensory Functions of the Skin in Primates.* Pergamon Press, Oxford, pp. 415–425.

Dickhaus, H., Pauser, G. and Zimmermann, M. (1985) Tonic descending inhibition affects intensity coding of nociceptive responses of spinal dorsal horn neurones in the cat. *Pain*, 23: 145–158.

Dobson, S.P., Quinn, J.P., Morrow, J.A. and Mulderry, P.K. (1994) The rat vasoactive intestinal polypeptide cyclic AMP response element regulates gene transcriptional responses differently in neonatal and adult rat sensory neurons. *Neurosci. Lett.*, 167: 19–23.

Eschenfelder, C.C., Benrath, J., Zimmermann, M. and Gillardon, F. (1995) Involvement of substance P in ultraviolet irradiation-induced inflammation in rat skin. *Eur. J. Neurosci.*, 7: 1520–1526.

Fiallos-Estrada, C.E., Kummer, W., Mayer, W., Bravo, R., Zimmermann, M. and Herdegen, T. (1993) Long-lasting increase of nitric oxide synthase immunoreactivity and NADPH-diaphorase reaction, and co-expression with the nuclear c-JUN protein in rat dorsal root ganglion neurons following sciatic nerve transection. *Neurosci. Lett.*, 150: 169–173.

Flor, H., Birbaumer, N., Backonja, M.M. and Bromm, B. (1994) Acquisition of chronic pain: psychophysiological mechanisms. *Am. Pain Soc. J.*, 3: 119–197.

Foerster, O. (1927) *Die Leitungsbahnen des Schmerzgefühls und die chirurgische Behandlung der Schmerzzustände.* Urban and Schwarzenberg, Berlin.

Gass, P. and Herdegen, T. (1996) Expression of AP-1 proteins in excitotoxic-neurodegenerative disorders and following nerve fiber injury. *Prog. Neurobiol.*, 47: 257–290.

Gillardon, F., Schröck, H., Morano, M. and Zimmermann, M. (1992) Long-term increase in CGRP levels in rat spinal dorsal horn following skin ultraviolet irradiation. A mechanism of sunburn pain? In: Y. Tache, P. Holzer and M. Rosenfeld (Eds.), *Calcitonin Gene-Related Peptide. Ann. N. Y. Acad. Sci.*, Vol. 657, pp. 493–496.

Gillardon, F., Beck, H., Uhlmann, E., Herdegen, T., Sandkühler, J., Peymann, A. and Zimmermann, M. (1994) Inhibition of c-Fos protein expression in rat spinal cord by antisense oligodeoxynucleotide superfusion. *Eur. J. Neurosci.*, 6: 880–884.

Gillardon, F., Vogel, I., Zimmerman, M. and Uhlmann, E. (1996) Inhibition of carrageenan-induced spinal c-fos activation by systemically administered c-fos antisense digodeoxynucleotides may be facilitated by local opening of the blood-spinal cord barrier. *J. Neurosci Res.*, in press.

Ginzburg, R. and Seltzer, Z. (1990) Allografting adrenal medulla of adult rats into the lumbar subarachnoid space suppresses autotomy: a model of deafferentation-induced pain. *Pain*, Suppl. 5: S462.

Goadsby, P.J. and Edvinsson, L. (1993) The trigeminovascular system and migraine: studies characterizing cerebrovascular and neuropeptide changes seen in humans and cats. *Ann. Neurol.*, 33: 48–56.

Goadsby, P.J., Edvinsson, L. and Ekman, R. (1990) Vasoactive peptide release in the extracerebral circulation of humans during migraine headache. *Ann. Neurol.*, 28: 183–187.

Gosh, A. and Greenberg, M.E. (1995) Calcium signaling in neurons: molecular mechanisms and cellular consequences. *Science*, 268: 239–247.

Handwerker, H.O., Iggo, A. and Zimmermann, M. (1975) Segmental and supraspinal actions on dorsal horn neurons responding to noxious and non-noxious stimuli. *Pain*, 1: 147–165.

Herdegen, T. (1996) Jun, Fos, and CREB/ATF transcription factors in the brain: control of expression under normal and pathophysiological conditions. *Neuroscientist*, 2: in press.

Herdegen, T. and Zimmermann, M. (1994) Induction of c-Jun and JunD transcription factors represents specific changes in neuronal gene expression following axotomy. In: F. Seil (Ed.), *Neural Regeneration. Progress in Brain Research*, Vol. 103. Elsevier, Amsterdam, pp. 153–171.

Herdegen, T. and Zimmermann, M. (1995) Immediate-early genes (IEGs) encoding for inducible transcription factors (ITFs) and neuropeptides in the nervous system: functional network for long-term plasticity and pain. In: F. Nyberg, H.S. Sharma and Z. Wiesenfeld-Hallin (Eds.), *Neuropeptides in the Spinal Cord. Progress in Brain Research*, Vol. 104. Elsevier, Amsterdam, pp. 299–321.

Herdegen, T., Kovary, K., Leah, J.D. and Bravo, R. (1991a) Specific temporal and spatial distribution of JUN, FOS and KROX-24 proteins in spinal neurons following noxious transynaptic stimulation. *J. Comp. Neurol.*, 313: 178–191.

Herdegen, T., Kummer, K., Fiallos-Estrada, C.E., Leah, J.D. and Bravo, R. (1991b) Expression of c-JUN, JUN B and JUN D in the rat nervous system following transection of vagus nerve and cervical sympathetic trunk. *Neuroscience*, 45: 413–422.

Herdegen, T., Tölle, T., Bravo, R., Zieglgänsberger, W. and Zimmermann, M. (1991c) Sequential expression of JUN B, JUN D and FOS B proteins in rat spinal neurons: cascade of transcriptional operations during nociception. *Neurosci. Lett.*, 129: 221–224.

Herdegen, T., Fiallos-Estrada, C.E., Schmid, W., Bravo, R. and Zimmermann, M. (1992) Transcription factors c-JUN,

JUN D and CREB, but not FOS and KROX-24, are differentially regulated in neurons following axotomy of rat sciatic nerve. *Mol. Brain Res.,* 14: 155–165.

Herdegen, T., Bastmeyer, M., Bähr, M., Stürmer, C.A.O., Bravo, R. and Zimmermann, M. (1993a) Expression of JUN, KROX and CREB transcription factors in goldfish and rat retinal ganglion cells following optic nerve lesions is related to axonal sprouting. *J. Neurobiol.,* 24: 528–543.

Herdegen, T., Brecht, S., Kummer, W., Mayer, B., Leah, J., Bravo, R. and Zimmermann, M. (1993b) Persisting expression of JUN and KROX transcription factors and nitric oxide synthase in rat central neurons following axotomy. *J. Neurosci.,* 13: 4130–4146.

Herdegen, T., Fiallos-Estrada, C.E., Bravo, R. and Zimmermann, M. (1993c) Colocalisation and covariation of the nuclear c-JUN protein with galanin in primary afferent neurons and with CGRP in spinal motoneurons following transection of rat sciatic nerve. *Mol. Brain Res.,* 17: 147–154.

Herdegen, T., Rüdiger, S. and Zimmermann, M. (1994) Increase of nitric oxide synthase and colocalization with Jun, Fos and Krox proteins in spinal neurons following noxious peripheral stimulation. *Mol. Brain Res.,* 22: 245–258.

Herdegen, T., Brecht, S., Fiallos-Estrada, C.E., Wickert, H., Gillardon, F., Voss, S. and Bravo, R. (1995) A novel face of immediate-early genes: transcriptional operations dominated by c-Jun and JunD proteins in neurons following axotomy and during regenerative efforts. In: T. Tölle, J. Schadrack and W. Zieglgänsberger (Eds.), *Immediate-Early Genes in the CNS: More than Activity Markers.* Springer-Verlag, Berlin, pp. 78–103.

Hunt, S.P., Pini, A. and Evan, G. (1987) Induction of c-fos-like protein in spinal cord neurons following sensory stimulation. *Nature,* 328: 632–634.

Igwe, O.I. and Ning, L. (1994) Regulation of the second-messenger system in the rat spinal cord during prolonged peripheral inflammation. *Pain,* 56: 63–75.

Jänig, W. and Schmidt, R.F. (Eds.) (1992) *Reflex Sympathetic Dystrophy: Pathophysiological Mechanisms and Clinical Implications.* VCH Verlagsgesellschaft, Weinheim.

Jasmin, L., Gogas, K.R., Ahlgren, S.C., Levine, J.D. and Basbaum, A.I. (1994) Walking evokes a distinctive pattern of Fos-like immunoreactivity in the caudal brainstem and spinal cord of the rat. *Neuroscience,* 58: 275–286.

Jenkins. R., McMahon, S.B., Bond, A.B. and Hunt, S.P. (1993) Expression of c-Jun as a response to dorsal root and peripheral nerve section in damaged and adjacent intact primary sensory neurons in the rat. *Eur. J. Neurosci.,* 5: 751–759.

Kantner, R.M., Goldstein, B.D. and Kirby, M.L. (1986) Regulatory mechanism for substance P in the dorsal horn during a nociceptive stimulus: axoplasmatic transport vs electrical activity. *Brain Res.,* 385: 282–290.

Kishore, K.R., Max, M.B., Schafer, S.C., Gaughan, A.M., Smoller, B., Gracely, R.H. and Dubner, R. (1990) Desipramine relieves postherpetic neuralgia. *Clin. Pharmacol. Ther.,* 47, 305–312.

Lanteri-Minet, M., de Pommery, J., Herdegen, T., Weil-Fugazza, J., Bravo, R. and Menetrey, D. (1993) Differential time course and spatial expression of Fos, Jun, and Krox-24 proteins in spinal cord of rats undergoing subacute or chronic somatic inflammation. *J. Comp. Neurol.,* 333: 223–235.

Leah, J.D., Cameron, A.A. and Snow, P.J. (1989a) Neuropeptides in physiologically identified mammalian sensory neurones. *Neurosci. Lett.,* 56: 257–263.

Leah, J., Herdegen, T. and Zimmermann, M. (1989b) Physiological and pharmacological induction of c-fos protein immunoreactivity in superficial dorsal horn neurones. In: F. Cervero, G.J. Bennett and P.M. Headley (Eds.), *Processing of Sensory Information in the Superficial Dorsal Horn of the Spinal Cord.* Plenum Press, New York, pp. 307–310.

Leah, J.D., Herdegen, T., Kovary, K. and Bravo, R. (1991) Selective expression of JUN proteins following axotomy and axonal transport block in peripheral nerves in the rat: evidence for a role in the regeneration process. *Brain Res.,* 566: 198–207.

Leah, J.D., Sandkühler, J., Herdegen, T., Murashov, A. and Zimmermann, M. (1992) Potentiated expression of FOS protein in the rat spinal cord following bilateral noxious cutaneous stimulation. *Neuroscience,* 48: 525–532.

Leah, J.D., Porter, J., De-Pommery, J., Menetrey, D. and Weil-Fugazza, J. (1996) Effect of acute stimulation on Fos expression in spinal neurons in the presence of persisting C-fiber activity. *Brain Res.,* 719: 104–111.

Lee, J.H., Wilcox, G. and Beitz, A. (1992) Nitric oxide mediates Fos expression in the spinal cord induced by noxious stimulation. *NeuroReport,* 3: 841–844.

Lenz, A.F., Tasker, R.R., Dostrovsky, J.O. et al. (1988) Abnormal single-unit activity and responses to stimulation in the presumed ventrocaudal nucleus of patients with central pain. In: R. Dubner, G.F. Gebhart and M.R. Bond (Eds.), *Proceedings of the Vth World Congress on Pain (Pain Research and Clinical Management),* Vol. 3. Elsevier, Amsterdam, pp. 157–164.

Levine, J.D., Clark, R., Devor, M., Helms, C., Moskowitz, M.A. and Basbaum, A.I. (1984) Intraneuronal substance P contributes to the severity of experimental arthritis. *Science,* 226: 547–549.

Levine, J.D., Fields, H.L. and Basbaum, A.I. (1993) Peptides and the primary afferent nociceptor. *J. Neurosci.,* 13: 2273–2286.

Liebermann, A.R. (1971) The axon reaction: a review of the principal features of perikaryal response to axon injury. *Int. Rev. Neurobiol.,* 14: 49–124.

Lima, D., Avelino, A. and Coimbra, A. (1993) Differential activation of c-fos in spinal neurones by distinct classes of noxious stimuli. *NeuroReport,* 4: 747–750.

Lombard, M.-C. and Besson, J.-M. (1989) Attempts to gauge the relative importance of pre- and postsynaptic effects of morphine on the transmission of noxious messages in the dorsal horn of the rat spinal cord. *Pain,* 37: 335–345.

Lombard, M.-C., Nashold, Jr., B.S., Albe-Fessard, D., Salman, N. and Sakr, C. (1979) Deafferentation hypersensitivity in the rat after dorsal rhizotomy: a possible animal model of chronic pain. *Pain,* 6: 163–174.

Lotz, M., Carson, D.A. and Vaughan, J.H. (1987) Substance P activation of rheumatoid synoviocytes: neural pathway in pathogenesis of arthritis. *Science,* 235: 893–895.

Maggi, C.A. and Meli, A. (1988) The sensory-efferent function of capsaicin-sensitive sensory neurons. *Gen. Pharmacol.,* 19: 1–43.

Mao, J., Price, C.D. and Mayer, D.J. (1995) Experimental mononeuropathy reduces the antinociceptive effects of morphine: implications for common intracellular mechanisms involved in morphine tolerance and neuropathic pain. *Pain,* 61: 353–364.

Mayer, D.J., Mao, J. and Price, D.D. (1995) The development of morphine tolerance and dependence is associated with translocation of protein kinase C. *Pain,* 61: 365–374.

Menkes, C.J., Renoux, M., Laoussadi, S., Mauborgne, A., Bruxelle, J. and Cesselin, F. (1993) Substance P levels in the synovium and synovial fluid from patients with rheumatoid arthritis and osteoarthritis. *J. Rheumatol.,* 20: 714–717.

Mikucki, S.A. and Oblinger, M.M. (1991) Corticospinal neurons exhibit a novel pattern of cytoskeletal gene expression after injury. *J. Neurosci. Res.,* 30: 213–335.

Molander, C., Hongpaisan, J. and Grant, G. (1992) Changing pattern of c-fos expression in spinal cord neurons after electrical stimulation of the chronically injured sciatic nerve in the rat. *Neuroscience,* 50: 223–236.

Moncada, S., Palmer, R.M.J. and Higgs, E.A. (1991) Nitric oxide: physiology, pathophysiology and pharmacology. *Pharmacol. Rev.,* 43: 109–142.

Morgan, J.I. and Curran, T. (1989) Stimulus-transcription coupling in neurons: role of cellular immediate-early genes. *Trends Neurosci.,* 12: 459–462.

Moskowitz, M.A. and MacFarlane, R. (1993) Neurovascular and molecular mechanisms in migraine headache. *Cerebrovasc. Brain Metab. Rev.,* 5: 159–177.

Nedivi, E., Hevroni, D., Naot, D., Israeli, D. and Citri, Y. (1993) Numerous candidate plasticity-related genes revealed by differential cDNA cloning. *Nature,* 363: 718–721.

New, H.V. and Mudge, A.W. (1986) Calcitonin gene-related peptide regulates muscle acetylcholine receptor synthesis. *Nature,* 323: 809–811.

Panerai, A.E., Sacerdote, P., Brini, A., Bianchi, M. and Mantegazza, P. (1987) Autotomy and central nervous system neuropeptides after section of the sciatic nerve in rats of different strains. *Pharmacol. Biochem. Behav.,* 28: 385–388.

Pretel, S. and Piekut, D.T. (1991a) ACTH and enkephalin axonal input to paraventricular neurons containing c-fos like immunoreactivity. *Synapse,* 8: 100–106.

Pretel, S. and Piekut, D.T. (1991b) Enkephalin, substance P, and serotonin axonal input to C-fos-like immunoreactive neurons of the rat spinal cord. *Peptides,* 12: 1243–1250.

Przewlocki, R., Haarmann, I., Nikolarakis, K., Herz, A. and Höllt, V. (1988) Prodynorphin gene expression in spinal cord is enhanced after traumatic injury in the rat. *Brain Res.,* 464: 37–41.

Rivot, J.P., Calvino, B. and Besson, J.M. (1987) Is there a serotonergic tonic descending inhibition on the responses of dorsal horn convergent neurons to C-fibre inputs? *Brain Res.,* 403: 142–146.

Ryseck, P. and Bravo, R. (1991) c-JUN, JUN B, and JUN D differ in their binding affinities to AP-1 and CRE consensus sequences: effect of FOS proteins. Oncogene, 6: 533–542.

Saito, Y., Collins, J.G. and Iwasaki, H. (1990) Tonic 5-HT modulation of spinal dorsal horn neuron activity evoked by both noxious and non-noxious stimuli: a source of neuronal plasticity. *Pain,* 40: 205–219.

Schaden, H., Stürmer, C.A.O. and Bähr, M. (1994) GAP-43 immunoreactivity and axon regeneration in retinal ganglion cells of the rat. *J. Neurobiol.,* 25: 1570–1578.

Schmidt, R.F., Schaible, H.-G., Meßlinger, K., Heppelmann, B., Hanesch, U. and Pawlak, M. (1994) Silent and active nociceptors: structure, functions, and clinical implications. In: G.F. Gebhart, D.L. Hammond and D.S. Jensen (Eds.), *Proceedings of the 7th World Congress on Pain. Progress in Pain Research and Management,* Vol. 2. IASP Press, Seattle, WA, pp. 213–250.

Seltzer, S., Marcus, R. and Stoch, R. (1981) Perspectives in the control of chronic pain by nutritional manipulation. *Pain,* 11: 141–148.

Seltzer, Z., Tal, M. and Sharav, Y. (1989) Autotomy behavior in rats following peripheral deafferentation is suppressed by daily injections of amitriptyline, diazepam and saline. *Pain,* 37: 245–250.

Shea, V.K. and Perl, E.R. (1985) Regeneration of cutaneous afferent unmyelinated C-fibres after transection. *J. Neurophysiol.,* 54: 502–512.

Terenius, L. (1979) Endorphins in chronic pain. In: J.J. Bonica, J.C. Liebeskind and D.G. Albe-Fessard (Eds.), *Proceedings of the Second World Congress on Pain. Adv. Pain Res. Ther.* Vol. 3, Raven Press, New York, pp. 459–471.

Vaeroy, H., Helle, R., Forre, O., Kass, E. and Terenius, L. (1988) Elevated CSF levels of substance P and high incidence of Raynaud phenomenon in patients with fibromyalgia: new features for diagnosis. *Pain,* 32: 21–26.

Vendrell, M., Pujol, M.J., Tusell, J.M. and Serratosa, J. (1992) Effect of different convulsant levels and proto-oncogene c-fos expression in the central nervous system. *Mol. Brain Res.,* 14: 285–292.

Verge, V.K.M., Xu, Z., Xu, X.J., Wiesenfeld-Hallin, S. and Hökfelt, T. (1992) Marked increase in nitric oxide synthase mRNA in rat dorsal root ganglia after peripheral axotomy: in situ hybridization and functional studies. *Proc. Natl. Acad. Sci. USA,* 89: 11617–11621.

Walker, J.B. and Katz, R.L. (1979) Neural plasticity and analgesia – aspects of the same phenomenon. [Letter]. *Lancet,* ii: 1307.

Wall, P.D. (1976) Plasticity in the adult mammalian central nervous system. In: M.A. Corner and D.F.C. Swaab (Eds.), *Perspectives in Brain Research. Progress in Brain Research*, Vol. 45. Elsevier, Amsterdam, pp. 359–379.

Wall, P.D., Devor, M., Inbal, R., Scadding, J.W., Schonfeld, D., Seltzer, Z. and Tomkiewicz, M.M. (1979) Autotomy following peripheral nerve lesions: experimental anaesthesia dolorosa. *Pain*, 7, 103–115.

Wall, P.D., Bery, J. and Saade, N. (1988) Effects of lesions to rat spinal cord lamina I cell projection pathways on reactions to acute and chronic noxious stimuli. *Pain*, 35: 327–339.

Weihe, E. (1990) Neuropeptides in primary afferent neurons. In: W. Zenker and W.L. Neuhuber (Eds.), *The Primary Afferent Neuron*. Plenum, New York, pp. 127–159.

Williams, S., Evan, G. and Hunt, S.P. (1991) C-fos induction in the spinal cord after peripheral nerve lesion. *Eur. J. Neurosci.*, 3: 887–894.

Xie, Y., Zhang, J., Petersen, M. and LaMotte, R.H. (1995) Functional changes in dorsal root ganglion cells after chronic nerve constriction in the rat. *J. Neurophysiol.*, 73: 1811–1820.

Xu, X.J, Wiesenfeld-Hallin, Z., Villar, M.J, Fahrenkrug, J. and Hökfelt, T. (1990) On the role of galanin, substance P and other neuropeptides in primary sensory neurons in the rat: studies on spinal reflex excitability and peripheral axotomy. *Eur. J. Neurosci.*, 2: 733–743.

Zajac, J.M., Lombard, M.C., Peschanski, M., Besson, J.M. and Roques, B.P. (1989) Autoradiographic study of mu and delta opioid binding sites and neutral endopeptidase-24.11 in rat after dorsal root rhizotomy. *Brain Res.*, 477: 400–403.

Zimmermann, M. (1979) Peripheral and central nervous mechanisms of nociception, pain, and pain therapy: facts and hypotheses. In: J.J. Bonica, J.C. Liebeskind and D.G. Albe-Fessard (Eds.), *Advances in Pain Research and Therapy*, Vol. 3. Raven Press, New York, pp. 3–32.

Zimmermann, M. (1991) Central nervous mechanisms modulating pain-related information: do they become deficient after lesions of the peripheral or central nervous system? In: K.L. Casey (Ed.), *Pain and Central Nervous System Disease: The Central Pain Syndromes*. Raven Press, New York, pp. 183–199.

Zimmermann, M. and Herdegen, T. (1994) Control of gene transcription by Jun and Fos proteins in the nervous system. *Am. Pain Soc. J.*, 3: 33–48.

Zimmermann, M., Koschorke, G.M. and Sanders, K. (1987) Response characteristics of fibers in regenerating and regenerated cutaneous nerves in cat and rat. In: L.M. Pubols and B.J. Sessle (Eds.), *Effects of Injury on Trigeminal and Spinal Somatosensory Systems*. Alan R. Liss, New York, pp. 93–106.

Subject index

α-2 Adrenergic receptors play a role in injured C fibres, 107
α-2 Receptors, 107
α-Adrenergic antagonists, 108
α-Adrenergic antagonists on the responses of C fibres to norepinephrine, 107
α_2-Adrenomimetics, 137
β-Endorphin, 33, 40, 96–98, 117, 143
β-Endorphin binding in cutaneous nerve fibres, 100
β-Endorphin content, 241
β-Endorphin release, by interleukin-1, 97
δ-Receptor antagonist ICI 174,864, 96
κ-Agonist drugs depress spinal reflexes, 142
κ Antagonist nor-binaltorphamine on spinal neuron, 171
κ-Opioid receptor, 96
κ-Opioid receptor ligand, 95
κ Opioid agonist, 171
κ Opioid system, 173
κ-Receptors, 99, 171
κ Receptor agonist, 142
κ-Receptor antagonist nor-binaltorphimine, 96
μ- and δ-opioid receptors on small cutaneous nerve fibres, 98
μ-Agonist opioids, 137
μ-Receptor antagonist CTOP, 96

Aδ- and C-fibres released burst-like discharges, 218
A-fibre induced pain, 88
AA in synaptic plasticity and long-term potentiation, 178
Aberrant sensory processing, 227
Abnormal functional properties of sensory neurons after a nerve injury, 105
Abnormal pain behaviour, 12
Abnormal pain states, 7
Abnormal posture, 17
Abnormal response to norepinephrine, 106, 110
Abnormal sensory symptoms, 7
Acetylsalicylate, 12
Acidic fibroblast growth factor (aFGF), 89
ACTH, 40
ACTH plasma levels following formalin s.c., 41
Acute pain and its modulation, 20
Acute to chronic pain, transition, 133
Adaptation to aversive situations and pain, 42
Adenosine, 229
    Acting at the inhibitory A1 receptor, 228
Adhesion molecules, 86
Adrenal medulla grafted to spinal cord, 241
Adrenaline excitatory effects on nociceptive afferents, 236
Adrenaline, 235
Adrenoceptor, 236
Adrenoceptors expressed on afferent fibres, 90
Adrenocorticotropin, 71
Affect as an intrinsic component of pain, 63
Affective central pain processes, 68
Affective component of pain contributes to learning, 65

Affective component of pain, 77
Affective dimension of pain, 63, 67
Affective processes in pain chronicity, 77
Afterdischarges, exaggerated, in dorsal horn neurones, 10
Allodynia, 5, 6, 8, 17, 88, 105, 110, 114, 133, 144, 151, 169, 177, 189, 226
    Following topical capsaicin, 152
    Pain evoked by a normally innocuous stimulus, 24
    Produced by interference with spinal GABA, 227
Alpha-2 receptors, 107
Alpha-2 adrenergic receptors play a role in injured C fibres, 107
Alpha-adrenergic antagonists, 108
Alpha-adrenergic antagonists on the responses of C fibres to norepinephrine, 107
Alternative splicing of NMDAR1 subunit controls NMDA receptor phosphorylation, 202
Alternative splicing of the NMDAR1 subunit gene
Alternative splicing of the NMDAR1 subunit mRNA, 194
Amnesia, 227
AMPA, 153, 179, 180, 181, 189, 194, 196
    And metabotropic glutamate receptors coactivation, 177
    And metabotropic glutamate receptors play a role in hyperalgesia, 186, 181
    And metabotropic glutamate receptors, interaction between, 187
AMPA receptor, 195, 225
AMPA receptor antagonist 6,7-dinitroquinoxaline-2,3-dione, DNQX, 179
AMPA receptor is involved in fast synaptic transmission, 187
AMPA receptor selective antagonist DNQX, 181
AMPA receptor subunit, 196–198
AMPA sensitive calcium channel, 188
AMPA subunit, 197–198
AMPA/NMDA components of wind-up, 225
Amygdala, 65–66
Analgesia, 151
    During stress, 72
    Through opioid peptides, 144
Analgesic effect of tricyclic antidepressants, 241
Analgesic potency, 12
Ancient philosophers considered pain an emotion, 63
Animal models
    Acute pain-like responses, 17
    Myositis, 127
    Neuropathic pain, 11, 17
    Neuropathy, 12
    Pain, 11
Animal pain sensation, 17
Antagonist drug administered by microdialysis, 154
Antagonists of the NMDA receptor, 26
Anterior cingulate cortex, 75
Anti-hyperalgesics, 226
Anti-inflammatory agents, 11
Antibody microprobes, 137, 139, 159

Antidepressants, tricyclic, analgesic effect of, 241
Antinociception, 100, 207, 242
  Induced by morphine, 242
Antinociceptive, 207
Antisense ODN against c-fos translation, 253
Antisense oligodeoxynucleotides, 220, 253
AP-1 complexes, 249
  Following chronic inflammation, 249
  FosB:JunD dimers, 249
AP-1 compositions and transcriptional operations, 250
AP-1 protein expression, 253
AP-1 response elements, 245
AP-1 transcription factors following noxious stimulation, 242
AP5, 157, 170
AP7, 154
APV, 182
Arthritic animals, 9, 12, 160
Arthritic pain, 157
Arthritis, 151, 159, 162, 171, 250
Arthritis, acute, 10
Arthritis, experimental, 157
Arthritis-induced expression of Fos, Jun and Krox in dorsal horn, 250
Arthroscopic knee surgery, 99
Aspartate, 152, 154
Aspirin-like drugs, 11
Assessment of
  Hyperalgesia in animal models, 25
  Hyperalgesia in animals, 17
  Nocifensive behavior, 24
  Pain in animals, 17
ATF-2, 243
Attention, 17
Autonomic arousal, 72
Autotomy, 240–241
  A behavioural model of neuropathic pain, 120
  A pain related behaviour, 239
  After sciatic nerve section, 241
  Reduced during treatment with amitriptyline, 241
  Sign of chronic neuropathic pain, 241
Avoidance behavior, 34, 67
Axonal elongation, 245
Axonal growth cones, 131
Axonal injury induces expression of transcription factors Jun and Jun D, 109
Axonal transport, 237
  Of opioid receptors, 98
  Of substance P is increased in inflammation, 251
Axotomized retinal ganglion cells, 244–245
Axotomy, 109

Baicalein on hyperalgesia, 185
Behavioural analgesic tests, 12
Biting and scratching behavior, 115, 180
Block of expression of immediate-early genes, 220
Block of gene expression by antisense oligodeoxynucleotides, 254
Blocks of spinal descending axons, 239

Blood flow, 47
Blood-brain barrier, 48
Body map, 252
Bradykinin, 95, 100, 117, 131, 235
Bradykinin receptors, 235
Bulbo-spinal inhibitory system, 241
Burn injury, 5
Bursting discharges in dorsal horn and thalamic neurons after dorsal root transections, 240

C-fibre, 107, 109, 117, 129, 151, 226–227, 229, 249
  Activity, 108, 250
  Afferents, 90
  c-Fos expression, 251
  Evoked potentials, 6, 217–218
  From injured nerve respond to norepinephrine, 108
  From the muscle, 127
  Neuropeptides, 76
  Nociceptors, regeneration of, 8
  Repetitive stimulation of, 90
  Stimulation, 26, 225
c-Fos, 143, 218–220, 222, 243–244, 249–250, 252–254
  Control transcription of dyn A gene, 254
  Dorsal horn neurons, 243
  ncl. gracilis, 247
c-Fos and c-Jun proteins dimerize and bind to promoter or enhancer elements of genes, 243
c-Fos and Jun B proteins, 219
  And increased sensitivity of spinal neurons following noxious stimulation, 251
c-Fos antisense ODN, 253–254
c-Fos expressing cells and modality of noxious stimuli, 250
c-Fos expressing neurons and stimulation intensity, 250
c-Fos expression, 25, 250
  Controlled by the antisense technique, 253
  Due to noxious stimulation, 253
c-Fos expression in mono-arthritic rats, 250
c-fos mRNA, 220, 253
c-Fos protein, 219–220, 247
  Instability of, 253
  Negative cis-autoregulation of, 253
c-Fos transcription factor, change gene transcription, 247
c-fos transcription, 253
c-Fos translation, block, 253
c-Fos/c-Jun transcription complex, 254
c-Jun, 220, 243–246, 249–250, 253–254
  Elevated levels of, 244
  In axotomized neurons, 244
  mRNA in juvenile rats, 245
  Synthesis, 244
  Re-induction of, 244
  Upregulation of in axotomized neurons, 244
c-Jun expression
  And axonal elongation, 245
  Regenerative propensity of injured neurons, 245
  Related to cell body response, 244
  Tempero-spatial pattern, 244
c-jun gene, persistent transcription of, 244

C-nociceptor, 152
Ca^{2+} channels, 156
Ca^{2+} influx leads to increase in cell excitability, 26
Ca^{2+} ions, 218
Calcitonin gene-related peptide (CGRP), 85, 138, 246
    In inflamed muscle, 127
    Upregulation of during inflammation, 172
Calcium binding proteins, 252
    Prevent activation of transcription factors, 253
Calcium coupled transmitter release, 90
Calcium influx, 252
Calcium ion influx, 34
Calcium-dependent gene expression, 252
Calmodulin, 253
Calmodulin-antagonists prevent c-fos mRNA, 253
cAMP, 252
Cancer patients with diffuse bone metastasis, pain, 77
Capsaicin, 86, 101, 106, 186
    Administered at birth, 10
    Injection causes a long-lasting increase in the excitability of STT neurons, 153
    Injection, 152
    Intradermal injection, 6, 151–153, 154
    Neonatal, 9
    Pretreatment decreases the severity of the acute inflammatory response, 157
    Response, 155
    Secondary hyperalgesia induced by an intradermal injection of, 110
    Topical application of, 152
Capsaicin-sensitive primary sensory neurones, 90
Carrageenan, 95, 127, 188
Carrageenan inflammation, 228
Carrageenan-induced myositis, 128
Carrier-mediated transport, 48
Cascade of gene expressions in neurons, 253
Cause of upregulation of alpha-2 adrenergic receptors, 110
CBF changes, 74
CBF differences between painful and nonpainful conditions, 75
CBF in patients with atypical facial pain, 75
CCK, 118, 120
    A physiological brake on opioid analgesia, 228
CCK antagonists in combination with opioids, a new approach for treating neuropathic pain, 120
CCK antagonizes the analgesic effect of morphine, 117
CCK is an opioid antagonist, 117
CCK may function as an endogenous opioid antagonist, 117
CCK-B receptor
    In the rat spinal cord, 117
    Interfere with morphine, 228
CCK-mediated enhanced opioid effects, 228
Cell body response, 234, 245
    After nerve transection, 234–235, 244–245
Cell death, 245, 253
Cellular and molecular mechanisms of nociception, 222
Central changes induced by noxious peripheral input, 8
Central excitability, 6
Central excitatory hypersensitivity, 230, 240

Central hypersensitivity in pain states, 226
Central inhibition, 9
Central pain states, 3
Central pain system, 243
Central sensitization, 116
Central sensitization, 6–7, 151–152, 193
    Increased activity in primary afferents by, 157
    Interaction between receptors, 163
Centrally acting analgesics, 137
Cerebellum, 250
Cerebral blood flow studies, 73
CGRP, 86, 89, 100, 116, 118–120, 172–173, 229, 236–237, 245, 250
    After skin UV-B irradiation, 239
    Co-administered with SP potentiated the effect of SP, 115
    Content in intraspinal primary afferent terminals, 239
    Facilitatory effects on spiral nociceptive processing, 172
    In dorsal root ganglia, upregulation of in inflammation, 173
    In perivascular nociceptive trigeminal fibres, 238
    In presynaptic terminals, 237
    Increase after inflammation, 131
    Increase in level, 237
    Inhibit the degradation of SP, 115
    Is involved in neurogenic inflammation of the joint, 172
    mRNA in sensory neurones, 88
    Synthesis, 88
    Triggers acetylcholine receptor, 246
CGRP-ir fibres, 128
CGRP-ir units, around arterioles or venules, 128
Chaos research, 221
ChAT activity, increased by cold stress, 42
ChAT in the hippocampal formation, 42
Chelerythrine, 179
    On hyperalgesia, 183
Chemical coupling between sympathetic neurons and afferent neurons, 105
Chemosensitive C fibres, 105
Chemosensitivity in regenerating nerve fibres, 235
Cholecystokinin (CCK), 117, 228
    On the flexor reflex, 117
    Reduces the ability of morphine to produce analgesia, 227
Cholinergic nerve terminals, 42
Cholinergic neurons, 68
Chronaxy, 217
Chronic inflammation, 11, 85, 167
Chronic inflammatory pain, 167
Chronic nerve constriction, 105–106, 110
Chronic pain, 89, 227
    Balances between excitatory and inhibitory events, 225
    Caused by inflammation and neuropathy, 225
    Mechanisms and hypotheses, 3
    Nociceptor discharge, 207
    Of acquired transcription failures, 255
Chronic pain states, 207
Chronicity of clinical pain, 233
Chronicity of neuropathic pain, 230
Ciliary neurotrophic factor (CNTF), 87
Cingulate cortex, 69, 74

Cingulum, 66
Clonidine, 70, 108–109
CNQX, 154, 170
Co-release of EAAs and SP in the dorsal horn, 154
Coactivation of spinal AMPA and metabotropic glutamate receptors produce an acute mechanical hyperalgesia, 184
Coefficient of dispersion in burst-like neuronal discharges, 211
Coexistence between galanin and VIP in dorsal root ganglion cells following axotomy, 120
Cognitions and emotions function interdependently, 64
Cognitive-evaluative component associated with nocifensive behavior, 28
Collagenase release, 237
Colocalisation
    Between c-Jun and GAP-43, 245
    c-Jun with CGRP, 246
    Of NOS and AP-1 proteins, 247
    Of transcription factor and target gene, 245
Colocalised peptides have a functional interaction, 120
Complete Freund's adjuvant (CFA), 200
Computer models of neuronal interactions, 221
Computer simulation of neuronal network, 215–216
Computerized image processing systems, 47
Conditioning of negative emotional associations, 67
Consensus sequences for phosphorylation by protein kinase C, 193
Consensus sequences for phosphorylation, 195
Contralateral excitatory inputs, 168
Cooperative neurotransmitter action central nervous sensitization, 151
Cooperativity between receptor subtypes on spinal neurons, 187
Coping reactions, 43
Corneal injury, 6
Corticosteroids, 71
Corticosterone, 40
Corticotrophin releasing hormone, 71
Corticotrophin-releasing factor to spinal cord produced a release of dynorphin, 142
Coupling to G proteins may depend on low pH, 102
COX (indomethacin) and LOX (baicalein), inhibitors of, 183
Cox cyclooxygenase, 178
COX enzyme, 11
COX inhibitors, 12
COX inhibitor indomethacin, 180
COX products, 189
COX-1 enzyme, 11
COX-2, 11
COX-3, 12
CP 96 345, a non-peptide antagonist of the NK_1 receptor, 115, 154, 156
CRE, 245
CRE or AP-1 response elements, binding sites for c-jun, 245
CREB, 243
Cross-correlograms of neuronal discharges, 208, 214–216
    Synchronous discharges, 215
Cumulative index of pain scores, 38
Cutaneous dysesthesias, 105
Cutaneous hyperalgesia, 106

Cutaneous mechanical hyperalgesia, 12
Cutaneous receptive fields, 215
Cutaneous secondary hyperalgesia, 8
Cyclooxygenase (COX), 178
Cyclooxygenase enzyme, 11
Cyclooxygenase products, 177
Cyclosporin A, 97
Cytokine production from monocytes, 86
Cytokines, 40, 86, 89
    And REM sleep, 42
    And sleep, 42
    In the opioid release, 97
    Inflammation, 40
    Participate in the modulation of animal behaviour, 40
Cytoskeleton proteins, 245

D_2 correlation dimension, 211, 214
    Discharges, 209
    Background activity, 209
Damage to peripheral nerves, 7
Deafferentated neurons
    Dynorphin, 247
    Enkephalin, 247
    Of spinal cord, 246
Deafferentation evokes a new pattern of c-Fos, 247
Deafferentation induced hyperexcitability, 240
Deafferentation, zone of, 240
Deep somatic pain, 125
Defensive threat behaviors, 72
Degrees of freedom of discharges of nociceptive neurons, 209, 221
Degrees of freedom of neuronal discharges, 207
Denervation supersensitivity, 110
2-Deoxyglucose, 48
    Autoradiographic images of pain related activity, 47
    Labelled with a radioactive isotope, 48
2-Deoxyglucose-6-phosphate, 48
Deoxyhemoglobin, 48
Depolarization of DRG neurones by substance P, 90
Descending alpha-2 activity, 228
Descending control from supraspinal centres, 8
Descending control of long-term potentiation, 218
Descending excitation of spinal neurones, 10
Descending inhibition, 10, 251
    Decreased efficiency, 240
    From the PAG, 241
    Lifting of, 10
    Of dorsal horn neurons, 240
Descending modulatory effects operating on spinal dorsal horn neurons, 22
Desensitization at the genetic and transcriptional levels, 251
Desynchronizations, 215, 221
Desynchronized discharges, 209, 215
Deterministic pattern of neuroanl discharges, 209, 211, 214
    Phase-space portrait, 209
Development and maintenance of heat hyperalgesia, 159
Development and maintenance of persistent pain, 10
Development of hyperalgesia following tissue damage, 25

Development of spinal hyperexcitability, 170
2-DG uptake, 50
Diabetes induced by streptozotocin, 88
Diacylglycerol (DAG) binding site on PKC, 200
Diclofenac, 12
Dietary supplementation of pain patients with L-tryptophan, 241
Differential sensitivity to endogenous opiates, 39
Dimers containing c-Jun, JunB, and c-Fos, 249
Diprenorphine in pain, 75
Dis-inhibition, 10
Discharge pattern, 209–210, 217, 221
    Random or highly complex, 209
Discharge properties of spinal neurons in chronic inflammation, 167
Discharge rates, 209, 221
    Nociceptive spinal dorsal horn neurons, 219
    Of nociceptive neurons, 207
Discharges of nociceptive neurons, 208
Diseases due to transcription deficiencies, 255
Disinhibition of spinal mechanisms cause pain, 239
DMSO, 180, 182
DNA binding affinity and transcriptional activity, 249
DNIC, 9
Dominant animal, 42
Dopamine, 139
Dopaminergic pathways from substantia nigra, 68
Dorsal columns, 50
Dorsal horn, 211
Dorsal horn excitability increase via disinhibition, 26
Dorsal horn neuronal responses to graded noxious heating, 24
Dorsal horn neurons, 6, 236, 240
    And withdrawal reflexes, 28
    Death of, 26
    Denervated, 9
    Enhancing of the excitability of, 90
    Impulse activity of, 128
    Response to noxious heating of hind paw and tail, 22
    Sensitized by exposure to an EAA and SP, 154
Dorsal horn nociceptive neurone, 229
Dorsal horn unit responses to noxious heat, 22
Dorsal horn, release of neurotransmitters, 151
Dorsal noradrenergic bundle, 67–68, 70
Dorsal rhizotomy, 9–10, 89, 242
Dorsal root ganglion, 236
    Abnormal nerve impulses in chemosensitive C fibres after nerve injury, 110
    In vitro, 109
    Neurons in culture, 141
Dorsal root ganglion cells, 89, 237
    With unmyelinated axons, 105
Dorsal root ganglion neurones express NPY after injury, 88
Dorsal root reflexes, 161
    In C-fibres, 161
Dorsal root sections, 240
Dorsal roots sectioning prevented the swelling and thermographic changes, 161
Dorso-lateral funiculus (DLF), 50
Dorso-ventral concentration gradients, 211

DRG cells, dissociated, 106
Dyn A, 254
Dynorphin A, 142, 253
    In the ventral horn, 143
    Release in the spinal cord, 142
Dynorphin A, basal and evoked release in the spinal cord
    Role in antinociception, 143
    Influence on motor behaviour, 143
Dynorphin, 25, 96, 137, 142, 173
Dynorphin B, 143
Dynorphin induced spinal cord necrosis, 144
Dynorphin synthesis upregulated under inflammatory conditions, 171–172
Dynorphins are endogenous ligands at κ opioid receptors, 142
Dynorphins in the spinal cord in inflammation, 144
Dysfunction of serotoninergic inhibitory systems, 241
Dysfunctional regeneration of nerve fibres, 245

EAA and SP release from the terminals of nociceptors in the dorsal horn, 157
EAA concentration, 160
EAA receptor subtypes involved in hyperalgesia, 185
Early gene expression, c-fos, c-jun, 89
Ectopic discharge, 8
Ectopic electrogenesis, 105
Ectopic generators at sites in injured primary afferent neurons proximal to a neuroma, 106
Ectopic spontaneous activity, 109
Eicosanoid inhibitor nordihydroguaiarate, 179
Electrical stimulation
    Associated with analgesia, 19
    In midbrain, 19
    Of A- and C-fibres in afferent nerves, 208
    Of dorsal roots, 219
    Of the sympathetic trunk, 107
EMG, 18
Emotion determines cognition, 68
Emotion, a crude intelligence, 67
Emotion, its functions and its expressions, 64
Emotional arousal, 72
Emotional dimension of pain, 64, 71
Emotional evaluation of cognitive events, 67
Emotional tone of sensory information, 68
Emotional well being of the patient a high priority goal of pain control, 78
Encoding of nociceptive information by multireceptive neurons, 221
Endogenous inhibitory systems, 233
Endogenous opioid antagonist, 120
Endogenous opioid peptides, 95
Endogenous opioid system, 241
β-Endorphin, 117
Enhanced pain sensations, 5
Enhanced perception of painful stimuli, 10
Enkephalin, 96–97, 143
    In immune cells infiltrating the inflamed tissue, 96
Enkephalinase inhibitors, 97
Epidural, 137

Ergot alkaloids, 87
Event-related potentials (ERP), 49
Excitability of neurons, enhanced, 154
Excitability, trigger for persistent increases, 26
Excitatory amino acids, 152
Excitatory amino acid glutamate, 225
Excitatory amino acid receptors in chronic inflammation, 169
Excitatory neurotransmission, 208
Excitotoxicity
    From glutamate release, 26
Exploratory behaviours, 38
Expression in dorsal horn
    c-fos, 247
    c-jun, 247
    JunB, 247
    Krox-24, 247
Expression of c-Fos or jun B, 220
Expression of glutamate receptor subunit genes following persistent pain, 203
Expression of glutamate receptor subunits, 197
Expression of IEG-encoded proteins, 249
Expression of immediate-early genes, 219–220
Expression of induced transcription factors, 252
Expression of substance P and NADPH diaphorase in dorsal horn, after nerve transection, 248
Extrasynaptic actions of SP, 128
Extrasynaptic mediators of long-term changes in spinal nociception, 209
Extrasynaptic neuromediators, 208, 222
Extrasynaptic neuropeptides, 210–211, 216, 219
Extrasynaptic spread of substance P, mismatch of release sites and binding sites for substance P, 209
Extrasynaptic substance P, 215–216, 219
Extrasynaptic transmission, 216
Extravasation of plasma, 85

Facilitation of the flexor reflex, 114, 119
    Reduced by blockade of the NMDA receptor, 116
Facilitation of the flexor reflex, 115
Fast synaptic transmission, 222
Fear conditioning, 66–67
Fentanyl, 95
Fibroblasts, 87, 236
Fibromyalgia, 75
    Altered central nervous system sensitivity in, 76
Fibromyalgia patients, 238
Fibromyalgia patients have elevated Substance P in the cerebrospinal fluid, 133
Flare response, 151
Flare, 8
Flare, hyperalgesia and pain, 9
Flat, 215
Flexion withdrawal reflex, 226
Flexor EMG recordings, 18
Flexor reflex, 117
    Evoked by noxious stimulation is correlated to painful sensations in humans, 113
    Facilitation of, 115

fMRI, 48
Formalin injection, 36
Formalin injection (s.c.), time course of nociceptive responses, 36
Formalin injection into skin, 49
Formalin test, 12, 26, 95, 218
    Central β-endorphin, 39
    Effects of different formalin concentrations, 35
    Exploratory activity, 37
    Familiarization, 38
    Food-hoarding, 37–38
    Hippocampal ChAT activity, 42–43
    Hormonal modifications, 40
    In rats, 33
    Novelty, 38
    Pain avoidance, 37
    Restricted diet, 38
    Self-grooming, 38
    Stress response, 41–42
Fos B, 220, 243–244, 247, 249–250
Fos, 243
Fra-2, 253
Freund's adjuvant, 95–96, 100, 139, 143, 167, 170, 172
Frontal lobotomies, 65
Functional connectivity of spinal neurons, 211
Functional heterogeneity of nociceptive afferents, 91
Functional imaging mapping techniques, 2-DG, PET, fMRI, 49
Functional imaging studies of the pain system, 49
Functional magnetic resonance imaging (fMRI), 48–49
Functional organization of the pain system, 48
Functional plasticity, 235, 252

G-protein, 90
G-protein coupled metabotropic receptors (mGluR), 193
GABA, 227–228, 241
GABA in spinal cord after nerve damage, 227
$GABA_A$ receptor antagonist (bicuculline), 159–161
Galanin, 86, 88, 115–116, 118, 120, 137–140, 143, 236, 245–246
    Analgesics for treatment of neuropathic pain, 120
    And dynorphin A release in the spinal cord, 137
    Antagonist M-35, 116–117
    Antagonist of excitatory neuropeptides, 116
    Binding, suppression of, 139
    Dorsal root ganglion neurones, 138
    Enhanced inhibitory role of following nerve section, 120
    Functional role in peripheral inflammation, 141
    Functions in the spinal cord, 138
    Inflammation, 139
    Inhibitory role, 119
    Inhibits the excitation of deafferented spinal neurons, 246
    Inhibits the facilitation of the flexor reflex, 116
    Inhibits the release of acetylcholine, 138
    Interaction with SP and CGRP, 116
    Intrathecal, depressant actions of, 138
    Intrinsic neurones of the spinal cord, 138
    Potentiates the analgesic effect of morphine, 117
    Produces hyperalgesia, 138
    Rate of degradation of, 140
    Reduces transmitter release, 141

Release in the spinal cord, 139
Silence afferents during regeneration, 141
Stores of releasable, 141
Superficial dorsal horn, 140
Upregulated in dorsal root ganglion cells following nerve section, 119
Upregulated in dorsal root ganglion neurons, 246
Galaninergic control of nociceptive input to the spinal cord is enhanced after nerve injury, 120
GAP-43 growth associated protein, 245
GAP-43/B-50, 131
Gating kinetics, 194
Gene expression
    And chronic diseases, 251
    And nociception, 222
    Of neuropeptides, 236
    Reactive, 252
Gene induction, 26, 229
Gene promoters, Calgcat, 87
Gene transcription, 87
GENESIS script language, 216
Glial cells, 91
    Release of inflammatory mediators, 91
GLU receptor agonists, 152
Glucose, 48
Glucose-6-phosphate, 48
GluR-A, 196, 197
GluR-A flip and flop, 196
GluR-C gene, 198
Glutamate, 90, 152, 154, 229
    Activated receptor channels, 193
    And SP, co-released upon C-fibre stimulation, 116
    Coexists with peptides, Substance P and CGRP, 225
    In dorsal root ganglion cells, 113
    In spinal and thalamic transmission of nociception, 225
    In the spinal cord, 113
    Receptor subunit mRNA, 193, 195, 197, 200
    Receptors for, 113
    Receptors, regulating the sensitivity of by protein phosphorylation, 203
    Release from glial cells, 91
Glutamatergic neurotransmission, 193
Glycine, 227
Glycine modulatory site, 12
GR82334, 154, 156
Grafting of peripheral nerve fibres, 245
Growth arrest, 253
Growth factors, 86–87
Growth-associated protein GAP-43, 109, 128
Guanylate cyclase (GC-S), 179

Habituation, 20
Harmonic oscillation in neuronal discharges, 211
Heat hyperalgesia, 157, 158
Heat sensitive nociceptors, 234
Heat stimulus, 8, 9
Heating of tail skin, 22
Hemidiaphragm-phrenic nerve preparation in vitro, 126

Hemoglobin, 188
Heterooligomeric assemblies of glutamate receptor subunits, 193
Hexokinase-catalyzed phosphorylation, 48
Hind limb withdrawal by intrathecal NMDA, facilitation of, 27
Hippocampal ChAT activity following injection of formalin, 42, 33
Histamine, 85, 88
Hormones released in response to average stimuli, 40
Hot-plate and tail-flick latencies, 220
Hot-plate and tail-flick tests, 138
HTM nociceptive units, 126
HTM receptors in vitro, 127
HTM receptors, 126
Human brain, functional activity levels of, 48
5-Hydroxytryptamine, 85
Hyperalgesia, 4, 5, 7, 12, 17, 86, 88–89, 95, 110, 114, 133, 144, 151–152, 167, 169, 177, 184, 188, 226, 229, 233–234, 236–237, 246–247
    After a burn injury, 6
    Associated with nerve injuries, 88
    By coactivation of AMPA and metabotropic glutamate receptors, 188
    Cellular mechanisms of, 25
    Evoked by local inflammation, 12
    Following intradermal injection of capsaicin, 26
    Following nerve damage, 25
    Following nerve growth factor treatment, 186
    Following ultraviolet irradiation, 236
    In animal models, 24
    In neuropathic pain models, 24
    In patients, 133
    In the chronic pain model, 17
    Increased pain evoked by a noxious stimulus, 24
    Induced by inflammation, 12
    Induced by release of prostanoids from sympathetic fibres, 90
    Mechanical stimulation of the skin, 105
    Mechanical, 177, 180, 182–183
    Mechanisms, 26, 185
    Persistent mechanical, 179, 185
    Pharmacological mechanisms of, 27
    Produced by injury, 7
    Punctate, 9
    Responses abolished by NMDA antagonists, 226
    Secondary and primary, 110
    Secondary mechanical, 151
    Secondary, 5, 6, 8, 25, 154
    Thermal, 177, 187, 189
    To heat stimuli during the formalin test, 34
    Unaffected by NMDA receptor antagonist, 186
Hyperexcitability, 26, 109, 118, 169
    During inflammation, 168
    Of dorsal horn neurons, 250
    Of partially deafferented neurons, 240
    Represents plasticity of the nociceptive system, 169
Hyperpathia, 10
Hyperreflexia, 28
Hyperresponsiveness following intradermal capsaicin, 26
Hypersensitive reflex, 228

Hypersensitivity to von Frey filaments, 152
Hypersensitivity, reduction due to local anaesthetics, 186
Hypervigilance, 71
Hypothalamic response to tissue injury, 72
Hypothalamic-pituitary system, 43
Hypothalamo-pituitary-adrenocortical axis (HPA), noxious stimulation, 71
Hypothalamus, 69, 72
Hypothalamus, direct stimulation of, 72
Hypothalamus, membrane preparations, 140

Ibuprofen, 12
IEG
   In the contralateral dorsal horn, 250
   Modality-specific expression, 250
IEG basal expression, 244
IEG encoded proteins, 244
IEG expression
   Dissociation of electrophysiological activity and, 251
   Following chronic stimulation, 250
   Following noxious stimulation, 243, 247
   Following peripheral stimulation, 244
   Following stimulation of C-fibres, 250
   In juvenile rats, 251
   Neuronal program, 250
   Pain behaviour, 250
   Protein synthesis, 250
   Stimulus duration, 250
IEG in
   Amygdala, 250
   Dorsal raphe, 250
   Habenular nucleus, 250
   Hypothalamus, 250
   Limbic system and hypothalamic areas, 243
   Locus coerulus, 250
   Parabrachial nucleus, 250
   Periaqueductal grey, 250
   Spinal cord, 247
   Thalamus, 250
IEG-encoded proteins in the brain, 249
IEGs in spinal neurons, 246
IEGs, temporal and spatial patterns in spinal cord following noxious stimulation, 249
IEGs, transynaptic expression of following noxious stimulation, 247
IEGs, visualize neuronal plasticity, 250
Image analysis of microprobes, 141, 137
Imaging of pain processes, 63
Imaging studies of the pain system, 47
Immediate-early genes, 163, 219, 220, 222, 242
Immediate-early gene expression, 219
Immune and nerve cell interactions, 86
Immune cells, 86, 97
Immune response, 236
Immune response to antigen challenge, 86
Immuno-suppressive procedures abolish stress-induced hyperalgesia, 102
Immunogenic inflammation, 237

In situ hybridization, 200
Increase in the responsiveness of a primate spinothalamic tract neuron, 154–155
Indicator for pathological (inflammatory) conditions, 209
Indicators for spinal nociception, 207
Indomethacin on hyperalgesia, 185
Induced gene transcription, 243
Induced transcription factors (ITF), 252
Induction of the c-fos gene, 247
Inflamed skeletal muscle, 127
Inflammation, 40, 95–96, 100, 173, 202, 215–216, 218–219, 221, 227–229, 234
   Acute, 167
   Chronic, synthesis of peptidergic transmitters is upregulated, 173
   Cytokines, illness and pain, relationship between, 42
   Enhances axonal flow of opioid receptors, 101
   In a joint, 237
   Increases activity of inhibitory systems, 227
   Induced binding of [^{3}H]PDBU in dorsal horn, 202
   Inducing a peripheral opioid site, 228
   Leads to increase in descending inhibition, 169
   Spinal levels of dynorphins are greatly increased, 142
Inflammation of skin, 209, 214
   At cutaneous receptive fields, 208
Inflammation of the knee joint, 9
Inflammation of the skin, 220
   Acute skin heating, 215
Inflammation related cells, 236
Inflammation-evoked hyperexcitability of spinal neurons, 169
Inflammation-induced changes in innervation density in SP-ir fibres, 131
Inflammation-induced hyperalgesia, 101
Inflammatory cells, recruitment of, 86
Inflammatory lesions, 171
Inflammatory pain and central sensitization, 162
Inflammatory pain promotes physiological inhibitory controls, 230
Inflammatory process, 7
Information transfer during skin inflammations, 221
Inhibitory actions on interneurons in the spinal reflex, 20
Inhibitory controls, 227
   Compensatory increases in, 228
   Decrease following neuropathic lesions, 239
Inhibitory influences more active in polyarthritic rats, 169
Inhibitory kappa opioid ligand(s), 171
Inhibitory neurons, loss of, 25
Inhibitory pathways originating in the brainstem, 114
Inhibitory systems in chronic pain, 227
Inhibitory systems related to monoaminergic transmitters, 241
Injury and inflammation, effect on peptidergic fibre functions, 87
Injury and pain, 3
Injury discharges by nerve damage, 226
Injury to a peripheral cutaneous nerve, 105
Injury-induced pain in humans, 49
Injury-induced sensitization of spinal nociceptive neurons, 26
Insula, 75
Insular cortex, 75

Integration of spinal and supraspinal centres, 34
Interleukin 1-β, 87
Interleukin-6, 97
Internal drives and the need to react to nociceptive stimulation, 33
Interspike interval distributions, 210–211, 221
Intraneural electrical stimulation, 26
Intrathecal administration, 137
    Of morphine, 242
    Of NMDA, 180
    Of quisqualate, 180
Intrathecal injections of NMDA, 219
Intrathecal neurokinin A, 115
Intrathecal NK1 receptor antagonists, 13
Intrathecal NMDA, 26
Intrathecal NMDA antagonists, 27
Intrathecal SP, 128
Intrathecal SP associated with a change in responsiveness of dorsal horn neurones, 132
Intrathecal stimulations, 219
Ionophoretically administered AMPA and NMDA, 157, 170
Ionotropic glutamate receptor, 194–195
    Structure, 195
Ionotropic glutamate receptor subunit mRNA, 194, 200
Iontophoresis, 151
ITF, 253–254

Joint afferent fibre, 157
Joint inflammation, 158, 237
Joint inflammation, acute, 157
Joint nociceptors, 9
Jun kinase, 243
Jun, 243
jun-B, 219
junB, 243–244, 249–250
JunB:c-Fos dimers, 249
junD, 243–244, 247, 249

KA-2 mRNA abundant in dorsal horn, 198
Kainate, 196
Kainate receptor transcripts, 198
Kainate receptors, 194
Kainic acid, 153
Kappa antagonist nor-binaltorphamine on spinal neuron, 171
Kappa opioid agonist, 171
Kappa opioid system, 173
Kappa receptor agonist, 142
Kappa receptors, 171
Keratinocytes, 87, 89
Ketamine, 157
    Effective in prolonged clinical pains, 227
Kinase activation, 247, 252
Krox-20, 244
krox-24, 243–244

L-NAME, 179, 182, 188
Lamina II neurons, 216
Laminae I and II of the dorsal horn, 196, 212
Latency of nocifensive responses, 24

Lateral reticular nucleus, 72
LC excitation a consistent response to nociception, 69
LC inactivation during REM sleep, 70
Learning, 254
Lenticular nucleus, 74
Lentiform nucleus, 75
Lesions of the spinal cord, 240
Leukaemia inhibitory factor (LIF), 89
Leukocytes, 88
Leukotriene C4, 88
Ligation of the sciatic nerve, 105
Limb flexion reflex, interneuronal connections, 22
Limb withdrawal
    Flexor EMG, 18
    Isometric force, 25
    Magnitude correlated with stimulus, 18, 21
    Midbrain suppression of, 21
Limbic brain, 66, 68
Limbic cortex, 75
Limbic forebrain, 72
Limbic function involves frontal and temporal cortex, 68
Limbic processes, 63
Limbic structures and functions, 66
Limbic structures in the distress of pain, 65
Limbic system, 65
Lipoxygenase (LOX), 178
Local anaesthesia of ectopic focus, 8
Local anaesthetics, 230
    In chronic neuropathic pain, 186
Local application of opioids, 99
Local glucose utilization rates, 47
Locomotion in the open field, 35
Locus coeruleus (LC), 68–69, 241
    And the dorsal noradrenergic bundle, 68
    Regulates attentional processes, 70
Long-lasting changes in nociception, 220
Long-term change in nervous system function, gene expression in nerve cells, 242
Long-term changes of nociception, 219
Long-term enhancement or decrease of EPSP amplitude, 202
Long-term plasticity, 243
Long-term plasticity in the pain system, 244
Long-term potentiation (LTP), 6, 188, 217–218
    Of C-fibre-evoked spinal field potentials, 218
    Of synaptic transmission, 217, 252
    Of synaptic transmission in spinal nociception, 221
Low threshold mechano-receptive fields, 208
Low threshold mechanoreceptors, 5
LOX inhibitor baicalein, 180
LOX, 183
Lymphocytes, 97

Macrophages, 97, 236
Magnetic resonance imaging (MRI), 48
    Of the limbic brain structures, pain experience, 74
Magnetoencephalography (MEG), 49
Malignant transformation, 253
Mammalian genome, 252

Mast cells, 88, 236
Mast cell degranulation, 85, 237
Medial forebrain bundle, 71
Medial pain system
  Involves emotional components, 73
  Pain chronicity, 73
Medial thalamus, 75
Mediator of enhanced excitability of neurons, 203
Mediators of inflammation, 85
Medullary reticular formation, 71
Membrane properties and membrane conductances, 221
Memory formation, 252
Memory research, 254
MEN 10207, a specific antagonist of the NK_2 receptor, 115
Mepacrine on hyperalgesia, 184
Mepacrine, 179, 188
Messenger RNA, 241
Met-enkephalin, 171
Metabolic changes, spatial map of CNS, 49
Metabolic patterns in the spinal cord related to pain, 49
Metabotropic EAA receptor agonist, trans-ACPD, 162
Metabotropic EAA receptors play a role in central sensitization, 162
Metabotropic glutamate receptors, 189, 203
Metabotropic receptor antagonist 2-amino-3-phosphonoproprionate, AP3, 179
Methylene blue, 179, 182, 188
N-Methyl-D-aspartate, see NMDA
N-Methyl-D-aspartate receptor, see NMDA receptors, 113, 225
Mg^{2+} block, 26
Mg^{2+} ion block of NMDA receptors is in part responsible for the central sensitization, 202
Microdialysis, 151, 159, 162
Microneurographic recordings, 151
Microprobe autoradiographs, 137
Microprobes, 140, 142–143
Migraine, 87, 238
  Etiology of, 87
Mitogenic effect of SP, 132
MK801, 226
Modelling single neurons and networks, 221
Models of allodynia, 226
Models of chronic pain, 49
Models of pain and hyperalgesia, 106
Models of pain processing, 4
Models of painful peripheral neuropathy, 13
Modulation at synaptic relays, 3
Modulation by analgesic brain stimulation, 24
Modulation by endogenous antinociceptive systems, 49
Modulation of pain, 22
Molecular cloning, 193
Molecular cloning studies have identified seven isoforms of PKC, 203
Molecular mechanisms
  Long term changes of spinal nociception, 207
  Spinal nociception, 219
Monoarthritis, 95, 194, 200
Monocytes, 97

Morphine, 18–20, 22, 95–96, 101, 117, 226, 242
  Effectiveness, 227
  Injected into the knee joint, 99
  Tolerance, 117
  Enhancement of the spinal effects of, 228
Morphological alterations in the dorsal horn, 151
Motivation, 17
Motivation to explore and react to the painful stimulus
Motor neurons, 200
mRNA, 200
mRNA expression of NMDA receptor subunits, 194
mRNA for prodynorphin, 142
mRNA splicing, 196
Multi-neuron recordings, 211
Multireceptive neurons, 211, 208, 214–215, 221
  Excitability of, 216
  In the superficial dorsal horn, 216
Multireceptive spinal dorsal horn neurons, 211
Muscle nociceptors, 131
Mustard oil, 6
Myositis, 128
Myositis by Freund's complete adjuvant, 127

NADPH-diaphorase, 246
NADPH diaphorase and galanin, after transection of nerve, 246
NADPH-diaphorase-labelled neurons, resistant against insults, 246
Naloxone, 18–20, 96
  Intraarticular injection in patients, 99
  On responses evoked by formalin pain, 39
Naltrexone, 96
Natural skin stimuli, 219
NBQX, a selective AMPA receptor antagonist, 225
NDGA, 183
NDGA on hyperalgesia, 184
NE can act directly on sensory neurons with injured C fibres, 110
NE responsiveness in C-fibres, 108
NE, shortage of, 110
Neomycin, 188
Neomycin on hyperalgesia, 183
Neonatal dorsal root ganglion neurons, 252
Neovascularization and proliferation of endothelial cells, 132
Nerve blocks, 152
Nerve constriction, 106
Nerve cut resulted in an up-regulation of alpha-2 receptors in cell bodies of neurons, 110
Nerve damage, inhibitory controls are lost, 230
Nerve growth factor (NGF), 128
Nerve injury by compression, 25
Nerve injury, 87, 88, 218–219
  Functional changes in primary sensory neurons after, 105
  Neurons develop ongoing discharge, 105
  Partial, 110
Nerve lesions, 88
Nerve lesions behavioral hyperalgesia, 241
Nerve ligation, 25, 88
Nerve sprouting, 87

Nerve transection, 89–90, 117
Nerve trauma, 236
Nervous activity alters gene transcription, 254
Nervous system plasticity, 233
Neurochemical changes occur in the dorsal horn, 159
Neurocircuity mediating a nocifensive behavior, 20
Neuroendocrine factors determining pain behaviour in animals, 33
Neurogenic component of chronic polyarthritis, 237
Neurogenic effects of the neurokinins, 91
Neurogenic inflammation, 86, 125, 161, 236–237
    Facilitate chronic inflammation, 237
    Sustained, 237
Neurogenic mechanisms in chronic pain, 85
Neurogenic (secondary) hyperalgesia, 105
Neurokinins, 86, 89–90
Neurokinin 1 or neurokinin 2 receptors, long term potentiation of, 217
Neurokinin 1 receptors, 157, 217
Neurokinin 2 receptors, 157, 217
Neurokinin A (NK-A), 85, 90, 157, 209, 211, 217–219, 229
    Tissue concentration in lumbar spinal cord, 211
Neurokinin precursor peptides, 87
Neurokinin receptor, 26
Neurokinin receptors, NK1, NK2, NK3, 87
Neurokinin receptors in the spinal cord involved in the induction and maintenance of heat hyperalgesia, 159
Neurokinins in plasma extravasation in the dura, 87
Neurological diseases, 7
Neuroma, 235
Neuroma C-fibres, 235
Neuromediators, 216
Neuromuscular transmission, re-establishment of, 246
Neuronal excitability, 25
Neuronal network, 217
Neuronal plasticity, 233
    In the dorsal horn, 177
Neuronal regeneration, 86
Neuropathic damage, 229
    Reduces opioid receptor numbers, 227
Neuropathic lesions, 240
Neuropathic pain, 7–8, 11, 24, 77, 89, 113, 227
    Functional plasticity of neurotransmitters, 113
    Patient, hypersensitivity in, abolished by NMDA receptor antagonist, 186
    Patients, 8
    Prevented by GMI gangliosides, 26
    Treatment, 121
Neuropathic states, 230
    Loss of inhibitions in, 227
Neuropathy, 10, 89, 227, 229, 242
Neuropeptide and receptor expression, changes following nerve section in dorsal root ganglion cells, 119
Neuropeptide containing afferent neurons, 85
Neuropeptide nerve function induced by nerve lesions, 88
Neuropeptide release, 25
    Pattern of, 91
Neuropeptide synthesis, 85
Neuropeptide Y (NPY), 85, 86, 88, 139

Neuropeptide-containing afferents, pathophysiological role, 87
Neuropeptide-mediated cell signalling, 85
Neuropeptides, 113, 125, 209, 211, 216, 218–219
    After deafferentation, 113
    And glutamate on spinal nociceptive mechanisms, 114
    And the orchestration of inflammation, 85–86
    Following nerve injury, 119
    Regulate lymphoid tissues, 86
    Released from primary afferents, 85
    Synergistic interaction, 115
    Transport rate, 131
Neuroplastic changes, 133, 207
Neuroplastic changes underlying pain, 246
Neuroplasticity, 233
Neuroplasticity after noxious stimuli, 254
Neurotransmitters, 207
Neurotrophin-3 (NT3), 87
Neurotrophins affect sensory neurones, 88
Neurotrophins, 87, 89
N^G-Nitro-L-arginine methyl ester, L-NAME, 179, 182, 188
NGF, 86–88
    Expression in the inflamed muscle, 131
    Increase tissue sensitivity and responsiveness to noxious stimuli, 88
    Infusions, 88
    Mediator of hyperalgesia, 88
NGF, increased, to induce sprouting of nerve fibres, 131
NGF-induced hyperalgesia, 88
NGF-induced sprouting, 87
Nitric oxide (NO), 177, 229
Nitric oxide synthase (NOS), 236, 245
Nitric oxide synthase expression following axotomy, 246
NK receptor antagonists, 90
NK-1, NK-3 and glutamate receptors
    On spinal neurons, in arthritis, 250
NK_1 and NK_2 receptors for SP and NKA, 115
NK1 antagonists, 156
    Producing analgesia in chronic pain, 91
NK_1 receptor antagonist CP-96,345, 116
NK1 receptor antagonist, CP99994-1
    Spinal cord microdialysis, 159
NK1 receptor antagonists in Phase 3 pain, 13
NK1 receptor antagonists, 12, 87
NK1 receptors, 12, 154, 156
    Expression of, 12
    In Phase 2 pain, 12
    On primary afferents, 90
NK1 tachykinin receptor, 12
NK1/NMDA receptor interactions
    On wide dynamic range dorsal horn cells, 90
NK2 receptors, 154
    After inflammation, 87
NK2 receptor antagonist, SR48968, 159
NK2 receptor antagonists, 87, 156
NK-A, 86–87, 115–116
NK-B, 87
NMDA, 152, 154–155, 179–180, 189, 194, 196, 227, 229
    And AMPA receptors, 193

And non-NMDA antagonists led to reduction in the receptive field size, 170
And non-NMDA receptors, 173
And SP, co-application of by iontophoresis, 155
NMDA antagonist, 114, 225, 227
  Prevent or block central hypersensitive states, 228
NMDA antagonist, AP7, 158
NMDA antagonist ketamine, 170
NMDA antagonist MK-801, 116, 118, 202, 242
  On wind-up of the flexor reflex, 118
NMDA antagonists reduce hyperalgesia in models of tonic pain, 202
NMDA glutamate receptor subtype, 12
NMDA mediated central events leading to hypersensitivity, 228
NMDA mediated wind-up, 226, 228
NMDA or QUIS, coadministration with substance P (SP) can result in enhancement of excitability, 154
NMDA receptor, 25, 26, 113, 169, 188, 194, 218, 222, 225, 228
  A role in neuronal plasticity, 202
  And the transmission of pain, 225
  Blockade, 12
  Channel, 226
  Consist of hetero-oligomeric assemblies of receptor subunits, 196
  In hyperalgesia, 186
  Induction of long term potentiation, 217
  Linked to the activation of NOS, 187
  Modulatory sites of, 12
  Of spinal dorsal horn neurons, 34
  Phosphorylation, 203
  Subunits, 198
NMDA receptor antagonist (AP7), 159
NMDA receptor antagonist (D-CPP), 217
NMDA receptor antagonist AP5, 170
NMDA receptor antagonist APV, 179
NMDA receptor antagonist, 12, 157, 242
  Attenuate hyperresponsiveness of dorsal horn neurons, 186
  On C fibre-evoked activity, 202
  Reduced spinal neuron receptive fields, 186
NMDA receptor proteins carry consensus sequences for phosphorylation by PKC, 202
NMDA receptor selective antagonist APV on hyperalgesia, 181
NMDA, AMPA and kainate subtypes of the ionotropic glutamate receptors, 202
NMDA-evoked hyperalgesia, 26
NMDA-maintained thermal hyperalgesia, 177
NMDA-mediated process of the second phase of formalin response, 35
NMDAR1 splice variant mRNA revealed by in situ hybridization, 196, 198
NMDAR1 subunit, 197
NMDAR1-1 splice variants mRNA, 197
NMDAR1-a, 196, 198, 200
NO, 187, 189, 222, 246–247
NO inhibitors, 35, 188
NO involved in expression of c-Fos, 247
NO release from vascular endothelium, 85
NO synthase (NOS), 109, 179

Nociception, 207
Nociception and central noradrenergic processing, 68
Nociceptive afferent barrage, 5
Nociceptive C-fibres, 250
Nociceptive ectopic activity, 6
Nociceptive input, 3
Nociceptive neurons, 220
  Nociceptive system, 219
Nociceptive processing in the spinal cord, 167
Nociceptive processing in wide dynamic range neurons with ankle input, 170
Nociceptive sensory function, 236
Nociceptive substances, 236
Nociceptive terminals, 151
Nociceptive thresholds, 219
  In the inflamed paw, 97
  Thermal and mechanical, 219
Nociceptive transmission is under tonic galaninergic inhibitory influence, 117
Nociceptive withdrawal reflexes, 178
Nociceptive, terminal region of, 101
Nociceptors, 234
  Thermal, 20
Nocifensive behavior, 18–19
Nocifensive behavioral response, magnitude of, 20
Nocifensive flexor reflex, 113
Nocifensive response, 17, 24
Nocifensive response threshold, 18
Non-damaging noxious stimuli, 209
Non-linear dynamics of neuronal discharges, 208, 221
  Of multireceptive neurons, 209, 214
Non-NMDA and NMDA receptors, 157
Non-NMDA receptors, 155
Non-NMDA receptor antagonist CNQX, 170
Non-NMDA, NMDA, NK1 and NK2 receptors, 159
Non-steroidal anti-inflammatory drugs, 230
Noradrenaline, 139
Noradrenaline and 5-hydroxytryptamine, 229
Noradrenergic fibres in spinal cord, hypothalamus, thalamus, hippocampus, 68
Noradrenergic pathway, links to negative emotional states, 68
Noradrenergic system, ascending, 71
Noradrenergic transmission in a primate brain, 70
Norepinephrine, 107
  Abnormal responsiveness of secondary neurons to, 105
Norepinephrine responsive C fibres, 107
NOS in cutaneous nerve fibres, 238
NOS inhibitor, L-NAME, 27
NOS metabotropic receptor linked to activation of adenylate cyclase and PLC, 187
NOS, 26, 253
  Coexpressed with c-jun, 246–247
  Increased in axotomized DRG neurons, 247
Novelty of stimuli provoke IEG expression, 251
Noxious heat stimuli, 18
Noxious mechanical skin stimulation, 218
Noxious skin heating, 208, 214–215, 221
  Phase-space portraits, 209

Noxious stimulation, 219, 221
    Activates limbic structures, 73
    Induced NOS expression, 247
    Motivation of animal to avoid, 28
NPY, 89, 118
    And galanin, nerve damage enhances the synthesis of, 141
    In sympathetic fibres, 88
    Reduces substance P release, 141
NSAIDs, 11
Nuclear level processes, 244
Nucleus basalis, 68
Nucleus cuneatus, 250
Nucleus gracilis, 250
Nucleus tractus solitarius, 72

Oct-2 transcription activators, 87
Oligodeoxynucleotides, 196, 220
Oncology, 253
Operantly conditioned response, 18, 20
    Suppressed by midbrain stimulation, 20
Opiate receptor binding after dorsal rhizotomy, 241
Opiate suppression of withdrawal reflexes, 22
Opiate system changes after damage, 11
Opiates, 11
Opiates on neuropathic pain, 242
Opioid action in the presence of inflammation, 99
Opioid analgesia following a nerve lesion, 242
Opioid effectiveness, 229
Opioid functions after nerve lesions, 241
Opioid inhibitory system in deafferentated and arthritic animals, 242
Opioid ligands exert peripheral antinociceptive effects, 100
Opioid peptide release
    From the inflamed synovia, 99
    Under clinical conditions, 102
Opioid peptides, 90
    Hydrophilic, 99
Opioid receptors, 75, 97
    Activation during the development of inflammation, 102
    Gated ion channel, 242
    Types, 11, 95
    Tertiary structure of, 102
    Upregulation of, 101
Opioid sensitivity, compensatory increases in, 228
Opioid suppression of the tail flick reflex, 19
Opioid system
    Activation by formalin s.c., 43
    Ineffectiveness, 242
    Insensitivity, 242
Opioid tolerance, 242
Opioid-induced inhibition of nociceptors, mechanisms involved in, during inflammation, 102
Opioids, 95, 229, 230
    Origin of the endogenous opioids, 96
    Peripheral antinociceptive effect of, 100
    Peripheral antinociceptive effect of, 101

Prevent the release of neuropeptides, 100
Reduced sensitivity to, 228
With high lipid solubility, Fentanyl, 101
Ornithine decarboxylase activity, 109

PAG stimulation, 21–22, 24
Pain and novelty, 38
Pain as an integrated rather than deterministic process, 33
Pain behavior
    Open-field device, 35
    Depends on stimulus strength, 35
Pain chronification, 233
Pain experienced by the animal, 34
Pain induced hypervigilance, 77
Pain is a language for some chronic pain patients, 77
Pain modulation, 233
Pain responses, exaggerated, 10
Pain responses, modulation of, 91
Pain sensation in human pain patients, 5, 12
Pain sensation, increased, 28
Pain, stress and chronicity, 73
Pain-related behaviour, 12
Painful diagnostic or treatment procedures, 67
Painful peripheral neuropathy, 8
Panic, 71
Parabrachial nucleus, 241
Parachlorophenylalanine (PCPA), 241
Paraventricular nucleus of the hypothalamus, 70
Patch-clamp recording
    From DRG cells after a constriction injury, 108
    Of a response to NE in a DRG cell, 109
    Recording techniques, 106
Paw pressure test, 95
Peltier heating thermode, 25
Peptide, 211
Peptide antagonists, 156
Peptide coexistence, 85
Peptide synthesis and release, 90
Peptidergic mediators of nociception, 88
Perceptual decision-making processes, 20
Periaqueductal gray (PAG), 19, 70, 75
Peripheral nerve graft, 245
Peripheral nerve section, 118
Peripheral neuropathy, 3, 6, 12
Peripheral opioid analgesia, 95
    Clinical aspects of, 99
Peripheral opioid receptors, 97
Peripheral sensitization, 5
Persistent pain, 12
PET, 74–75
Pharmacology of pain, 11
Phase 1 pain, 3
Phase 2 pain, 4, 10, 12
Phase 2 pain central mechanisms, 5
Phase 3 pain, 7, 10–11
Phase-space portrait, 209, 214
    Discharges, nonlinear dynamics, 210
    Of neuronal discharges, 211

Phases of pain, 3
Phases of pain-related behavioral changes, 49
Phenotypic changes in sensory neurones, 89
[^{3}H]Phorbol-12,13-dibutyrate ([^{3}H]PDBU), 200
Phorbol esters, 203
Phospholipase A_2, 177–178
Phosphorothioate antisense oligodeoxynucleotide, 253
Phosphorylation at NMDA-receptor, 242
Phosphorylation of ATF-2, 243
Phosphorylation of CREB, 243
PKC, 187
    A cellular link in neuronal plasticity, 193
    Activation of, 25
    Activation phorbol esters, 163
    Distribution of in the spinal dorsal horn, 194
    In the substantia gelatinosa of the spinal cord, 203
    Increases the responses of dorsal horn neurons to excitatory amino acids, 203
PKC inhibitor, chelerythrine, 188
PLA_2, 187
    Activation, 188
PLA_2 inhibitor mepacrine, 182
Plasma ACTH, 33
Plasma protein extravasation, 90, 237
Plastic changes in nociception, 207, 219
Plastic changes of the excitability of CNS, 49
Plastic changes, 215, 221
Plastic features of afferents, 91
Plastic modifications in the central nervous system, 34
Plasticity, 219
    And stability of gene expression, 252–253
    At the genetic level, 251–252
    Controlled at the level of gene induction, 252
    In primary afferent nociceptive neurons, 234
    Of messenger function in primary afferents following nerve injury, 113
    Of neuropeptide expression, 236
    Of neuropeptide function, 118
    Of neuropeptides in primary afferent neurons, 237
    Of nociceptive neurons, 234
    Of spinal glutamate function, 118
    Of the nervous system, 233
    Response of the peripheral nervous system, 237
    Responses in neuropathic pain, 244
    Transynaptic processes, 246
PLC inhibitor neomycin, 179, 182
Pleurisy, 87
Polyamine-sensitive site on the NMDA receptor, 188
Polyarthritis, c-fos expressing spinal neurons in, 251
Polymodal nociceptors responsive to NE, 110
Positive feedback loop between spinal cord and brain stem, 10
Positron emission tomography (PET), 47, 49, 73
Post injury pain, 88
Post-operative pain, 225, 230
Post-synaptic receptors for neurokinins, 90
Post-transcriptional modifications of subunits, 196
Postganglionic sympathetic C-fibres, 236
Potentiation by NMDA agonists, 251
Potentiation by PKC, 194
Potentiation of c-Fos expression by coincident painful events, 251
Potentiation of IEG expression, 252
Potentiation of the analgesic effect of morphine, 120
Prazosin, 107, 108
Pre-emptive analgesia, 6, 144, 230
Presynaptic actions of neurokinins, 90
Primary afferent, 216
Primary afferent C-fibres, 208, 217
Primary hyperalgesia, 5, 151
Primate spinothalamic tract, 151–152, 155
Principal component analysis, 211
Pro-enkephalin (PENK), 97–98
Pro-enkephalin A, 171
Pro-opiomelanocortin (POMC), 97–98
Prodynorphin, 97–98, 142–241
    In spinal cord neurons, 171
    Synthesis of, 143
Proliferation of synoviocytes, 237
Prolonged (tonic) pain, 49
Prolonged activation of C-fibres, 49
Prolonged changes in spinal cord excitability, 114
Prolonged electrical stimulation of Aδ and C fibres, 34
Prolonged facilitation of the flexor reflex, 115
Prolonged neurogenic inflammation, chronicity, 237
Promoter elements of the c-fos and c-jun genes, 243
Promoters of the *galanin* genes, 245
Propriospinal, 207
Prostaglandin release, 237
Prostaglandins, 11, 95, 100
    Synthesis of, upon inflammation, 11
Prostanoids, 229
    Derivatives, 88
    Receptors, 189
    In hyperalgesia, 189
Protective behaviour, 34
Protein kinase C (PKC), 26, 162–163, 179, 193, 200, 243
    Binding of in the spinal dorsal horn, 200
    Translocation of, 177
Protein phosphorylation in central sensitization by noxious stimuli, 203
Protein phosphorylation in secondary hyperalgesia, 193
Protein phosphorylation of ligand-gated ion channels, 193
PRV, 22
PRV-labeled spinal neurons, 22

QA, 180
Quantal effect of morphine, 23
Quaternary opioids with alkaloid structure, 99
Quisqualate, 154, 179
Quisqualic acid, 152

Rapid eye movement (REM) sleep, 70
Ratings of pain sensation, 18
Receptive fields, 217
    Enlargement, 151

Expansion, 26, 144, 169
Expansion depends on the continuous activation of non-NMDA and NMDA receptors, 171
Increase in size, 6
Increase in size during spinal superfusion, 129
Of individual STT cells, 152
Of spinal neurons were larger in inflammation, 168
Receptor mapping studies, 48
Recuperative behaviour, 34
Reflex hyperexcitability, 119
Reflex interneurons, 23
Reflex modulation, 22
Reflex sympathetic dystrophy, 105
Regenerated nociceptors, 234
Regenerated sensory receptors, 8
Regenerating C-fibre, 235
Regenerating nociceptive nerve fibres, 234
Regenerating sensory fibres, chemical responsiveness of, 235
Regenerating sprouts, sensitivity of, 8, 236
Regeneration after injury, 91
Regeneration, 8, 234
    And collateral sprouting, 109
    Prevented by ligation of nerve stump, 244
Regeneration-associated genes, 244–245
Regenerative efforts of damaged neurons, 244
Regional blood flow rates, 47
Regional cerebral metabolism, 47
Regional glucose consumption, 48
Regional glucose metabolism in rat and chronic pain, 49
Renewal process or stochastic noise in neuronal discharges, 209
Responsiveness of dorsal horn neurons in neuropathic rats, 25
Restitution of function of nociceptors, 234
Retrograde transneuronal transport of pseudorabies virus (PRV), 21, 23
Reuptake of 5-HT, 241
Reverberant neuronal circuits, 215
Reverberant pathways in the spinal cord, 221
Rheumatoid arthritis, 87
Rheumatoid arthritis patients, 75

Sacral dorsal horn neurons, 22
Sacral spinal cord, 24
Schwann cells, 87
Sciatic nerve, 217
    Ligated to create a neuropathy, 24
Sciatic nerve section, 119
Second messengers
    Calcium ions, 222
    Cascades, 252
    Protein kinase C, 222
    Systems, 162
Secondary and primary hyperalgesia, 110
Secondary hyperalgesia, 5, 6, 8, 25, 154
    Induced by an intradermal injection of capsaicin, 110
Secondary somatosensory cortex, 75
Self-mutilation, 240
Sensitivity, increase, 234
Sensitization, 5, 151, 233–234

Sensitization by capsaicin, 156
Sensitization is prevented by CNQX, 155
Sensitization of central neurons, 154
Sensitization of chemosensory response, 235
Sensitization of dorsal horn neurons, 157
Sensitization of neurons by arthritis, 251
Sensitization of nociceptors, 4–5, 7, 177
    And hyperalgesia, 100
    By inflammation, 235
    In the inflamed skin, 237
Sensitization of regenerated cutaneous nociceptors, 234
Sensitization of spinal neurons, 237
Sensitization of spinothalamic tract cells, 110
Sensitization of the bradykinin response by adrenaline, 236
Sensitization of the flexor reflex by C-afferents is altered after axotomy, 119
Sensitization of the nociceptive afferents, 10
Sensitization of the nociceptive system due to deficits in inhibition, 239
Sensitization process depends on the release of NO, 163
Sensitization produced by intradermal capsaicin, 162
Sensory aspect of pain, 74
Sensory features of patient complaint can distract from the affective dimension of pain, 77
Sensory function, loss of, 7
Sensory nerve regeneration, 89
Sensory neurones and growth regulators, 89
Sensory neuropeptides, 85, 86, 236
    Involved in the sensitizarion of the flexor reflex, 114
    In the spinal cord, 90
    Provide chemical interface for neuroimmune regulation, 91
    Substance P, 236
Sensory plasticity, 234
Sensory stimulation, 217
Septo-hippocampal neurons driven by noxious stimulation, 42
Septum, 65–66
Serotonergic fibres and raphe nuclei, 68
Serotonin synthesis, 241
Short term potentiation, 6
Signal transduction events in the sensitization of dorsal horn neurons, 162
Signal transduction linked to glutamate receptor activation, 187
Signal-to-noise ratio of neuronal discharges, 207
Signaling capacity increased by nerve sprouting, 90
Silent neuron activation, 157
Silent nociceptors, 5, 7, 235
Single photon emission computerized tomography (SPECT), 47
Skeletal muscle, pathological alterations of, 127
Skin heating, 209, 218
    D_2 correlation dimension, 209
Skin inflammation, non-linear dynamics of neuronal discharges, 209
SOM, 118, 120
Somatosensory cortex, 74–75
Somatosensory system, damage to, 7
Somatostatin (SOM), 86, 90, 116, 118, 120, 138, 209, 218–219, 236
Somatotopic representation, 252

SP, 118–120
   A powerful excitant of receptors, 131
   And afferent fibres in muscle, 125
   And NKA applied on the spinal cord excite dorsal horn interneurons, 114
   And NKA release, 159
   Antagonist, 126
   Concentration used for intrathecal administration, 130
   Fibres in the wall of larger arteries, 128
   In volume transmission, 128
   Interaction with other neuropeptides in primary afferents, 115
   Is colocalized with calcitonin gene-related peptide (CGRP), 115
   Release, 125
   Superfused on the spinal cord, 129
SP, CGRP and galanin are colocalized, 116
SP-induced discharges of primary afferent, 131
SP-induced increase in RF size of dorsal horn neurones, 130, 132
Spantide, 126
Spantide II, antagonist of tachykinin receptors, 114
Species differences in the reaction of nervous tissue to injury, 121
Specificity of pain mechanisms, 3
Specificity of transcription control, 255
Spinal analgesic studies, 211
Spinal control of inflammation, 161
Spinal cord, 12, 20, 200
   Influences the degree of joint inflammation, 161
   Plasticity, 203
Spinal dorsal horn after UV-B irradiation, 237
Spinal dorsal horn neurons, 22, 209, 251
   Excitability, 211, 219
Spinal dorsolateral funiculus, 241
Spinal field potentials, 217
Spinal hyperexcitability after nerve injury, 117
Spinal interaction of SP and glutamate, 116
Spinal monoamine levels, 241
Spinal neuronal network, 208
Spinal neurons remain in a state of hyperexcitability, 169
Spinal NMDA play a role in thermal hyperalgesia, 177
Spinal nociception, 220
Spinal nociceptive mechanisms, 113
Spinal nociceptive neurons, excitability of, 25
Spinal opioid system, responsiveness of, 242
Spinal pharmacology of pain, 229
Spinal plasticity, 193
Spinal prostaglandins are important in hyperalgesia, 189
Spinal reflex circuitry, 28
Spinal reflexes, 6
Spinal release of galanin in inflammation, 139
Spinal release of ir-dynorphin A by compression of the inflamed ankle region, 143
Spinal release of neuropeptides, 137
Spinal sensitization, 237
Spinal transection, 142, 218
Spinalization removes the tonic descending control, 114
Spinothalamic cells become sensitized, 157
Spinothalamic cells, responses of identified, 156–157
Spinothalamic pathways, 68
Spontaneous pain and maintenance of, 5, 7
Sprouting capacity, 245
Sprouting sensory C-fibres, 235
Stimulation in midbrain PAG, 21
Stimulation in the lateral midbrain reticular formation, 21
Stimulation induced IEG expression, 250
Stimulation of receptors in deep somatic tissues, 128
Stimulus-response functions for paw withdrawal force, 25
Stochastic discharges of nociceptive neurons, 207
Stress, 73
Stress responses, 73
Stress response is nonlinear, recursive, and highly complex, 78
Stress-induced activation of endogenous opioids, 95
Stress-induced analgesia, 95–96
Stress-induced antinociception, 97
Stress-induced peripheral antinociception, 97
Stressor, 73
Stroking hyperalgesia, 9
Substance P, 12, 85–90, 100, 113, 125, 127, 132, 139, 189, 209, 215, 217, 219, 229, 236–237, 250
   Administration into the spinal dorsal horn, 90
   And CGRP, release from trigeminal afferent fibres, 239
   And myositis, 125
   c-Fos labelled spinal neurons, 251
   Central release of, 138
   Changes in excitability of afferent nerve terminals, 91
   Cross-correlograms of neuronal discharges, 216
   Did not activate the dorsal horn neurons, 128
   Efflux of glutamate, 211
   Fibres show a close relationship to mast cells, 132
   Has a weak excitatory action, 125
   In inflamed muscle, 127
   In nociceptive endings, 125
   In perivascular nociceptive trigeminal fibres, 238
   In synovial fluid from patients with osteoarthritis, 237
   Increased the degree of input convergence, 129
   Induce c-fos in spinal cord, 251
   Neuromodulator, 211
   Neuropeptides, skin inflammation, 211
   New response to C-fibre input during superfusion with, 129
   On dorsal horn neurones, 128
   On primary afferent neurones, 126
   Pro-inflammatory effects of, 132
   Receptor antagonists, 12
   Release from sensory fibres, 86
   Strengthens synaptic connections with afferent fibres, 132
   Synthesis is upregulated, 131
Substantia gelatinosa, 22, 23
Sumatriptan, 87
Sunburn reaction, 236
Superficial spinal dorsal horn, 20, 50, 90, 217
Superfusion of spinal cord, 208, 211, 215–216, 218–220
   SP, 128–129
   Substance P, long term potentiation, 218
Supersensitivity, 233
Supraspinal components of Phase 2 and Phase 3 pain, 8
Supraspinal mechanisms, 10
Supraspinal, descending, 207

Sural nerve and electrical stimulation, 218
Sympathectomy, 110
Sympathectomy on joint swelling thermographic changes, 161
Sympathetic blockade, 105
Sympathetic nerves in the generation of chronic pain, 89
Sympathetic nerves interactions with afferent fibres, 89
Sympathetic nervous system, 105, 236
Sympathetic neurons and afferent fibres, 89
Sympathetic reflex dystrophy, 236
Sympathetically controlled sensitization, 236
Sympathetically maintained pain, 105
Synaptic efficacy increase of afferent fibres, 129
Synaptic pathways, strengthening, 215
Synaptic plasticity, 188
Synaptic plasticity in the spinal cord and phosphorylation, 202
Synaptic re-arrangements, 252
Synaptic re-modelling, 245
Synaptic strength, 208, 211, 217, 221
Synaptic transmission between primary afferent C-fibres, 219
Synchronization of discharges, 211, 214–216, 221
    Multireceptive neurons, 215
    Skin inflammation, 211
Synergistic interaction, 115
Synovitis, 99

Tachykinins, 172
    Have potent immunomodulatory actions, 132
    Lose their mediating spinal reflex hyperexcitability following nerve section, 119
    May mediate spinal cord hyperexcitability, 115
    Nociceptive neurons, 217
Tachykinins SP, 115
Tachykinins substance P, 218
Tactile sensations, 5
Tail flick force, 19, 22
    Slope of the stimulus-response function for, 22
Tail flick force vector, 19
Tail flick latency, 19
Tail flick magnitude, 19
Tail flick reflex, 18
Tail flick reflex arc, 20
Tail flick reflex magnitude, suppression by morphine, 19
Tail flick reflex pathway, 21–23
Tail withdrawal thresholds, 184
Target genes of the c-Jun transcription factor, 245
Target-derived inhibitory factors, 88
Target-derived NGF, 89
Target-derived trophic factors, 91
Temporal correlation of discharges, 208, 214–216
Tenderness, 114
Tetanic stimulation of sciatic nerve, 217
Thalamic neurones, 12
Thalamocingulate, 65
Thalamus, 69, 74, 208, 211
Therapy for Phase 2 and Phase 3 pain, 12
Thermal stimulation, graded noxious, 18
Tocainide relieves hypersensitivity, 186
Tonic descending inhibition, 9, 239–240

Tonic descending pathways, 218, 221
Tonic descending systems, 219
Tonic inhibitory control, 240
Tonic opioidergic inhibition of spinal neurons, 171
trans-ACPD, 179, 181
Transactivation of target genes
    Transcription factor proteins c-fos and c-jun, 243
Transcription control, 242, 245
    De novo protein synthesis, 252
    Of the dynorphin A gene, 254
    Therapeutic possibilities, 255
Transcription factors, 245, 252
Transcription factor expression, transient changes in and lasting functional alterations, 253
Transcription factor proteins, 249
Transcription of c-Fos, 251
Transcription of ITF, 253
Transcriptional processes, 234, 253
Transcription processes in lesioned neurons, 244
Transcriptionally operating proteins, 252
Transection of nerve fibres, 244
Transection of sciatic nerve, 242
Transforming growth factor α (TGFα), 89
Transmitter systems, 11
Tricyclic antidepressant drugs in treating neuropathic pain, 241
Trigeminal neurones, excitability of, 6
Trk receptors, trkA, trkB, trkC, 87
Trophin, 87
Tubulins, 245
Tumor necrosis factor, 97
Tumor necrosis factor -α, 87
Tyrosine kinase receptor (trk A), 87

U50,488H, 171
Ultraviolet irradiation, 102
Unmyelinated dorsal root fibres, 106
UV-B irradiation
    Human skin, elevated blood flow, 236
UV-irradiation, 236

Vasoactive intestinal peptide, 88
Vasoactive intestinal polypeptide on flexor reflex hyperexcitability, 116
Vasodilatation, 85, 237
    Extravasation, 85
    Extravasation in the dura mater, 87
Ventral horn, 211
Ventral noradrenergic bundle, 71
Ventrobasal complex of thalamus, 250
Vicious circle
    Inflammation nociceptor excitation, 237
Vigilance, 71
VIP, 116, 118, 120, 236, 252
    Is upregulated after nerve injury, 119
Visceral nociceptors, 5, 7
Visceral spinal pathways, 10
Voltage and ligand gated channels, 109
Voltage dependent Mg^{2+} block, 194

Volume transmission, 210, 222
von Frey hair, 216
von Frey thresholds, 220

Wallerian degeneration, 234
Wide dynamic range cells, 34, 113, 208
Wide dynamic range response profile, 168
Wide dynamic range spinothalamic tract cell, 158
Wide dynamic range, class 2 or multireceptive neurons, 221
Wind-up, 26, 114, 116, 118, 202, 225, 229
    A key to central hypersensitivity, 226

    Blocked by metabotropic glutamate receptor antagonists, 186
    Of a spinal nociceptive neurone, 226
    Produced by mustard oil, 186
Withdrawal
    Enhanced by intrathecal NMDA, 26
    Thresholds, 179–180

Y2 receptors located on sympathetic fibres, 89
Yohimbine, 107, 108

Zymosan, 188–189

Other Volumes in PROGRESS IN BRAIN RESEARCH

Volume 87: Role of the Forebrain in Sensation and Behavior, by G. Holstege (Ed.)—1991, ISBN 0-444-81181-8.
Volume 88: Neurology of the Locus Coeruleus, by C.D. Barnes and O. Pompeiano (Eds.) — 1991, ISBN 0-444-81394-2.
Volume 89: Protein Kinase C and its Substrates: Role in Neuronal Growth and Plasticity, by W.H. Gispen and A. Routtenberg (Eds.) — 1991, ISBN 0-444-81436-1.
Volume 90: GABA in the Retina and Central Visual System, by R.R. Mize, R.E. Marc and A.M. Sillito (Eds.) — 1992, ISBN 0-444-81446-9.
Volume 91: Circumventricular Organs and Brain Fluid Environment, by A. Ermisch, R. Landgraf and H.-J. Rühle (Eds.) — 1992, ISBN 0-444-81419-1.
Volume 92: The Peptidergic Neuron, by J. Joosse, R.M. Buijs and F.J.H. Tilders (Eds.) —1992, ISBN 0-444-81457-4.
Volume 93: The Human Hypothalamus in Health and Disease, by D.F. Swaab, M.A. Hofman, M. Mirmiran, R. David and F.W. van Leeuwen (Eds.) — 1992, ISBN 0-444-89538-8.
Volume 94: Neuronal-Astrocyte Interactions: Implications for Normal and Pathological CNS Function, by A.C.H. Yu, L. Hertz, M.D. Norenberg, E. Syková and S.G. Waxman (Eds.) — 1992, ISBN 0-444-89537-x.
Volume 95: The Visually Responsive Neuron: From Basic Neurophysiology to Behavior, by T.P. Hicks, S. Molotchnikoff and T. Ono (Eds.) — 1993, ISBN 0-444-89492-6.
Volume 96: Neurobiology of Ischemic Brain Damage, by K. Kogure, K.-A. Hossmann and B.K. Siesjö (Eds.) — 1993, ISBN 0-444-89603-1.
Volume 97: Natural and Artificial Control of Hearing and Balance, by J.H.J. Allum, D.J. Allum-Mecklenburg, F.P. Harris and R. Probst (Eds.) — 1993, ISBN 0-444-81252-0.
Volume 98: Cholinergic Function and Dysfunction, by A.C. Cuello (Ed.) — 1993, ISBN 0-444-89717-8.
Volume 99: Chemical Signalling in the Basal Ganglia, by G.W. Arbuthnott and P.C. Emson (Eds.) — 1993, ISBN 0-444-81562-7.
Volume 100: Neuroscience: From the Molecular to the Cognitive, by F.E. Bloom (Ed.) —1994, ISBN 0-444-01678-x.
Volume 101: Biological Function of Gangliosides, by L. Svennerholm et al. (Eds.) — 1994, ISBN 0-444-01658-5.
Volume 102: The Self-Organizing Brain: From Growth Cones to Functional Networks, by J. van Pelt, M.A. Corner, H.B.M. Uylings and F.H. Lopes da Silva (Eds.) —1994, ISBN 0-444-81819-7.
Volume 103: Neural Regeneration. by F.J. Seil (Ed.) — 1994. ISBN 0-444-81727-1
Volume 104: Neuropeptides in the Spinal Cord, by F. Nyberg, H.S. Sharma and Z. Wiesenfeld-Hallin (Eds.) — 1995, ISBN 0-444-81719-0.
Volume 105: Gene Expression in the CNS, by A.C.H. Yu et al. (Ed.) — 1995, ISBN 0-444-81852-9.
Volume 106: Current Neurochemical and Pharmacological Aspects of Biogenic Amines, by P.M. Yu, K.F. Tipton and A.A. Boulton (Eds.) — 1995, ISBN 0-444-81938-X.
Volume 107: The Emotional Motor System, by G. Holstege, R. Bandler, C.B. Saper (Eds.) — 1996, ISBN 0-444-81962-2.
Volume 108: Neural Development and Plasticity, by R.R. Mize and R.S. Erzurumlu (Eds.) — 1996, ISBN 0-444-82433-2.
Volume 109: Cholinergic Mechanisms: from Molecular Biology to Clinical Significance, by J. Klein and K. Löffelholz (Eds.) — 1996, ISBN 0-444-82166-x.

PROGRESS IN BRAIN RESEARCH

VOLUME 110

TOWARDS THE NEUROBIOLOGY OF CHRONIC PAIN

Millikan
RB127 .T682 1996
Towards the neurobiology of
chronic pain